DISASTERS ──────────

An Analysis of Natural and Human-Induced Hazards

Fourth Edition

D1408578

Charles H. V. Ebert
University at Buffalo
State University of New York

KENDALL/HUNT PUBLISHING COMPANY
4050 Westmark Drive   Dubuque, Iowa 52002

ACKNOWLEDGEMENT - ◎

The author wishes to express his gratitude to all who assisted in the many efforts that led to the completion of this book.

I want to thank Mr. James A. Ulrich of the Educational Communication Center of our University for producing the new photographs and maps of this edition, and additional thanks go to Betsy Abraham for typing the final version of the manuscript.

Last, but certainly not least, I want to thank my wife, Ilse, for her constant encouragement and for her patience during the time the work was in progress.

C.H.V.E.

Copyright © 1988, 1993, 1997, 2000 by Kendall/Hunt Publishing Company

Library of Congress Catalog Card Number: 00-103529

ISBN 0-7872-7073-3

Printed in the United States of America

10 9 8 7 6 5 4 3 2 1

ABOUT THE AUTHOR – ⊚

Professor Charles H.V. Ebert is a Distinguished Teaching Professor of Physical Geography at the State University of New York at Buffalo. In 1963, he became the first chairman of the newly established Department of Geography. He headed the department until 1970, then moved on to serve as University Dean for Undergraduate Education until 1977.

He lectured to thousands of enthusiastic students in subjects such as physical geography, soils, land development, and environmental impact studies. Indeed, the excellence of his teaching is reflected in having received two Chancellor's Awards for Excellence in Teaching (1974 and 1976), the Milton Plesur Memorial Teaching Award (1988), and in his Promotion to Distinguished Teaching Professor (1989). He has also received the Award for Distinguished Achievement in the Arts and Sciences (1989), and the Distinguished Teaching Achievement Award from the National Council for Geographic Education (1990)

CONTENTS

PREFACE

Disasters: An Analysis of Natural and Human-Induced Hazards is written for all who wish to acquire a thorough understanding of the causes, the nature, and the geographical occurrence of natural and human-induced disasters.

No single book that deals with this complex topic can cover all types of disasters, nor can it be exhaustive about all their causal, technological, geographical, and societal aspects. Therefore, this volume is a comprehensive sampler. It is designed to probe thoroughly into a wide range of potential hazards as well as into actual disasters, and to sensitize the reader to the impact and consequences of such events.

This interdisciplinary text combines a rigorous presentation of natural science principles with a humanistic approach to demonstrate and to strengthen the critical relationships between the physical and the social sciences. This endeavor makes it necessary and desirable to cross some of the barriers that so often keep academic disciplines artificially apart.

The material is presented not only to serve students of geography, of the earth sciences, and those of environmental studies, but also to assist any person who needs to obtain a solid but not excessively technical insight into the workings and consequences of disasters and hazards facing humankind.

All critical processes involved in natural hazards, as well as those in which humans have an ever-deepening involvement, are systematically explained. Case studies, where helpful to the understanding, are offered for selected types of disaster. *Detailed references, key words* printed in *boldface and italics,* and a *glossary* containing essential terms and concepts are used to facilitate the search for follow-up readings and the learning of background vocabulary. Moreover, the open-ended structure of the book allows the instructor to introduce additional case studies and to expand, or reduce, the total coverage. The text is designed to serve either as a main reader or function as a supplementary source book in more specialized courses; its scope and style make this textbook especially suitable for use in science-oriented general education.

Newly recognized facts and quickly changing theories mark our contemporary world of sciences. This trend makes it difficult to keep a book of this nature and scope current in all aspects. However, past events offer us a great amount of knowledge. Moreover, a systematic understanding of the causes and effects of disasters will help humankind to cope more effectively with the future.

Natural scientists usually concentrate on the *trigger mechanisms* and on the physical processes that lead to disasters. Social scientists, on the other hand, focus primarily on the impact of disasters on humans, on property, and on institutions. We must use both approaches, the trigger mechanism and the impact on people, to understand the complexities of both natural and human-induced disasters. This book uses a cross-disciplinary approach. Moreover, we must understand both causes and effects.

This book, while purposefully integrating the aspects mentioned above, uses a clear and helpful way to group and to assess disasters. This method places the respective events into general *arenas of action,* or within the subdivisions of the total earth system: the *lithosphere,* the *hydrosphere,* the *atmosphere,* and the *biosphere.* Clarity of organization helps the student to understand.

Some disasters readily lend themselves to this method of classification; for example, earthquakes are primarily linked with the lithosphere, whereas hurricanes and tornadoes are unquestionably atmospheric phenomena. However, it is critical to realize—as will be demonstrated repeatedly in the subsequent chapters—that both *energy* and *matter* move across the boundaries of these subdivisions and become integrated in the dynamics of the entire earth-system matrix. Accordingly, this approach emphasizes the dynamic interrelations of the causes and effects of disasters on our planet.

The first four parts of this book center on those natural processes in the lithosphere, in the hydrosphere, in the atmosphere, and in the biosphere that may cause harm. However, in many instances the boundaries between the arenas of action are hazy; one disaster may cause other changes. It will also be shown that in many cases humans' activities interfere with natural processes. The impact of humans on the planet is increasingly significant. This interference, or the human factor in disasters, can further cloud attempts to classify disasters.

Part five of the presentation focuses on the specific *role of humankind* in causing environmental stress and disasters. Humans' role is by no means unilateral and involves complex interactions. But humans must accept a growing share of responsibility for making many parts of the world more disaster prone. Where useful to the understanding of the processes under discussion, case studies and shorter examples will illustrate the unwise and often reckless interaction between humans and nature that frequently lead to events that exact a high toll of casualties and cost.

Ultimately, we will focus on a devastating phenomenon that derives almost entirely from human aggression; it has reached, over the course of history, such actual and potential dimensions that it could destroy the vital base on which life on earth depends. This phenomenon is *war.*

The historical evolution of war and weapon systems, their ominous progression along a *J-curve,* and their increasingly violent impact on land and people concludes the analysis of disasters. The rationale for including war in the discussion of disasters is not only to underscore the devastating potentials of warfare—as illustrated by the obliteration of Hiroshima and Nagasaki by atom bombs, by the devastating firestorms that erased cities in World War II, and the burning oil wells during the recent Gulf War—but also to show that humankind is uniquely responsible for these truly human-made cataclysms and must find a way to eliminate the threat of war.

In concluding the presentations of natural and human-induced disasters, it is necessary to assess the questions how people of different backgrounds react to and deal with hazards which may turn into disasters.

Hazards occur in many forms and magnitude; however, it appears that hazards can exist without involving humans and present merely a potential. *Risks,* on the other hand, confront people and undermine their existence, their welfare, their property, and their activities. Risks must be assessed by fully understanding the nature of hazards. This understanding makes it possible to meet such threats by risk management and, if possible, by efforts to lessen their potential impact. The latter attempt demands prediction, planning, and an appropriate level of preparedness.

The Learning Environment

The coverage and style of this book—blending natural sciences with a humanistic theme—encourages non-science students to explore the realm of natural sciences in a practical and realistic fashion. At the same time this book gives the science-oriented person the opportunity to develop a meaningful understanding of an insight into societal implications of hazards and disasters.

A conscious effort is made to make this text readable and to avoid overspecialization without sacrificing the thorough treatment of scientific principles and processes. This book presents the scope and working methodologies of various science disciplines so that the reader may grasp each within its unique setting and yet may recognize the links to other topics. Each chapter presents some *principles* and *examples* and frequently concludes with a case study of a particular disaster's causes and effects. These case studies and special references, which range from the obscure beginnings of humankind's existence to modern nuclear disasters, lend an atmosphere of *"real life"* and reality to the discussions. This approach helps to instill curiosity as well as a feeling of personal involvement on the part of the reader. Unlike many books, this study presents *theories* and *practical cases* in close relationships, making both easy to understand.

Benefits to the Student

The integrated and systematic treatment employed in this volume offers students of varied backgrounds and interests a comprehensive and stimulating overview of the past and present events on our planet.

Parochial narrowness in language and scope is avoided while giving the serious student the opportunity to acquire a sound vocabulary in the sciences. Moreover, the student will comprehend the intricate *cause-and-effect* relationships between nature and humankind. Such understanding prepares the mind to see more clearly the often delicate lines that separate *stress* and *threshold* situations from breakdowns of natural and human systems.

The awareness of nature's sensitivity to humans' unwise and, at times, irresponsible actions will promote a sense of concern. This personal involvement may generate a desire to translate concern into responsible action.

INTRODUCTION: PERSPECTIVES OF DISASTERS

The geologic history of planet Earth, marked by the formation of the solid crust, probably began about 4.6 billion years ago. From its early existence on, our planet suffered violent upheavals and incessant changes. These changes, customarily referred to as *natural processes,* are still taking place and involve energy flow and dynamic interactions between the total physical environment and all living things.

The energy that is responsible for these ongoing changes derives primarily from two sources, namely (1) radiation from the Sun, called *solar insolation,* and (2) *heat flow* emanating from the interior of the planet. Activities associated with *plate tectonics,* or the dynamic movements of lithospheric segments, manifest themselves in earthquakes and in volcanism. Processes of the atmosphere may lead to violent storms such as hurricanes, tornadoes, and thunderstorms. These natural phenomena, and many others, are the reflections of the ongoing adjustments and dynamic changes endemic to our restless planet.

The complex processes that shape the *lithosphere,* the *hydrosphere,* the *atmosphere,* and the *biosphere* guided the evolution of living things as well as their adaptations to the ever-changing conditions on Earth. Of course, our knowledge about the earliest forms of life within the larger framework of natural forces is limited; such knowledge rests for the most part on deductions and inferences. However, sophisticated methods and new technologies have filled many gaps in our understanding of the past as well as that of the present.

In a compressed time model of our planet's history, where a time span of 30 minutes represents the 4.6 billion years of geologic history, the emergence of humankind and its subsequent development would occupy only the last few seconds. Humans, just as any other living creatures, were subjected to the violence of nature and had to learn how to cope with such forces. Earthquakes, violent volcanic eruptions, landslides, floods, droughts, dust storms, insect plagues, and diseases threatened prehistoric humans; these hazards still threaten our existence in modern times.

For example, massive volcanic activities ravaged the Eifel region in what is now Germany around 8300 B.C. Devastating famines plagued Egypt around 3500 B.C., and several cataclysmic earthquakes leveled numerous cities in central Italy in 1450 B.C. Scars of these upheavals still can be seen today in the form of a vast *caldera* (a cauldron-shaped volcanic depression) that is occupied by ancient Lake Ciminius, or Lago di Vico as it is called today. Almost concurrently, a powerful volcanic eruption occurred in the Aegean Sea to the south of Greece. Here, the island of Stronghyle, now known as Santorini, was torn apart by volcano Thera. This violent act of nature reputedly led to the destruction of the Minoan culture on Crete. Are these disasters "Acts of God" or are they just natural processes?

Literally, the term "disaster" means "bad star" and, in a wider sense, implies a bad omen, a calamity, a misfortune, or a harmful impact. These words immediately raise a fundamental question: Is a volcanic eruption, or an earthquake, in an unpopulated and remote area a disaster? Must humans be hurt or killed, and their property destroyed, before a violent natural event can be called a disaster? There is no simple answer to this question. The definition of the word *disaster* varies with one's personal perception as well as with the purpose and application of the definition.

For example, on June 6, 1912, a violent eruption tore apart Mount Katmai on the Alaskan Peninsula. The force of this explosion was estimated to be equivalent to 200,000 H-bombs; more than 7 cubic miles (30 km³) of pumice and ash overwhelmed the landscape. Rivers were blocked, thousands of acres of forest virtually disappeared, acids spilled into the coastal ocean, and all wildlife was killed within a vast area. There were no known human casualties, but such a devastating event must be viewed as a true disaster.

Another question arises: What is the role of humankind in creating hazards and calamitous events? There is a general view that it is possible to separate disasters into two basic categories: natural disasters and human-made ones. While it is true that many disasters are triggered by natural processes, it is equally evident that many types of disasters are not only made worse by human activities and miscalculations, but are caused directly by human mismanagement.

This conclusion does not only pertain to human-induced stress on the environment—deforestation, urbanization, soil erosion, release of toxic materials, contributing to global warming, and other detrimental actions—but also to many hazards anchored in the realm of nature. It has been stated that "the line between natural and man-made geohazards is finely drawn and may be blurred," and that many catastrophic events within the environment are man-induced or made worse by the intervention of humankind (McCall et al.; 1992, p.2)

The role of humankind on planet Earth was at one time that of an integral part of the living world; humans did not exert a noticeable influence on *ecosystems.* In prehistoric times our ancestors' fate was more or less dominated by the forces of nature which they often attempted to appease with offerings and rituals. In later days, Christianity viewed nature and disasters in its way. The Black Death, as mentioned in chapter 12, was interpreted as a divine visitation, while others took refuge in satanic rites and macabre practices.

In contemporary times, advanced technology and dramatic advances in science gave us powerful capabilities to protect ourselves from many hazards; however, our vulnerability to both natural and human-induced disasters seems to be on the increase. This increase, which parallels our never-sated desire for higher standards of living, can at times be linked directly to the very measures designed to shield us from dangers. A good example is given by Robert L. Kovach: "We build dams and levees to offset the cyclical effects of floods, but as the 1993 deluge in the Midwest showed, our elaborate flood-control systems may increase the magnitude of a flood by obstructing and channeling the wild river waters too rigidly" (Kovach; 1995, p.1)

As will be expanded upon in chapter 13 (Major global concerns), the influence of humankind on our planet has become comparable to, and possibly exceeding in many ways, many natural processes. Violence of nature found its match in the growing threats by humans. Many factors contribute to this worsening situation: (1) the seemingly unstoppable population explosion which now passed the 6-billion mark, (2) a widespread lack of adequate education and experience, (3) the fact that more people move into *disaster-prone environments,* (4) spreading poverty in many parts of the world, and (5) the inefficient infrastructure which prevails in many societies.

The last chapter of this book investigate how vulnerable humankind is to hazards, and how people differ in their ability, and willingness, to cope with disasters of various magnitudes. Societies in general and some communities in particular, respond differently to calamities and may develop highly non-traditional behavior. Some groups engage in spontaneous *self-help* while others appear to be incapable of making decisions on their own.

It will be shown that the ultimate success in the struggle to stem world-wide hazards and violent disasters depends not only on science and technology. Humankind must be willing, trained, and creative to apply scientific knowledge and sound judgment to make global survival possible. We have the responsibility to care for the only home we have: *Planet Earth.*

DISASTERS INVOLVING THE LITHOSPHERE AND SURFACE MATERIALS

Earth, our *dynamic planet,* is constantly affected by natural processes that occur deep in the interior or in the outer crust, in the oceans, and in the envelope of gases, the atmosphere. Some of these processes, such as the gradual erosion by rivers and glaciers, as well as the slow motion of the earth's crustal plates, take place almost unnoticeably if viewed on a day-by-day basis. Other processes, as illustrated by the sudden and violent motions of earthquakes, or by the cataclysmic eruptions of volcanoes, may radically change our landscapes within a short time span. But we must remember that stress buildup, leading later to sudden energy release, may extend over long time periods.

Stress buildup, in response to energy input, will eventually upset a system's state and cause changes. Such a change could be a momentary adjustment, such as a flood in a river system, or it may occur over a much longer period to reach a new state of *equilibrium.* The rebound of continental masses after the melting of the icecap is a good example of the latter.

The time interval that is needed for a system to find a new state of equilibrium is called its *response time.* The new equilibrium does not have to be the same condition that existed before; it rarely is. This is especially so for processes involving the lithosphere and the earth's surface features.

Disasters of this category are linked primarily to energy releases in the earth's **crust,** or **lithosphere,** leading to earthquakes and volcanic activities. Associated events can be triggered by disturbances in the crust but occur on the earth's surface. These include **land-slides,** mudflows, soil failure, snow and *rock avalanches,* and other surface processes.

EARTHQUAKES

Key Terms		
seismometers	epicenter	land subsidence
seismology	earthquake waves	soil liquefaction
faults	body waves	floodplain
elastic rebound	surface waves	seismic sea wave
subduction	primary waves	tsunami
mantle	secondary, shear waves (S-waves)	seismographs
asthenosphere	Moho discontinuity	perihelion
basaltic	Rayleigh waves	conjunction
volcanism	Love waves	opposition
magma chambers	intensity	perigee
Circum-Pacific Ring	magnitude	shears
rift	Mercalli scale	The Big One
transform faults	Richter scale	liquefaction
focus	amplitude	ground failure

Geologic evidence shows that major earthquakes have occurred throughout geologic history. These events are perfectly normal processes that will take place again and again as our planet undergoes dynamic changes.

Some of these earth movements are quite minor and can be detected only by sensitive *seismometers.* It is possible that, on a worldwide basis, about one million tremors occur each year; however, only about 10 percent are felt by people. Whether or not an earthquake is noticed depends not only on the amount of energy released but also, to a large measure, on the earth materials through which the earthquake energy propagates.

Modern *seismology,* the study of earthquakes, is a key science that helps us to understand the nature of earthquakes, their effects, and possibly their more accurate prediction. Moreover, by tracing the behavior of earthquake waves, geoscientists have learned a great deal about the interior structure of the earth that cannot be revealed by direct means. The crucial nature of our knowledge is reflected in the staggering loss of life connected with some of the worst quakes in recorded history: The Shensi Province earthquake of the 16th century in China led to more than 800,000 casualties, the 1755 devastation of the Lisbon area in Portugal killed nearly 60,000 people; and—the worst in modern times—that which leveled the Chinese industrial center of Tangshan in 1976, which caused over 300,000 deaths. Casualties like these, in addition to the massive destruction of property, make earthquake studies, and their prediction, a high priority.

Causes of Earthquakes

Most earthquakes originate in the earth's lithosphere where, through geological forces, sufficient strain builds up to cause fracturing of rock formations. These fractures are

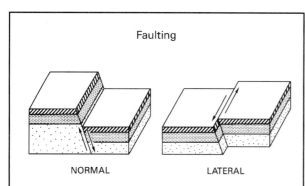

Figure 1.1. Rock formations may fracture along fault lines. This process results in vertical (normal) or horizontal (lateral) slippage. Faults can also move simultaneously in both ways.

referred to as *faults.* In fault systems rocks may be displaced vertically, or laterally, or both (fig. 1.1).

Movement along such fault lines is not smooth and continuous but occurs in sudden jolts as the pent-up stress is released. The sudden slippage occurs because an enormous amount of friction between the rock faces must first be overcome.

Some fault systems are quite active, and in such areas minor tremors are very frequent. Other faults may remain *frozen* for thousands of years, allowing the stress to build up to very dangerous levels. Rocks, when subjected to continuous strain, are capable of being slowly deformed similar to the gradual bending of a piece of wood. As the stored-up stress becomes excessive, individual fibers in the wood begin to tear. In a comparable fashion the crystalline structure of a rock begins to lose strength and, at a given level of strain, the structure disintegrates. It may, subsequent to fracturing, spring back to a position of prebreak equilibrium. This is the basic concept of *elastic rebound.*

Faulting in the earth's crust may occur universally; yet, there are regions where fault activities are particularly frequent. These active zones are mostly associated with *plate tectonics.* In these areas plates may collide, or slide past each other, are subject to *subduction* (downbending of a crustal plate and its reincorporation into the underlying *mantle*), or they break apart forming rifts (fig. 1.2). Faulting may also occur far away from plate boundaries within the *shield* masses of the continents. These granitic shields are also affected by stress which may originate within the shield because of postglacial rebound. This type of adjustment happens after the extensive masses of continental ice sheets have melted and the continent slowly rises after having

been depressed by the weight of the ice. Other forces are exerted by heat flow from the *asthenosphere* underlying the lithosphere. Here the semiplastic *basaltic* masses of the asthenosphere push upward against the roots of the continental shields.

Other causes of earthquakes are associated with *volcanism.* Before the mechanics of plate tectonics were understood, the idea prevailed that most earth tremors were caused by cave-ins of volcanic *magma chambers* underneath volcanoes or were mainly caused by magmatic materials shifting underground. Yet, as will be discussed later in this chapter, there is a close correlation between major zones of volcanism and earthquakes. Longtime monitoring of both phenomena in Japan has shown that there is not necessarily a direct connection between volcanic activity and earthquakes (Longwell et al.; 1948, p. 393). Moreover, it is now generally accepted that tremors, often preceding volcanic eruptions in considerable numbers, are of low intensity and do not result in major earthquakes.

Geographical Occurrence of Earthquakes

Although earthquakes may take place in just about any location on earth, there are very distinct zones of marked earthquake concentration. Probably the most-mentioned one is the so-called *Circum-Pacific Ring,* a relatively narrow band around the Pacific Ocean where, for the most part, oceanic plates are being subducted (fig. 1.3). This zone, which includes the Japanese Islands, the Kuril Islands, the Aleutians, the Alaskan coastal regions, the west coast of North and South America, as well as entire Central America,

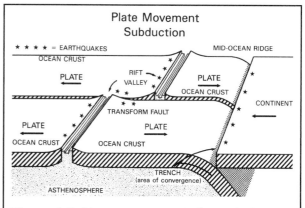

Figure 1.2. New ocean crust is formed where plates diverge along rift valleys, while old plate material is destroyed in subduction zones where plates converge.

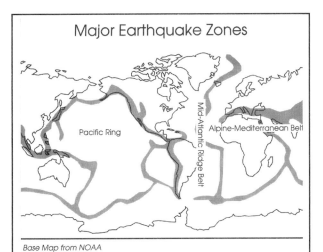

Figure 1.3. Major earthquake zones are located along active plate borders and major rifts. About 70 percent of the world's earthquakes occur in the Circum-Pacific Ring, a zone which is also known for widespread volcanic activity.

Base Map from NOAA

accounts for about 70 percent of the world's earthquakes. It is also well known for a great concentration of volcanoes.

In addition to the Circum-Pacific Ring several areas show a marked concentration of earthquakes, including a great number of *submarine quakes.* Submarine disturbances are associated with *oceanic ridges,* such as the *Atlantic Ridge.* Here, new oceanic crust material wells up and slowly spreads away from the central *rift.* The older basaltic rock, located further away from the rift, becomes brittle and in many places fractures along so-called **transform faults,** which tend to make 90-degree angles with the rift. A great number of submarine earthquakes originate in these fault zones.

Another very active seismic zone, marked by plate fractures and collisions, extends in a general west-to-east direction from the Straits of Gibraltar through the northern Mediterranean into Asia Minor (fig. 1.3). This earthquake belt, which coincides with considerable volcanic activities, has produced more than 60 major events since the 1750s just in the area involving Greece, the Aegean Sea, the Caucasus region, Turkey and Iran. In Iran, for example, a very destructive earthquake, estimated at a magnitude of about 7.4 on the Richter scale, struck the Gilan Province on June 21, 1990, causing casualties estimated as high as 50,000 (NCEER Bulletin: 1991, p.1).

A recent event occurred in Turkey on August 17, 1999, when a very strong earthquake with a magnitude of 7.4 struck near Izmit, about 55 miles (88km) south-southeast of Istanbul. The epicenter was associated with a major movement along the North Anatolian Fault, which has caused hundreds of disturbances in modern times. A case study of this event is presented later in this chapter.

The Anatolian Fault system is a major segment of the west-to east seismic zone mentioned above. This zone extends into Asia where several other plates engage in collisions and subduction. On July 16, 1990, a major earthquake with a magnitude of 7.3 took place in the northern part of the Philippine Island of Luzon. This quake was triggered by a slippage in the very active Philippine Fault, which is part of the subduction zone discussed in connection with the eruption of Mount Pinatubo presented in the following chapter.

Earthquakes, as well as volcanism, are also concentrated in continental fracture zones such as the one represented by the East African Rift, where both tremors and active volcanoes are encountered.

Earthquake Energy and Waves

We have learned that rock masses along fault systems, as a result of dynamic geological forces, build up to high stress levels. When the stored energy exceeds the frictional resistance between rock faces, a slippage occurs. Usually this takes place where the friction is lowest. This location is the point of energy release or the earthquake *focus.* The point at the earth's surface, directly above the focus, is called the *epicenter* (fig. 1.4). Earthquake foci may be located a few miles beneath the surface or may occur at depths of up to 430 miles (690 km). The majority of foci lie at depths of from 40 to 100 miles (64 to 161 km). Deep-focus earthquakes, ranging from 180 to 400 miles (290 to 645 km) typically occur on the northern segments of the Pacific Ring to the subduc-

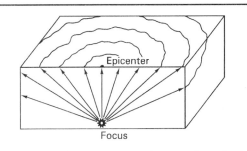

Figure 1.4. The epicenter of an earthquake is located at the earth's surface directly above the point of energy release, or focus.

tion zones along the Kuril and Aleutian Islands (Foster; 1971, pp. 116–117).

The breaking of rock masses and the resulting rebound take place over time periods ranging from fractions of a second to several minutes. Thus the formerly stored-up potential energy is converted into *earthquake waves,* a form of *kinetic energy* or energy of motion. The wave energy disperses from the focus in all directions, but at different velocities and in different wave types.

Basically, there are two wave types: so-called *body waves,* which travel through and within the earth, and *surface waves,* which propagate on the earth's surface. The relative difference in their travel rates and their behavior as they encounter different layers makes it possible to obtain a much more detailed picture of the earth's interior than is possible by direct investigations. The Deep Sea Drilling Project of 1968 is one example of such direct probing.

The two major forms of body waves are the *primary waves* (P-waves) and the *secondary* or *shear waves (S-waves).* The primary waves, as implied by their name, travel fastest, or about 1.7 times faster than the S-waves. The primary waves are able, very much like sound energy, to move through solids and liquids. The wave energy moves forward and backward in the direction of wave travel, comparable to the compression and expansion of a steel coil spring. The S-waves, on the other hand, make particles move up and down at a right angle to the direction of wave travel. They behave in a fashion of a rope tied to a post and shaken at the free end (Wyllie; 1976, p. 90). These secondary waves cannot propagate through liquids. For this reason they do not pass through the liquid core of the earth.

Earthquake waves are refracted and deflected as they encounter earth layers of different density. Abrupt changes in wave velocity and patterns of directional changes in wave travel indicate the location and thickness of these layers. For example, the velocity of P-waves in the outer *granitic* continental rock mass is about 3.8 mi/sec (6.1 km/sec) whereas the speed increases to more than 5 mi/sec (8 km/sec) in the denser deep-seated *basaltic* crustal layer of the oceans' basins and underlying the lighter continental masses. An additional and very marked change in wave velocity reveals the location of the so-called *Moho discontinuity,* named after A. Mohorovicic, 1909 (Bath; 1973, p. 23). This discontinuity separates the lithosphere from the earth's mantle.

The surface waves are the longest waves. They impart motions to earth particles similar to the orbital paths followed by water particles in ocean waves; but in contrast to ocean waves they tend to be much lower.

Surface waves are extremely destructive to buildings, highways, bridges, and other structures because they make the ground heave, and buildings sway like ships in a stormy sea. This heaving motion is caused mainly by the **Rayleigh waves** (R-waves), a subclassification of surface waves. Sideways motions, at right angles to the direction of surface wave travel, result in highly destructive shear waves called *Love waves* (L-waves).

In addition to revealing the interior structure of the earth, earthquake waves also make it possible to locate the points of origin of such disturbances. The difference in velocity between P-waves and S-waves results in the primary waves arriving faster at a seismographic station than the S-waves. The difference in arrival time is proportional to the distance from the earthquake's point of origin.

Seismic wave graphs indicate the distance but cannot show the direction from where the waves came. For this reason a minimum of *three* seismographic stations is required to identify the location of the quake origin. Although the location of the epicenter can be found in a fairly simple manner, the pinpointing of the focus, in principle following the same method, is more complicated (fig. 1.5).

Intensity and Magnitude

Earthquakes can be described in a variety of ways: casualties, property damage, type of structural damage, cost, duration, depth of origin, and many other

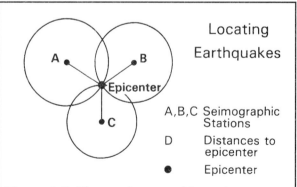

Figure 1.5. Three seismographic stations are needed for locating the epicenter. The radii of the circles represent the distances between the individual stations and the epicenter. The distance is proportional to the time difference in the arrival of the P- and S-waves.

ways. But it is necessary to apply other means to assess earth shocks and to use a scale of measurement. Two aspects are commonly identified: *intensity* and *magnitude.*

The *intensity* of an earthquake is a more descriptive term and is represented by the *Mercalli scale.* This scale, employing categories I through XII, reflects what is felt by persons subjected to the quake and what effects an earthquake has on structures and natural earth features. Diverse geologic materials respond in very different ways to earthquake waves. Consequently, the Mercalli scale, which qualifies the effects, cannot be used to actually measure the severity of an earthquake. The measurement of severity is achieved by the *Richter scale,* expressing magnitude.

This scale uses the common logarithm of the size (*amplitude*) of the various earthquake waves as they are recorded at the seismic station. Earthquake waves lose energy over distance traveled, and this factor must be taken into account. The scale begins at a set zero point determined by the smallest tremor that can be recorded by a seismometer. The upper scale is open, although quakes above a magnitude of 9 are quite rare.

It is important to understand that because of the *logarithmic* nature of the scale, the amplitude grows ten times between a magnitude of 1 and 2 on the Richter scale. An advantage of this scale is that there is a correlation between the energy release at the focus and the amplitude of the resulting waves. Thus, in general terms, the energy release grows by a factor of about 30 between units of magnitude (Bolt; 1980, p. 4).

The damage caused by earthquake waves does not depend only on the energy involved but is also affected to a large extent by the earth material through which the waves pass. Many misconceptions exist concerning this factor. For example, it may appear reasonable to think that *sediments,* such as sand, mudflats, volcanic ash layers, river floodplains, and even artificial fill areas act as a cushion and absorb energy. On the contrary, sediments constitute a much greater damage potential than solid *bedrock,* especially when they contain a large amount of groundwater.

Sediments and ash layers, as demonstrated during the Mexico City earthquake of September 1985 contributed to violent surface undulations and partial subsidence. Mexico City, the old Aztec settlement of Tenochtitlan, is largely built on an ancient lake surface that readily propagates surface waves. Severe shaking also occurred in connection with the 1886 earthquake that struck Charleston, South Carolina,

which is located on unconsolidated marine sediments of the coastal plain. These examples reveal that the study of ground materials is a critical factor in the assessment of potential earthquake risk zones and should be used in urban and industrial planning as well as in the development of transportation systems.

The Loma Prieta Earthquake in California (October 17, 1989) further illustrates the role of unconsolidated ground materials in seismic events. This earthquake is discussed in detail at the end of this chapter.

Earthquake Hazards

The overwhelming impact of violent destruction and general upheaval caused by earthquakes, so intensely covered by the news media, tends to overshadow the fact that some of the *side effects* and *postquake hazards* cause more suffering and damage than the earthquake destruction itself. For example, the direct damage to structures and property during the April 18, 1906, San Francisco earthquake was estimated to be only about 5 percent of the total destruction. The subsequent fire that gutted the heart of the city accounted for the remainder. Similarly, uncontrollable fires, following the 1923 Tokyo earthquake, killed the majority of the 140,000 victims.

Other serious subsequent hazards that are associated with earthquakes, in addition to the primary effects, include extensive *land subsidence,* often combined with flooding, the triggering of rock, snow, and ice avalanches, landslides, mudflows, and *soil liquefaction.* Sand liquefaction became a major factor in the June 1964 earthquake that caused heavy damage to the Japanese city of Niigata. This city is built on flat *floodplain* deposits of the Shinano River. During the severe shaking the waterlogged sediments subsided and groundwater emerged at the surface. The quake coincided with the high tide, which intensified the flooding. Although many modern buildings survived the immediate quake effect, several sank into the unstable sands. Some of the buildings were said to have leaned to an angle of 40 degrees (Butler; 1976, p. 15).

Another devastating side effect of earthquakes, especially with those originating on the ocean floor, is the forming of a *seismic sea wave,* or *tsunami.* These waves, triggered by intense faulting and large-scale submarine landslides, will be discussed in chapter 4.

In addition to the physical hazards there are also the problems of sanitation, pestilence, and lack of food and water. The stress produced by these problems can become the ingredients for riots and political unrest. The latter is illustrated by the fact that the 1972 Managua earthquake, later presented as a case study, contributed to the fall of the Samoza regime, and the 1978 earthquake in Iran led to severe criticism of the Shah's government (Wijkman, Timberlake; 1984, p. 125). The disasters may reveal a government's inadequate ability to respond. In some cases the natural disasters may be regarded as "God's will," leading to a change in government or political orientation.

Earthquake Prediction

It is obvious, with more people living on the earth and with more of them moving into seismic hazard zones, that the prediction of earthquakes must receive increasing attention. Efforts to predict earthquakes, based on sound scientific principles, had to await the development of accurate and reliable instrumentation as well as the basic understanding of the dynamic earth processes. Sufficiently accurate *seismographs* were not available until shortly before 1900, and the general acceptance of the concepts of plate tectonics was not realized until after the International Geophysical year of 1957. However, even though impressive progress has been made in technological aspects dealing with research methodologies, data processing, and in the understanding of the geophysical processes leading to earthquakes, it is not possible to predict the precise moment for an earthquake to occur; yet, seismologists are capable of indicating its potential location.

Earthquake prediction takes place at various levels of generalization and involves very different approaches. The broadest category of prediction dwells on the recognition that seismic activities concentrate in *four earth zones:* (1) the areas where plates slide past each other, (2) the collision regions between continental masses and oceanic plates, as illustrated by the Nazca Plate and the South American Plate, (3) the boundary zone between the two active continental plates as in the region between the Indian Subcontinent and Asia, and (4) the midoceanic ridges, with their associated *transform faults,* as typified by the Mid-Atlantic Ridge. Transform faults cut across midoceanic ridges and offset them horizontally.

The second type of prediction effort centers on a wide range of instrumentation and techniques mainly developed in the United States, Japan, the Soviet Union, and China. Most of the techniques concentrate on precursory phenomena that may, or may not, result in earthquakes.

The following is a brief summary of some of the accepted monitoring methods used:

1. Persistent changes in the elevation of given topographic survey points
2. Changes in the attitudes of rock surfaces as indicated by *tiltmeters*
3. *Variations* in the *magnetic field* of large rock formations that are undergoing severe stress
4. *Strain-* and *stressmeters* measuring compression of bedrock
5. *Variation* in the *velocity* of earthquake waves in swarms of tremors that frequently precede earthquakes
6. Relatively rapid changes in the *electric conductivity* in rock formations partially following the development of microfissures in rocks and indicated by a resistivity gauge
7. An increase in the emission of a radioactive gas, *radon,* from rocks as monitored in deep wells
8. The use of *creepmeters,* basically wire strands extending across a fault trace, to indicate stress and movement (Scholz et al.; 1973, pp. 803–809).

Long before the introduction of these sophisticated methods, both Chinese and Japanese observed and recorded the abnormal behavior of certain animals before impending earthquakes: rats and mice leaving their burrows, dogs barking frantically, skittish behavior of horses, and the restiveness of fowl. The Chinese especially, in addition to modern methods, have engaged large segments of their population to observe and report any unusual patterns of animal conduct. In China, four out of five destructive earthquakes that happened during 1975 and 1976 are said to have been predicted (Rikitake; 1982, p. 40).

One of the successful predictions was the one preceding the devastating earthquake of Haicheng on February 4, 1975. Had it not been for the massive evacuation of the population the number of casualties (90,000) would have been considerably greater. However, about a year later a disastrous earthquake completely destroyed the industrial city of Tangshan, reputedly killing more than 300,000 people.

Another factor that may affect earthquake prediction, as far as cyclic occurrence is concerned, is the possible *correlation* of the occurrence of earthquakes with *tidal pull* on the earth's crust. Evidence exists that *moonquakes* occur at times of maximum tidal pull (Menard; 1974, p. 199). Whether this also pertains to our planet is not fully clear because of the much more dynamic nature of the earth as illustrated by plate tectonics.

It may be interesting to correlate maximum tidal phases with earthquake activities. The absolute largest tidal pull on the earth's crust occurs only once every 1,600 years with the next peak phase projected by the year 3300. At such times the sun is at *perihelion* and in *conjunction,* or in *opposition,* with the moon and earth, with the moon being at *perigee,* and at zero declination (Thurman; 1985, p. 252).

Another consideration in potential earthquake prediction is a series of human-made factors that are capable of triggering earth tremors. They include the building of large reservoirs (Kariba Dam, Zimbabwe; Koyna Dam, India; Hoover Dam, U.S.A.) where the sheer weight of the water mass, in addition to the infusion of water into rock crevices and fault planes, may trigger earthshocks. Other potential triggering agents are major conventional and nuclear underground explosions as well as the release of large amounts of liquids into underground chambers of abandoned mines.

ⓢ The Managua Earthquake (Nicaragua, December 23, 1972): A Test Case for Interacting Factors

Nicaragua, in Central America, is located in one of the most active seismic zones plagued by earthquakes and volcanic eruptions. The intense seismicity, for which Central America is notorious, is related to the activities of three plates: the Cocos Plate moves in a general northeasterly direction toward the Middle American Trench where it pushes underneath the Caribbean Plate that, in turn, appears to move toward the West Indies Arc. The third unit, the North American Plate, **shears** in a general west-northwest direction along the northern boundary of the Carribean Plate (fig. 1.6).

Managua, the capital of Nicaragua, is situated in a structural lowland, or graben, which extends from the northwest, near the southern border with Honduras, to the southeast toward the Caribbean Sea. This lowland is the result of a number of faults that spread from the shear zone between the Caribbean and North American Plates (Ebert; 1981, p. 60). Several smaller cross-faults cut across the lowland and a number of these fractures are located within the city boundaries of Managua.

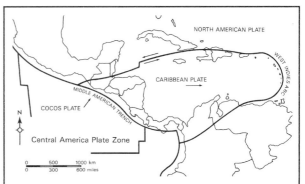

Figure 1.6. The intense seismic activity of Central America is the result of the interaction between several crustal plates including the North American Plate, the Caribbean Plate, and the Cocos Plate.

Between the years 1520 and 1972 Nicaragua experienced more than 450 earthquakes, of which 99 had a destructive magnitude of 6, or higher, on the Richter scale (Leeds; 1973, p. 27). Since the turn of this century Managua suffered about 10 earthquakes, of which 3 reached severe levels. These quakes occurred in March, 1931; January, 1968; and the most devastating one on December 23, 1972.

The 1972 earthquake, during which at least three major faults became active, destroyed the heart of the city of about 400,000. Shortly after midnight the quake struck and within about 12 seconds turned the center of Managua into rubble [Plate 1]. The casualties amounted to nearly 10,000 killed and more than 20,000 injured. One of the most prominent faults running through the city is the Tiscapa fault which moved up to 15 inches (38 cm) during the earthquake along a surface break about 3.7 miles (6 km) long. This fault system is clearly visible within a volcanic crater lake, the Laguna de Tiscapa. The rim of this crater lake shows both vertical and horizontal displacement [Plate 2]. The 1972 Managua earthquake was especially destructive because of at least five factors involved:

1. The focus, near or within the city proper, was very shallow having an estimated depth of less than 4 miles (6.4 km). This accounts for the fact that the three major seismic waves (P-waves, S-waves, and surface waves) arrived almost simultaneously, resulting in extreme motions.
2. At least three faults, less than 0.62 mile (1 km) apart, became active.
3. The largest portion of the city is built on or in thick volcanic ash layers so that high-intensity readings (VII to IX) occurred.

Plate 1. Utter devastation marked the center of Managua (Nicaragua) after the severe earthquake of December 23, 1972. Only a few modern high-rise buildings maintained their structural integrity. (Photo by C.H.V. Ebert).

4. Many of the houses were of the so-called taquezal construction (heavy-tiled roofs, wooden framework, adobe fill and plaster) leading to total disintegration [Plate 3].
5. The use of low-quality concrete that contained sandlike volcanic cinder material instead of high-integrity sand. This inferior concrete quickly failed and pulverized [Plate 4].

General conditions in the city became chaotic soon after the earthquake had struck. The government, to avoid the spread of diseases and to maintain control over the situation, initially refused to send food into the devastated center of the city. Water and assistance was offered only in the undamaged outer sections of Managua. Angry and desperate mobs rioted, set fires, and engaged in widespread looting. There is no question that the consequences of this earthquake severely undermined the Samoza government and contributed to its downfall.

⊚The Loma Prieta Earthquake

The Loma Prieta earthquake of October 17, 1989, with a magnitude of 7.1 on the Richter scale, had its epicenter about 10 miles (16 km) to the northeast

of Santa Cruz, California (fig. 1.7); it originated at a depth of about 12 miles (19 km). This disaster was the most severe seismic event since the San Fernando earthquake in 1971 and ranks right behind the 1906 San Francisco earthquake. Unless indicated otherwise, most of the information about the Loma Prieta earthquake was obtained from a special report on this event issued by the Department

Plate 2. The very active Tiscapa Fault (see arrow) cuts across a crater lake, the Laguna de Tiscapa, in the center of Managua (Nicaragua). Both vertical and horizontal slippage marks this fault. (Photo by C.H.V. Ebert).

Plate 3. The older houses of Managua (Nicaragua), with their wooden framework (taquezal construction) and heavy tiled roofs, have a very low resistance to earthquakes. In many cases they suffered total destruction. (Photo by C.H.V. Ebert).

Plate 4. The use of low-quality concrete and inadequately reinforced joints led to the complete destruction of this apartment building during the December 1972 earthquake in Managua, Nicaragua. (Photo by C.H.V. Ebert).

of Conservation, Division of Mines and Geology, in Sacramento, California.

The earthquake resulted from a movement along the San Andreas fault where such occurrences have taken place at irregular intervals over millions of years and shaped the geologic features in its vicinity. Geologists Mason L. Hill and T. W. Dibble proposed in 1953 the possibility that the total accumulated horizontal right-lateral movement of this fault amounted to 350 miles (560 km) since Jurassic time, or about 135 million years (Tank; 1983, p. 105).

Fortunately, the Loma Prieta earthquake was centered in an underpopulated area; had such a strong quake taken place near San Francisco, about 56 miles (90 km) to the north, the physical damage and human losses would have been much greater. The relatively low death toll was 63, with about 3,760 injuries reported.

The major shocks of this earthquake took place over a time period of about 10 to 15 seconds and were felt by millions of people; however, the intensity varied considerably in different regions.

In the epicentral area the intensity reached VIII on the Modified Mercalli scale, while an intensity of VII, or greater, was observed within an area of 1660 mi^2 (4,300 km^2). In some places the intensity of IX was indicated, leading to the collapse of the elevated section of Interstate 880 in Oakland [Plate 5]. The same intensity damaged Highway 480 and

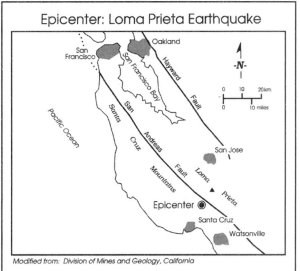

Modified from: Division of Mines and Geology, California

Figure 1.7. The epicenter of the 1989 Loma Prieta earthquake was located at the San Andreas Fault about 10 miles (16 km) to the northeast of Santa Cruz.

Plate 5. Aerial view of the collapsed Interstate 880 double-decked highway structure. Forty-one motorists were killed in this event. (Photograph credit: E. V. Leyendecker, U.S. Geological Survey).

raised havoc in the Marina district of San Francisco. Again, unstable sediments explain the damage caused by this earthquake [Plate 6].

Strong-motion ground accelerometers, operated by the California Division of Mines and Geology, recorded about 10 to 15 seconds of shaking stronger than 0.1 g, while the highest response site

Plate 6. A view along Jefferson Street in the San Francisco Marina District where tenants try to salvage belongings from the rubble. (Photograph credit: D. Perkins, U.S. Geological Survey).

was 0.64 g horizontal. Readings from accelerometers reflect the motion of the ground during shaking; the values are based on the well-known gravitational acceleration of 9.8 m/s^2. Accelerations in excess of 1.0 g would destroy most structures (Tank; 1983, p. 82).

The direct damage from the Loma Prieta earthquake was estimated at near $6 billion. About 234,000 homes were damaged and another 1,000 completely destroyed. Moreover, 366 business establishments were destroyed and 3,530 damaged. Damage to 1,544 public schools amounted to $81 million; 1,500 bridges were partially weakened and three collapsed. There was loss of electricity to 1.4 million customers, and considerable damage was done to the water supply system. Ground failure, especially in areas where soil liquified, was the major cause of the destruction of water mains.

A question that is on the minds of many Californians is, of course, whether additional earthquakes can be expected in the near future, or whether the massive release of energy during the Loma Prieta earthquake will give the region a respite for some time to come. Geologists are not in accord about this basic question. Some believe—and they seem to be in the majority—that earthquakes along major faults occur at longer intervals, which are needed to build up the seismic energy of the system. Others, including David D. Jackson and Yan Y. Kagan, challenge the so-called seismic gap hypothesis. Whether another magnitude 7 earthquake will strike the Loma Prieta area soon, or only in the far future, cannot be predicted with accuracy. However, the underlying danger—the San Andreas fault system—remains a constant threat; only the future will tell the full story.

The government agencies of the State of California, as well as local government bodies, are maintaining a realistic stance in view of the ever-present threat of earthquakes. The first more recent steps to launch a preparedness program in California were undertaken in the early 1980s by creating the Southern California Earthquake Preparedness Project (SCEPP) and the establishment of a Governor's Task Force on preparedness that brought together private groups, finance, and the media to assess aspects of risk and strategies. This action was followed by developing the Bay Area Regional Earthquake Preparedness Project (BAREPP). Many other civic efforts, educational programs, and new laws have further strengthened California's drive toward preparedness for future seismic events, including "The Big One" that one day—as many believe—will take place.

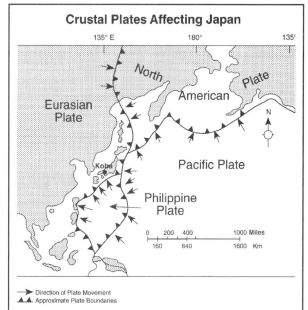

Crustal Plates Affecting Japan

Figure 1.8. The intense seismic activities and volcanism on and around the Japanese islands result from the complex interaction of four crustal plates: (1) the North American Plate, (2) the Pacific Plate, (3) the Eurasian Plate, and (4) the Philippine Plate. (Based on: W. Hamilton; U.S.G.S.)

◎ The Kobe Earthquake of January 17, 1995

On an average day, seismographs in Japan record more than twenty significant earth tremors. This activity is not surprising when one considers the geologic setting of the Japanese islands; they were born in a region known for very active plate tectonics involving four major plate units: (1) the North American Plate, (2) the Pacific Plate, (3) the Eurasian Plate, and (4) the Philippine Plate (fig. 1.8).

The area of southeastern Honshu, the largest island of Japan, borders on Osaka Bay and is referred to as the Yodogawa Basin. This basin contains two of the largest urban centers of Japan, Kobe and Osaka, and has a total regional population of more than 10 million. Kobe, the great port city, has a population of more than 1.5 million; it suffered the worst earthquake in Japan since the devastating event that destroyed Tokyo in 1923. The author visited Japan in May 1995 and obtained much of the subsequent

information from the Technical Research Institute of Pacific Consultants, Ltd., in Tokyo.

The violent earthquake which struck the Kobe area at 5:46 a.m. on January 17, 1995, has been given different names: the Hyogo-ken Nanbu, the Great Hanshin Earthquake, the Kansai Earthquake, or simply the Kobe Earthquake because of the major impact it had on that city (Seible; Priestly, 1995). This destructive earthquake was estimated at about 6.9 to 7.0 on the Richter scale. The quake and hundreds of aftershocks were triggered by a slippage of the Nojima Fault with its main focus a little more than 10 miles (16 km) underneath the northern end of Awaji Island on the southwestern margin of Osaka Bay (fig. 1.9). This fault system lies within the Eurasian Plate and to the west of the subduction zone of the Philippine Plate where it descends underneath the Eurasian Plate.

The Japanese Meteorological Agency rated the earthquake's intensity at VII on the Mercalli scale (NCEER; Special Report, 1995, p. 1), but the variable ground conditions resulted in markedly different levels of damage. Most severely damaged were

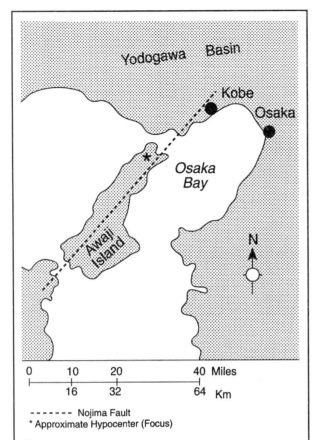

Figure 1.9.
Slippage of the Nojima Fault caused the devastating Kobe Earthquake of January 17, 1995. The main focus was located underneath the northern end of Awaji Island. (Based on: Earthquake Research Institute, University of Tokyo).

- - - - - - Nojima Fault
* Approximate Hypocenter (Focus)

Plate 7. The older sections of central Kobe suffered the worst destruction and accounted for the greatest number of casualties. (Photograph credit: Pacific Consultants, Ltd., Tokyo).

victims were trapped under collapsed low-rise buildings.

In summary, it can be said that the Kobe earthquake was so severe because of four major factors: (1) the focus—as in the case of the devastating earthquake which destroyed Managua, Nicaragua, in December 1972—was at a relatively shallow depth, (2) the exceedingly high building density which is so typical of Japanese cities, (3) the widespread ground failure induced by soft sediments and artificial fill, and (4) the fact that many older structures did not conform to the more recently established building code [Plate 8].

areas associated with unconsolidated sediments where strong ground motion and liquefaction prevailed. Here, as well as on the artificially created islands in Kobe Harbor, vertical ground displacement measured 8 feet (2.4m), and the ground motion lasted as long as 100 seconds. In the sections underlain by solid rock formations, the shaking lasted only about 20 seconds and the ground displacement was much less. The duration of the main shock was about the same as in the Loma Prieta event of 1989 and lasted somewhat longer than in the Northridge Quake of 1994 (Hays; 1995, p. 6).

The greatest damage, and also the largest number of the total casualties—estimated at 5,500—occurred in the older sections of Kobe [Plate 7]. Within minutes, intense fires broke out in these densely populated parts of the city, and many

⊚ The Izmit Earthquake of August 17, 1999

The Izmit earthquake occurred early in the morning on Tuesday, August 17, 1999. It had a magnitude of 7.4 and caused widespread devastation in many communities in northwestern Turkey. This region has suffered many major quakes which were associated with the North Anatolian Fault system.

The quake under discussion took place in a portion of the North Anatolian Fault that had not moved in seismic events between 1963 and 1967 including the severe earthquake of 1967. This temporarily dormant section of the fault was first identified in a study presented in 1997. It was then speculated that this fault sector has a 12 percent probability to

Plate 8. Even a considerable number of newer buildings in suburban Kobe showed severe damage because their construction did not conform to the more recent building code. (Photograph credit: Pacific Consultants, Ltd., Tokyo).

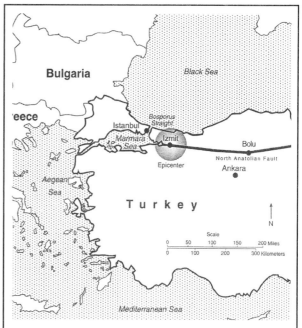

Figure 1.10.
The epicenter of the Izmit Earthquake is located under the North Anatolian Fault about 55 miles (88 km) south-southeast of Istanbul.

(Bruneau; 1999, p.3) who pointed out that thousands died in collapsed buildings. These buildings were constructed of reinforced concrete frames with unreinforced masonry infills [Plates 9 and 10].

Plate 9. Poorly constructed buildings, composed of reinforced concrete frames with unreinforced masonry infills, were typical in many sections of Izmit and usually suffered total destruction. (Photograph credit: Michel Bruneau; Multidisciplinary Center for Earthquake Engineering Research, SUNY/Buffalo, New York).

experience an earthquake during the period from 1996 to 2026 (Papageorgiou; 1999, p. 1).

The epicenter was about 6.8 miles (11km) southeast of the industrial city of Izmit below the fault line which runs from the Gulf of Izmit eastward past Bolu (fig 1.10). The initial shocks lasted for about 45 seconds and were felt over thousands of square miles of the most densely populated parts of Turkey.

Several factors contributed to the massive damage inflicted by this earthquake: (1) at least two subevents took place within the fault system and they were spaced only 19.6 miles (30 km) apart (Papageorgiou, 1999, p. 2), (2) the high population density of the region, (3) the early morning hours which found most people still in their homes, and (4) the poorly constructed buildings. The latter factor was studied by Professor Michel Bruneau

Plate 10.   Many houses in Izmit had heavy concrete floors. Such buildings are very top-heavy and lack the structural strength to withstand seismic shocks. (Photograph credit: Michel Bruneau; Multidisciplinary Center for Earthquake Engineering Research, SUNY/Buffalo, New York).

This situation, as well as the apparent inadequacy of government response to the disaster, led to severe criticism and political unrest. The press was highly critical of the lack of search and rescue operations and the tardy reaction of the military in rendering assistance (Mitchell; 1999, p.5).

VOLCANOES

Key Terms		
volcanic vent	lapilli	volcanic bombs
magma	nuée ardente	fissure eruption
Mid-Atlantic Ridge	glowing cloud	island arc
hot spot	crater eruption	fumaroles
cone volcano	flank eruptions	ash fall
shield volcano	pahoehoe	pyroclastic flow
stratovolcanoes	mudflows	greenhouse effect
pyroclastic	avalanche	global cooling
tephra		

Volcanoes represent only one of the many phenomena that are listed under volcanic activities or are covered by the science of *volcanology*. Moreover, most peoples' perception of volcanoes calls forth beautifully symmetrical cone volcanoes, as typified by Mount Fuji in Japan. Yet, there are great variations in the shapes, heights, structure, and behavior of volcanoes.

In addition to stereotyping the image of volcanoes many people think of volcanoes only in connection with catastrophic paroxysms: glowing rivers of lava burying villages, volcanic ash darkening the skies, and fiery fragments raining down to spread fire and destruction.

The main reason for viewing volcanoes as violent destroyers of people, towns, and nature probably is the fact that some volcanic eruptions are, indeed, spectacular. It is on these that public attention tends to focus. Second, in several instances volcanoes killed thousands of humans, and such killer volcanoes stand out in the memory of humankind: Mt. Pelée (1902), on the Caribbean island of Martinique, within a few minutes wiped the town of St. Pierre off the map and killed more than 30,000 people; volcano Tambora (1815), in the East Indies caused 56,000 casualties; Thera (1450 B.C.), on the island of Santorini, is said to have brought about the fall of the Minoan culture on Crete and buried the settlements on the Aegean island of Santorini; and, more recently in 1985 the giant volcano Nevado del Ruis obliterated the town of Armero in Colombia killing more than 20,000.

Are, then, volcanoes major killers? In some cases they certainly are; yet, over the past 500 years it is believed that volcanoes caused only about 200,000 fatalities, averaging 400 victims a year. By comparison, in this country we kill about 100 persons a day in car accidents alone! Thus, although volcanic activities may cause extensive property damage and destruction over wide areas, the number of human casualties, in a relative sense, is quite low. A partial explanation is that in many cases volcanoes give warning signs of imminent eruptions in time for the population to leave. Another factor is that many volcanoes are located in rural or in uninhabited zones. Also, if the eruption involves mainly lava flows it is possible to evacuate the threatened areas.

Formation of Volcanoes

The most basic requisite for volcanoes to form is the presence of a molten rock reservoir, or *magma chamber,* which is under sufficiently high pressure to force various forms of volcanic materials to the surface. The passage of this ejecta can take place through zones of structural weakness, or fissures, or through a more central conduit generally referred to as a ***volcanic vent.***

The source of such molten rock masses, which also contain large amounts of gases, is usually within 60 miles (100 km) from the earth's surface. Underneath the ocean, in the vicinity of midoceanic ridges, the ***magma*** may be at a depth of 15 miles (25 km) beneath the relatively thin basaltic crust of the ocean floor (fig. 2.1).

The composition of magmatic material varies considerably, and this difference is a major factor in what shape volcanoes attain and what type of activity they reveal. Other elements, such as gas content and pressure, are also critical in determining whether a lava eruption is of the quiet outflow type or whether it will take place in a sudden violent explosion.

Although *volcanic gases* vary somewhat in composition, both in absolute terms as well as during the various stages of the eruption, the most abundant gas is water vapor. Part of this water is actually a component of the molten rock and is only released, in the form of steam, when the pressure is vented. It is now generally accepted that the bulk of planet earth's surface and atmospheric water is of volcanic origin. Water makes up the bulk of volcanic gases, usually about 70 percent. Other constituents include carbon dioxide, carbon monoxide, hydrogen chloride, sulfur trioxide and sulfide, hydrogen fluoride as well as small amounts of methane, ammonia, hydrogen thiocynate, and some rare gases (Rittman; 1962, p. 63).

The exact nature of the origin of *earth heat,* the energy source behind plate tectonics and volcanism, is still being debated. The more acceptable explanations include:

1. Large amounts of *residual heat* that go back to the molten state of our planet
2. The *decomposition* of *radioactive* materials in magma generating zones
3. *Tidal friction,* caused by the decline of the earth's rotational speed, may generate some heat energy
4. *Pressure* and *friction* in the zones of plate subduction are believed to result in the melting of rock materials and in changes in the mineralogy (Bolt; 1980, p. 100).

Geographical Distribution of Volcanoes

We do not know precisely how many volcanoes exist on our planet, but the total number must be tens of thousands. It is estimated that there are at least 10,000 volcanoes in the Pacific, and this ocean constitutes about half of the total world ocean. Based on this estimate there could be more than 20,000 volcanoes on the earth's ocean bottom (Gross; 1982, p. 48). Of course, not all these volcanoes are active; they may be either *dormant* or *extinct.* At this time the world's active volcanoes number over 500. Nevertheless, though we do not know the total number of volcanoes, the basic pattern of their distribution is clear.

First, it can be stated that the distribution pattern of volcanoes correlates well with regions of plate subduction, with midoceanic ridges, and with continental rift valleys and fracture zones. Thus it is not surprising that a major region of volcanism is the *Circum-Pacific Ring,* as well as that of the ***Mid-Atlantic Ridge,*** the *East African Rift Valley,* and the *Atlas-Alpine-Caucasus* mountain-building zone extending from North Africa through the Mediter-

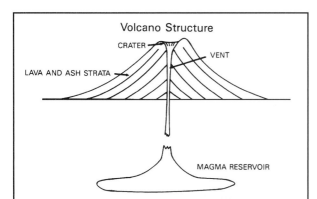

Figure 2.1. Volcanoes are composed of lava and/or pyroclastic materials. The lava derives from deep-seated magma reservoirs and rises through the vent to the center. The shape of a volcano is determined primarily by the type of the ejecta.

ranean realm and eastward. In addition a considerable number of volcanoes occur along extension and transform faults already discussed in connection with earthquakes.

The dominant role of the Circum-Pacific Ring is illustrated by the fact that out of an approximate 511 above-sea level volcanic eruptions recorded in historical times, 422, or 82 percent occurred in this zone (Krafft; 1975, p. 163). One glance at the map of the Pacific reveals that several island chains exist with a general alignment from northwest to southeast. One of these is between Midway and the Hawaiian islands. Hawaii is the youngest island while the age of the remainder increases away from Hawaii. It is only Hawaii itself that still has active volcanoes (Kilauea and Mauna Loa). The only other active volcano is the submarine volcano of Loihi immediately to the southeast of Hawaii.

One of the possible explanations is that the Pacific Plate slowly moves across a so-called **hot spot** located in the upper mantle that facilitates the upward motion of magma. Eventually, as Hawaii moves away from this location, the presently active volcanoes, too, will become extinct.

Another interesting fact involves the *latitudinal* distribution of volcanoes. Two-thirds of the world's volcanoes are located in the northern hemisphere and only 18 percent are found between latitude 10° S and the South Pole (Simkin et al.; 1981, p. 7).

Volcano Types and Activities

Volcanoes can be classified in various ways: by their shapes, their structures, their type of ejecta, and their eruptive activities.

Basically, there are two major shapes: the ***cone volcano*** and the ***shield volcano.*** Well-known cone volcanoes include Mt. Shasta, Mt. Rainier, Mayon (Philippine islands), Momotombo (Nicaragua), and the towering Cotopaxi in Ecuador with an elevation of 19,600 feet (5975 m) making this the highest active volcano in the world [Plate 11]. On the other hand, the best-known shield volcanoes are associated with the Hawaiian islands.

What type of volcano develops depends for the most part on the type of ejecta. For example, *basaltic lava* of low *viscosity,* remaining fluid at relatively low temperatures, usually engages in relatively tranquil outpourings and tends to build smooth-contoured shield volcanoes. On the other hand, lava that is high in silica (SiO_2) content, called *acidic lava,* tends to solidify quickly and shows greater structural strength. Cone volcanoes are mostly **stratovolcanoes** where lava flows alternate with cinder-type ejecta.

Plate 11. One of the active volcanoes of central Nicaragua, Momotombo, is a fine example of a stratovolcano; it is characterized by its conical, symmetrical shape. (Photo by C.H.V. Ebert).

This layering results in the building of high volcanic mountains of considerable structural strength. Many of the world's highest volcanoes are stratovolcanoes.

Neither volcanoes, nor lava, can be classified into two simple categories. There are many intermediate types in both with corresponding transitional characteristics. For example, Mt. Etna on Sicily is primarily built up by successive lava flows, but its summit is a **pyroclastic** structure composed of angular lava blocks.

There are also small volcanoes that are entirely made up of fine-textured **tephra** and small lava fragments called **lapilli.** These volcanoes may spring up along radial fissures near large volcanoes. They seldom grow higher than about 1,000 feet (300 m) and are referred to as *cinder cones.* Generally, they pose no threat.

Volcanoes are also classified by their activity patterns or phases. Some volcanoes engage in constant activity releasing gases and various types of ejecta. Stromboli, a member of the Lipari Islands in the eastern Tyrrhenian Sea, regularly erupts about four times an hour. Other volcanoes may remain inactive for long periods of time, building up enormous pressure between eruptions. They may erupt violently without much warning. The cataclysmic eruption of Mount St. Helens in 1980 is a good example of this characteristic.

Possibly the most awesome and disastrous behavior of a volcano is the so-called *Peléan phase,* a name that became accepted after the May 1902 explosive eruption of Mt. Pelée on the Caribbean island of Martinique. Here, a superheated gas and incandescent ash cloud, the **nuée ardente** (Fr. **glowing cloud**), burst out of a side opening of Mt. Pelée and totally destroyed the town of St. Pierre in a way similar to a nuclear blast.

Volcanic Hazards

Throughout human history volcanoes have been viewed with an awe instilled by their spectacular eruptions. For a visitor such stupendous events can be a once-in-a-lifetime experience, as documented by many scientific films and color slides [Plate 12]. Yet, for the population living in the immediate areas of volcanic hazards, the specter of pending destruction is a daily companion.

There are many examples, especially in tropical lands, that illustrate this situation:

1. *Mt. Vesuvius* buried Pompeii and Herculaneum in A.D. 79 and had erupted 83 times by 1944. Yet,

Plate 12. El Fuego is one of the most active volcanoes in Guatemala and frequently ejects both lava and ash materials (Photograph credit: G. Feucht).

villages and towns are thriving on and near its slopes, and the great city of Naples is located a few miles from the volcano

2. *Taal,* on the Philippine island of Luzon had erupted at least 33 times by 1977, and its companion, Mayon, 46 times by 1978. Again, agricultural communities dot the landscape around these volcanoes

3. *Irazu,* a massive volcano in Costa Rica, erupted 19 times by 1974. During the 1963 to 1965 eruption this volcano deposited an estimated 80 million tons of ash on the agricultural heartland of that country.

Similar situations could be cited for numerous other locations plagued by active volcanoes. Accordingly, the question arises why people are willing to live with such dangers.

This question is not easy to answer because of many variables involving culture, economics, and the individual willingness to take risks. In most cases it is just a matter of finding fertile land. This is especially the case in wet tropical lands where soils are easily leached of nutrients, and some of the volcanic soils are among the most productive and permit multiple cropping.

In other circumstances it is the attitude of the people. They tend to worry more about the loss of property than fear the loss of life. In a study of a hazard area around Puna, Hawaii, 41 percent of the respondents to an inquiry stated that they do not worry when they learn that an eruption of Kilauea is imminent. Ninety percent believed that there are greater advantages in living in that area than disadvantages (Murton; Shimabukuro, 1974, p. 156). This view is held despite the history that Kilauea has erupted more than 70 times since 1750!

Many kinds of volcanic hazards exist, ranging from the initial violent impact of an explosive eruption to ash and mudslides and all-engulfing lava flows. Moreover, the level of danger of a volcanic eruption is also determined from what point it issues. In the case of a **crater eruption** most energy is directed upward. While ash and tephra falls may be affected by wind direction, the heavier pyroclastic materials often fall back into the crater. **Flank eruptions,** directing the blast sideways, may occur if the vent of the main crater is plugged by solidified lava. Each eruption type has its own dangers and involves different ejecta. The main hazards are:

1. *Lava flows.* Depending on the lava type such flows may vary considerably. They may consist of the smooth-flowing **pahoehoe** lava of low viscosity. This lava may cover great distances and develops a smooth crust under which the lava continues to flow. The so-called *Aa* flow is made up of jagged, broken blocks that tumble forward along a slowly advancing lava front. This type of flow can be very destructive to structures because of its massive impact

2. *Volcanic* **mudflows.** These **avalanche**-type mass movements usually consist of a mixture of new, and old, wet volcanic ash. The high water content can be the result of heavy precipitation, of rapidly melting snow or glacial ice near the crater, or even the bursting of a crater lake. Mudflows, therefore, may occur independently from volcanic eruptions but they frequently follow them. This was the case when a massive mudflow descended the flank of volcano Irazu, in Costa Rica, and destroyed the town of Cartago. Heavy ash and rock debris had accumulated during the 1963 eruptions and torrential rainfall had saturated the slopes [see Plate 20 in chapter 3]

3. *Glowing avalanches* (*nuée sardentes*) (fig. 2.2). This phenomenon gained notoriety after the destruction of St. Pierre in 1902. The glowing cloud represents a density flow composed of gases, ash, and glowing rock fragments. As it rolls down the flank of a volcano, buoyed up by the superheated gases within, and by the air trapped below, it may attain velocities in excess of 60 mph (100 km/h) (Costa; Baker, 1981, pp. 102–103)

4. **Volcanic bombs** and *tephra*. The ejection of *volcanic bombs,* solidified hot lava fragment thrown high up into the air, and the massive ejection of

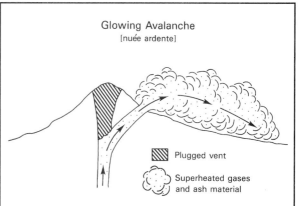

Figure 2.2. Glowing avalances (nuée ardente) may erupt from the flank of a volcano if solidified lava forms a plug that blocks the vent that leads to the crater.

tephra present another hazard capable of doing great damage. The bombs may smash through roofs and set buildings on fire. The tephra may bury entire towns (Pompeii) causing roof tops to collapse and choking agricultural lands. The case study of Heimaey, following this section, will illustrate these hazards

5. *Toxic volcanic gases.* Volcanic gases may vary somewhat in composition, but they always contain toxic constituents, which may prove deadly if encountered in sufficient concentrations. Some of the more dangerous gases are hydrogen sulfide, carbon dioxide, carbon monoxide, and sulfur dioxide. By virtue of being heavier than air these gases tend to move along topographic depressions and collect in low-lying areas. In some old volcanic craters, such as Kawah Upas, on the island of Java, carbon dioxide fills the bottom and has killed animals and people (Krafft; 1975, p. 133). On August 21, 1986, a disastrous outburst of toxic gases from an old crater lake (Lake Nyos), in Cameroon, asphyxiated nearly 2,000 people and killed thousands of livestock.

Prediction of Volcanic Eruptions

It may be recalled that many volcanoes give early warning signs preceding their eruptions. Swarms of tremors, sometimes heavier shocks, steam, and ash eruptions are frequent *precursors* of more serious activities. There are also a variety of scientific methods employed to monitor suspect volcanoes. Some of the instruments are similar to those used for predicting earthquakes.

Laser beams are trained on reflector targets placed in critical spots on the volcanoes. Changes in shape can be recorded by this method. The huge bulge, developing on the flank of Mount St. Helens preceding its explosive eruption, is a good illustration of this type of precursory behavior.

Sensitive portable seismometers, as well as special microphones, may reveal abnormal activities inside volcanoes. A flurry of tremors preceded Mount St. Helen's eruption and was recorded by seismometers. Some volcanic eruptions were heralded by actual earthquakes.

Other precursors of volcanic eruptions include changes in the temperature and composition of volcanic gases. In almost all such cases an increase in water vapor content was noted. Furthermore, changing *magnetic properties* in rocks, heated by rising hot gases or magma, may assist in monitoring dangerous volcanoes. However, as it is true for earthquakes, no single method is adequate to predict volcanic eruptions (Macdonald; 1972, p. 418).

◎ The Heimaey Eruption (Iceland; January 23, 1973)

The island of Heimaey is the largest of the Vestmann Islands, a group of volcanic islands offshore south of Iceland. These islands were formed in association with the volcanic activities of the Mid-Atlantic Rift.

Heimaey, just as the newly formed neighboring island of Surtsey, was created by a series of lavaflows and was covered by a veneer of tephra from the now dormant cone of ancient Helgafell. It is estimated that Helgafell has not erupted for the past 5,000 years. Thus the **fissure eruption,** which started just east of Helgafell early in the morning of January 23, 1973, came as a total surprise to the 6,000 inhabitants of Vestmannaeyar, the main settlement on Heimaey (fig. 2.3).

This fissure eruption, so typical for Iceland, pushed through a fracture about one mile (1.6 km) long just outside the town. Billowing clouds of dusty tephra, steam, and eventually a spattering of glowing lava fragments rose like a fiery curtain against the dark winter sky. Fortunately for the people the wind blew away from Vestmannaeyer and carried most of the volcanic ejecta out to the ocean.

Over the following day the eruption continued to intensify and seemed to concentrate on a site about half a mile (0.8 km) to the northeast of Hel-

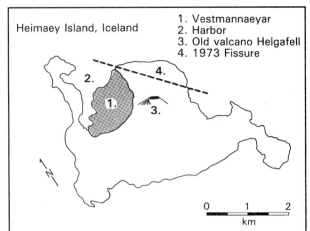

Figure 2.3. The 1973 fissure eruption on Heimaey (Iceland) occurred just to the northeast of the fishing town of Vestmannaeyar. Subsequent lava flows enlarged the island by about one square mile (2.5 km²).

gafell. Then the winds shifted and masses of glowing tephra descended on the town so that the population had to be evacuated to the mainland. The efficient and orderly evacuation was executed in six hours.

Within a few days a new volcanic cone began to grow over the fissure and lava poured out in ever-increasing quantity [Plate 13]. Many houses caught on fire or collapsed under the growing weight of the black tephra, but the greatest danger to the community was the steady flow of lava that crept toward the harbor entrance [Plate 14]. Unchecked, this flow could have cut off the harbor from the sea. This

Plate 13. On January 23, 1973, a new volcano erupted on the small Icelandic island of Heimaey, one of the many islands that is associated with volcanic activity along the Mid-Atlantic Rift. (Photo by C.H.V. Ebert).

Plate 14. A massive lava flow from the newly formed volcano on Heimaey threatened to close off the only entrance to the harbor of the fishing town of Vestmannaeyar. Huge quantities of seawater were sprayed on the advancing lava front and succeeded in diverting the flow. (Photo by C.H.V. Ebert).

would have spelled total economic doom for the island whose people depended almost entirely on the fishing industry. Heimaey contributes nearly 20 percent of Iceland's fish export. This dangerous lava flow, according to the Icelandic geologist Professor Sigidur Thorarinson, reached an output of about 3,500 cubic feet (100 m^3) per second.

A most unusual and very successful effort was then started to halt and to divert the lava flow. Bull-dozers threw up massive barriers of lava blocks, and huge quantities of seawater were sprayed on the slowly advancing lava front. The idea was to cool and solidify the outer crust of the lava so that it would act as a containment wall. In about two weeks, using a battery of 47 large U.S. Navy diesel pumps, the temperature of the lava was reduced from about 1830° F (1000° C) to 212° F (100° C).

By mid-June the eruptions had died down, although sporadic emissions of ash and lava bombs still occurred. It was estimated that a total of 191 million cubic yards (250 million m^3) of volcanic material was ejected. The lava flow enlarged the island by nearly one square mile (2.5 km^2). More than 300 houses were totally destroyed by fire, were buried by the lava, or collasped under the load of tephra. But it is remarkable that not a single casualty occured. This achievement must be ascribed to the timely evacuation and the remarkable discipline of the population.

With typical Icelandic determination a massive clean-up operation began. Nearly two-thirds of the people eventually returned to their island to salvage and to rebuild. Airplanes were used to drop grass seed and fertilizer onto the fresh tephra to stabilize it and to reclaim new pastures for the future [Plate 15].

Plate 15. Airplanes were used on Heimaey for dropping grass seed and fertilizer to stabilize the fresh tephra and to reclaim pasture land that had been buried by the volcanic ejecta. (Photo by C.H.V. Ebert).

How a government responds to a disaster can make a difference.

⑨ The 1991 Eruption of Mount Pinatubo in the Philippines ———

Mount Pinatubo is an **island arc** volcano and is a member of a chain of volcanoes that is associated with the subduction zone of the Philippine Plate. The volcano is located on the Philippine Island of Luzon about 62 miles (100 km) northwest of Manila (fig. 2.4).

Mount Pinatubo lay dormant for over 635 years; however, the violent eruption on June 15, 1991, made up for the long period of peace. The majestic mountain, with an elevation of 5,725 feet (1,745 m), was the highest of three volcanic centers, but the disastrous sequence of eruptions in 1991 lowered the volcano by about 200 feet (61 m).

On April 2, 1991, Mount Pinatubo awoke after months of ominous rumblings, and a line of new **fumaroles** developed (NOAA/739-A11-007; 1993, p. 1). Four days later a lava dome formed on the

Plate 16. Increasing eruptive activities of Mount Pinatubo on June 12, 1995, resulted in the evacuation of thousands of people within an area of 7.5 miles (12 km) from the volcano. (Photograph credit: U.S.G.S.; R. S. Culbreth, U.S.A.F.).

mountain and exuded glowing lava flows. These precursors of a pending eruption made it advisable to evacuate Clark Air Base which was located about 7.5 miles (12 km) from the volcano. On June 12, a towering cloud of ash rose up from the crater and reached a height of 12.5 miles (20 km). In view of

Location of Mt. Pinatubo

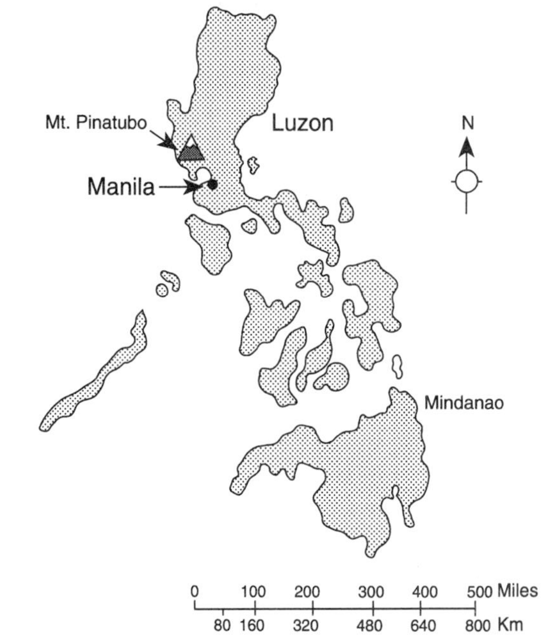

Mt. Pinatubo

Luzon

Manila

N

Mindanao

| 0 | 100 | 200 | 300 | 400 | 500 Miles |
| 80 | 160 | 320 | 480 | 640 | 800 Km |

Figure 2.4. The active volcano Mount Pinatubo is located on the Philippine island of Luzon about 62 miles (100 km) northwest of Manila. Volcanism in the Philippines is associated with the subduction of the Philippine Plate under the Eurasian Plate. (see Figure 1.8).

Plate 17. On June 15, 1995, the most powerful explosive eruption of Mount Pinatubo sent a plume of ash and steam to a height of 25 miles (40 km) into the atmosphere. (Photograph credit: U.S.G.S.; Karin Jackson, U.S.A.F.).

the increasing eruptive activities, thousands of people were evacuated within a distance of 7.5 miles (12 km) from the volcano [Plate 16].

Three days later, on June 15, Mount Pinatubo exploded in a cataclysmic eruption which was considered the most powerful of this century (Rondal; Yoshida; 1994, p. 1). The blast caused the death of over 700 people and sent a plume of ash and steam to a height of 25 miles (40 km) into the atmosphere [Plate 17]. The massive **ash fall** extended 93 miles (150 km) from the volcano, and the total area covered by at least 0.4 inches (1 cm) of ash was estimated at 1.48 million acres (600,000 hectares). Closer to Mount Pinatubo the ash layers ranged from 3 to 6 feet (1 to 2 m) in thickness.

According to the Philippine Bureau of Soils and Water Management more than 49,400 acres (20,000) of prime agricultural land have been rendered totally unproductive because of ash fall and subsequent mudflows. The cost to Philippine agriculture exceeded U.S. $32 million. Part of the damage resulted from the silting up of wells and canals which made the production of irrigated crops impossible or severely reduced the yields.

After June 16, the volcanic activities declined both in frequency and magnitude and averaged about three eruptions a day in July. Only three minor eruptions were recorded in September; however, a new threat arose: a severe hurricane struck Luzon on July 18. Heavy rainfall triggered extensive mudflows along river valleys to the east of the volcano. Some of these flows reached a height of over 16 feet (5 m), and by July 26, 100,000 houses had been destroyed. One **pyroclastic flow** dammed up a river and created a temporary lake. Later, the barrier ruptured and a massive flash flood destroyed 800 houses and killed numerous people.

In addition to the impact on land, towns, and people, Mount Pinatubo also affected the atmosphere. The force of the eruption carried large quantities of gases and dusty ash to heights in excess of 18 miles (30km) into the upper atmosphere. The role of aerosols in the so-called **greenhouse effect** will be discussed later (see Chapter 14); let it suffice to mention here that the ejecta of Mount Pinatubo reflected a considerable amount of sunshine before it could reach the Earth's surface. This reflection led to a 20 to 30 percent decline in solar energy that normally strikes the ground (Hoppe; 1992, p. 37). According to the National Geophysical Data Center in Boulder, Colorado, the maximum **global cooling** was 2.7 F (1.5 C); however, the cooling declined during the following three years.

To what extent the eruption of Mount Pinatubo had an effect on the chemical composition of the upper atmosphere, and especially on ozone levels, is more controversial. Initial readings that were obtained by NASA satellites indicated that a predicted decrease in upper-atmosphere ozone was not as severe as expected. Later data, which was collected by balloon-borne instruments, showed an all-time low of Antarctic ozone concentration (Monastersky; 1992, p. 278).

The Mount Pinatubo disaster has faded from the headlines of the world, but the people of Luzon still have to face hardships and the threat of more eruptions. Each heavy rainfall tends to loosen the thick layers of tephra and trigger new mudflows and more flooding in the ash-choked river valleys. Are there any positive aspects associated with this grim scenario? Eventually the volcanic ash will enrich the soils, but the massive destruction and suffering will dominate the situation for years to come.

Landslides and Avalanches

Key Terms		
mass wasting	colluvium	hoar frost
shear stress	creep	powder avalanche
hydrology	rockslide	turbulence
urban sprawl	joints	ground avalanche
shear strength	karst	slab avalanche
rockfall	sublimation	cornice
bedding planes	vapor pressure	orography

As stated in the introduction, natural systems will try to reach a *steady state,* or *equilibrium.* The action of gravity on all parts of the earth system could be called an ever-present equalizing force that assists in attaining such equilibrium.

To the casual observer almost any landscape appears stable; no motion is perceived. In rugged mountain environments an occasional rock- or ice-fall can be heard, but a hilly landscape seems peaceful and inactive. This general impression may lead to the rash conclusion that earth materials, unless disturbed by natural processes or by actions of man, are motionless. However, even the most gentle hillside is subject to external and internal forces that, over long time periods, will change the slopes. Some of these changes, whether they occur in rock, soil, or snow, are imperceptibly slow. But, if we were to view a time-lapse film, covering hundreds or even thousands of years, many of these changes in slope angles, elevations, or in general shape would be dramatic.

Downslope movements of earth materials may take place in a sudden catastrophic way, or they may occur in a slow internal creeping motion as revealed by leaning telephone poles, cracking walls, trees pointing in different directions, or by soil bulges at the bottom of a hill. Such movements, also referred to as *mass wasting,* are natural earth processes. When disasters occur in connection with landslides it is usually because of the interference by humans with such natural processes. Many actions may delay or accelerate landslides and avalanches; roads, towns, factories, or power lines may be placed in the path of an inevitable event.

In addition to rendering earth materials unstable in response to natural factors, there are many situations where *shear stress* in such materials is increased by human activities. Such actions include the removal of vegetation cover, excavations, dumping of mine waste, and also by changing the *hydrology* of an area. The latter can be effected by watering lawns and golf courses as well as by irrigation of agricultural land.

One of the best-known disasters which involved the dumping of coal mine waste struck the small town of Aberfan in South Wales. Heavy rainfall had saturated the waste

pile which stood almost 200 feet (61m) above Aberfan. On October 21, 1966, the steep slope gave way, and a mass flow buried parts of the community killing 144 people. The casualties included over 100 children in a schoolhouse which was buried by the slide.

One way in which human activities can increase landslide hazard potentials is demonstrated in Japan. There, a number of physical factors—heavy precipitation, frequent seismic disturbances, volcanic eruptions, and steep slopes—conspire in creating a fragile natural environment, and the situation worsened because of human action.

In response to a rapidly growing economy, land prices in cities accelerated **urban sprawl.** More and more land in landslide-prone regions was developed for urban or industrial use. By moving into such high-risk areas, the population became more vulnerable to disasters. In recognition of this danger, the Japanese authorities promoted intensive studies of all landslide phenomena and started a systematic preventive program (Japan Society of Landslide; 1988).

Basic Mechanics of Slides

Many variable factors are involved in producing the physical state at which a particular type of earth material, or even snow, becomes unstable and fails. The following explanations basically underlie all types of *material failure;* however, *trigger mechanisms,* the materials themselves, and the specific sequence of events vary greatly from case to case.

In the discussion of fault systems it was pointed out that rocks may build up stress over long time periods. Rocks, depending on their makeup and on the rate with which the stress force is applied, will either slowly deform or may fracture suddenly. In a similar way many other forces are at work and affect all earth materials as illustrated by the following simple example.

Place a rock on a level concrete slab and then slowly raise one end of the slab. Eventually, after the slab attains a sufficiently steep angle, the rock suddenly begins to slide. At that moment the friction between the rock and the concrete slab was overcome by the gravitational pull (weight) on the rock. Had a smooth object of the same weight replaced the rock, and a sheet of plate glass replaced the concrete slab, the downslope action would have started at a much shallower angle.

Slope, no doubt, is a very significant control factor in rendering a given mass unstable. It cannot always be predicted at what specific angle a mass tends to fail because this is also affected by its physical properties. In a study in the San Francisco Bay area it was noted that 75 to 85 percent of the landslides associated with built-up urban areas occurred on slopes greater than 15 percent (Keller; 1979, p. 116). Yet, it is important to realize that both *earthflows* and certain *snow avalanches* may release on much gentler slopes. Thus one can say that some form of failure will take place if the resisting forces, in this case friction, become less than the driving forces. The ratio between these opposing forces can be used to assess the stability of materials. Stated simply it can be said that the resisting forces, although more complex than the driving forces, are mainly the *frictional component* between surfaces and **shear strength** within the material.

Shear is the lateral deformation within a given mass in response to pressure, and shear strength is the material's resistance to such deformation. While somewhat different in solid rock, compared to soils and other unconsolidated material, basically two components determine shear strength: (1) *internal frictional forces,* which are primarily the result of particles interlocking, and (2) *cohesion* between particles, or their ability to stick together. The relative magnitude of these forces, especially in soils, varies primarily with particle size (Hausenbuiller; 1985, p. 516). The more varied the particles' sizes are, and the more angular their shape, the greater the shear strength.

The role of water in shear strength is not so much the lubricating effect between particles, which appears like a plausible explanation, but centers more on adding weight to the total mass. Water also exerts pressure inside the pore spaces and thereby reduces cohesion.

The significance of water content in earth materials was underscored during the earthquake in California on October 17, 1989, referred to as the Loma Prieta earthquake. This event was discussed in detail in the chapter on earthquakes.

Throughout an area of about 5,460 mi^2 (14,000 km^2) a variety of rock falls and landslides occurred. Several million dollars in damage occurred along the coastal region where steep bluffs are located within 6 to 12 miles (9 to 20 km) from the epicenter.

Scientists speculate that considerably more landslides and much greater damage would have resulted had the quake taken place in midwinter, the rainy season in California (California Department of Conservation Report; 1990, p. 59). At that time the unconsolidated materials composing the marine terraces, underlain by weak bedrock, would have been in a moist state. Earthquake amplitudes are greater and shaking is prolonged in unconsolidated deposits, and the danger of slides is much greater under these conditions (Tank; 1983, p. 82).

Neither the driving factors nor the resisting forces can be viewed as being constant. Earth materials are subjected to chemical as well as mechanical *weathering processes* that progressively weaken rock and soil materials. The rapid addition or removal of water, often associated with the quick melting of snow, or with heavy precipitation, change weight relationships and may also change the physical properties of earth materials and snow, and thus induce instability. When the driving forces and the resisting forces are just about in balance any destabilizing factor can trigger failure. In some instances, in addition to the infusion of water or other loading factors, *vibrations* can act as the critical agent. Such vibrations can be the result of earth tremors, heavy traffic or nearby highways, blasting at a construction site, as well as the sonic booms caused by supersonic planes.

Major Types of Earth Material Failures

The names for the various types of rock and soil failures differ in the literature and, unfortunately, are not always consistent. The expression "landslide" seems to be only loosely defined and could be viewed as an umbrella term that covers a great variety of processes and circumstances (Jones; 1992, p. 118). Therefore, it appears to be more practical to stress the common factors of such events and also to point out the significant differences between them. Moreover, it is very difficult to predict which of these downslope movements are the most dangerous ones and are prone to do the greatest amount of damage. Complex equations vary so much under different circumstances that any generalization is out of place. Also, in many instances several types of failures are intermixed, or causes and effects may be widely separated in a geographical sense. For example, an *ice-* and *rockfall* near the summit of one of Peru's highest mountains, Huascaran, in 1970 triggered an *ice-*

and *debris avalanche* that totally destroyed the town of Yungay, 9 miles (15 km) away, causing a total of about 25, 000 casualties.

The simplest failure, in a mechanical sense, is a rockfall off a cliff. Some rock formations have wide overhangs, which attest to the strength of that particular material. The famous Niagara Falls tumble over a highly resistant rock ledge, the Lockport Dolomite, which juts out over the more easily erodable layers of softer rock. At undeterminable intervals the *cap rock,* weakened by undercutting, will tumble into the river gorge.

How strong a rock formation is depends on its internal structure and on whether it has weak zones such as fissures or **bedding planes,** which are layers of different materials. There is less cohesiveness between such layers than in the layers themselves.

Other failures and downslope motions occur in unconsolidated materials ranging from coarse **colluvium,** gravel, sands, volcanic ash, mud, and soils [Plate 18]. If the movement is relatively rapid it is generally called a *flow,* while a very slow motion is referred to as a **creep.** Obviously, such downslope movements do not have a constant speed because of variable water content, particle size, and terrain.

The term *slide,* as used in landslides and rock and debris slides, indicates a massive failure within a large body of earth materials [Plate 19]. Slides of catastrophic dimensions are the events that make the headlines and find their place in the literature because of the large number of casualties, the immense damage done to property, or the way in

Plate 18. Unstable soil conditions, as well as strong vibrations from a nearby highway, caused repeated soil failures and landslides near Ithaca, New York. (Photo by C.H.V. Ebert).

Plate 19. On April 29, 1903, about 90 million tons of rock slid down Turtle Mountain and buried part of the coal mining town of Frank in Alberta, Canada. The momentum carried the slide up the opposite valley slope to an elevation of nearly 395 feet (120 m) above the valley bottom. (Photograph credit: A. Abrahams).

which they radically changed the landscape of the area affected [Plate 20].

The single largest *rockslide* ever to occur in recent history of the United States took place in the Gros Ventre River Valley in Wyoming. Steeply inclined rock formations, interbedded with clay layers, had been weakened by weathering and by frequent earth tremors in that region. In response to above-average precipitation, in addition to water derived from melting snow, about 39 million cubic yards (39 million m^3) slid down the mountainside and blocked the Gros Ventre River. The momentum of this slide carried part of the mass 348 feet (106 m) up the opposite valley side. Fortunately this event did not result in human casualties.

A truly disastrous rockslide, killing about 2,500 people, occurred in northern Italy on October 9, 1963. Ironically, the killer was not the rockslide itself but a gigantic floodwave of water displaced by a rockslide that plunged into a large reservoir just above the town of Longarone.

The waters of the artificial lake were held back by a high, thin-arch concrete dam. Steeply inclined rock beds, which rested on top of clay and soft marl, had engaged in minor slides during the construction phase of the dam. Monitoring of the suspect slope indicated mass movements up to 9 to 12 inches (24 to 30 cm) per week (Butler; 1976, p. 104).

Late in the evening of that fateful day in October a mass of about 311 million cubic yards (238 million

Plate 20. Subsequent to the extensive 1963 ash eruption of Volcan Irazu in Costa Rica, a massive wet ash avalanche obliterated a large part of the town of Cartago and left behind a wasteland of mud and boulders. (Photo by C.H.V. Ebert).

m³) detached and roared into the reservoir with a speed estimated at nearly 59 miles per hour (95 km/h) at impact. A huge wave of water, at least 295 feet high (90 m) rose over the dam and rushed down the valley wiping away the town of Longarone. Critics asserted that the dam never should have been built in this location well known for its geologic instability. The final irony was that the Vaiont Dam remained unscathed by this event.

Landslides in Southern California

A growing population and intensive commercial development increase the pressure on land resources in southern California. More land surfaces are being modified to accommodate new or extended human activities. Reservoirs are being built, terraced land replaces natural hillslopes to allow the construction of homes or to start new orchards and vineyards. More water is forced into the soils to support agriculture or to make lawns and gardens more attractive. All these changes result in new and somewhat artificial landscapes and conspire to render many tracts of land more unstable.

The general *instability* of the land is not just the consequence of *seismicity,* which is blamed in most events involving mass wasting; however, a great part of land instability can be traced to particular soil characteristics, excessive watering, changes in vegetation cover, as well as unusually severe rainfall events associated with the occurrences of El Niño (see chapter 10; The El Niño-La Niña Phenomenon).

The soils in the area of Santa Barbara, as typical for that part of California, are mostly a suborder of *Vertisols,* namely *Xererts,* which develop from volcanic materials as well as from old marine deposits, under a climate typified by long dry summers and mild rainy winter seasons. These soils may have up to 30 percent clay of the *Montmorillonite* type which has a high *shrink-swell potential* upon drying and wetting, respectively (Miller; Donahue; 1990, p. 607).

Two aspects make these soils highly unstable: (1) upon irrigation, or during heavy downpours—as under El Niño conditions—water enters the deep surface cracks which can extend to a depth of 40 inches (1 m) or more. When the water encounters a less permeable stratum underneath the surface layer it will spread horizontally and downslope. This process lubricates the interface between the sub-stratum and the surface layer and thus destabilizes the soil mass; (2) repeated wetting may flush out saline

electrolytes associated with old marine material. The salt remnants normally help the soil to maintain structural integrity. The removal of this "electrolytic glue" increases the clay-rich soil's sensitivity to shock and other disturbances (Tank; 1983, p.148).

Massive landslides occurred around Montecito near Santa Barbara during the disastrous *El Niño* rains of 1995 (Plate 21). The damage to homes, orchards, and roads was extensive and required costly remedial work. After the removal of the landslide debris the main efforts centered on making the scarred slopes more resistant to future instability problems. In most instances, the steeper slopes on which the failures had occurred were made gentler in gradient and less variable in slope degree. This work, at first glance, seems to be quite reasonable; however, a shallower gradient, while reducing the volume and the speed of water runoff, will increase water *infiltration* into the soil surface (Johnson; Lewis; 1995, p. 94). This increased infiltration not only adds to the weight of the soil mass but also lowers the shear strength of the subsoil which induces renewed slope instability in the future.

Another effort consists of establishing a more coherent vegetation cover on the endangered slopes. Unfortunately, not all vegetation has necessarily a beneficial effect on *slope hydrology.* A study made by R.H. Campbell, as discussed by Edward Keller, shows that *chaparral* vegetation tends to increase water infiltration into slopes and thereby raises slide risks. Furthermore, it has been seen that grass-covered slopes experience 3 to 5 times more soil slips than those under chaparral (Keller; 1979, p.118).

Plate 21. Massive landslides and soil failures occurred around Montecito near Santa Barbara, California, as the result of the excessive El Niño rains in 1995. (Photograph credit: Werner Herrman, Montecito).

The key factor in stabilizing slide-prone slopes is to alter the hydrology to prevent upslope water from wetting the lower slope sections. Upslope water must be intercepted and diverted into erosion-proof conduits which descend the entire slope length. Such diversion allows excess water to be safely controlled and transported into natural or constructed drainage channels. A very successful slop stabilization project in Montecito, using this technique of water interception and diversion, was studies by the author on location (Plate 22).

Land Subsidence

In addition to the types of earth material failures discussed so far there are also other concerns, including a phenomenon called *land subsidence.* Land subsidence is not necessarily tied to slope instability. It may occur both naturally or as a result of humans' activities. The basic principles of stress and resistance also apply but under rather different circum-

stances. For example, a land surface may give way under the load of a building or a bridge support. If the soil material has low shear strength the weight of the structure may cause at first some *compression,* and subsequently *lateral flow* away from the compressed area (Hausenbuiller; 1985, p. 515).

Land subsidence may also be the result of natural *underground solution* of limestone. Limestone, a carbonate rock, is easily dissolved by rain and soil water, which contain a variety of acids. This water seeps through natural fissures, or *joints,* in the limestone until the water table is reached, then the water will flow horizontally. In this manner solution forms large voids which may induce the overlying rock ceiling to cave in. *Solution holes* of this kind are referred to as sinkholes and eventually lead to the creation of a *karst* landscape.

Human-induced factors in land subsidence include the massive extraction of water from water-bearing rock formations, or *aquifers,* and pumping natural gas and oil from underground fields. Prob-

Plate 22. After establishing a gentler slope gradient the slide-prone area was further stabilized by intercepting upper-slope runoff and by diverting the excessive waterflow with long pipes leading to appropriate drainage facilities. (Photo by C.H.V. Ebert).

lems of surface collapse are also associated with many other mining operations. In view of a growing population and increasing urbanization one must ask whether landslides and land subsidence can be predicted or possibly be controlled.

It was pointed out that the surface of the earth is constantly changing in response to a variety of dynamic processes. These are natural processes with which we must learn to live. Thus, it is the responsibility of the people to study such potential hazards, to avoid them wherever possible, and to encroach upon them once they are recognized. Geologists, civil engineers, and geographers have the knowledge to understand the danger signs and to give ample warnings. The case of the Vaiont Dam illustrates the fact that such warnings are not always heeded. Ignoring the knowledge, led to disastrous results. Moreover, man's activities which promote mass wasting and land subsidence are well known and must be curtailed, or even prevented, where feasible.

Finally, several techniques can be applied to slow down mass wasting processes or to effectively modify them. All such techniques require a basic knowledge of how earth materials behave and an understanding of the cause-and-effect relationships. Some of the preventive, or remedial, measures include: artificial reduction of slope gradients; installing drainage systems to carry water away from unstable slopes; planting vegetation cover; inserting rock bolts into hazardous rock units; avoiding excessive watering of soils near precarious slopes; infusion of chemicals that bind unconsolidated materials into porous layers; building retaining and diversion walls. In some selective soils a relatively new process, *electro-osmosis,* can be used to dewater subsoils. Pore water is driven out by passing an electrical current through the materials (Coates; 1981, p. 357).

Snow Avalanches

In the preceding sections we investigated the basic mechanics leading to failures in earth materials. A number of these principles also pertain to the development of snow avalanches. However, a major difference is that a given mass of snow may undergo rapid changes in its physical makeup and can accumulate, or diminish, within short time periods.

Avalanche Requisites

There are *four basic requisites* for an avalanche to develop: (1) the accumulation of a critical mass of snow, (2) structural changes within the snow that reduce the stability of the mass, (3) an adequate slope angle to permit a gravity flow, and (4) a triggering mechanism. These conditions are most typically associated with high mountain regions such as the Rocky Mountains, the European Alps, the New Zealand Alps, and Himalayas, and any other mountainous areas high enough to collect large amounts of snow.

How much snow must fall to produce a *critical mass?* This cannot be stated in absolute terms because snow is not of constant density. Snowflakes, that is ice crystals, attain different shapes and weights depending on the conditions in the atmosphere under which they form. They further change as they strike the ground and are buried under additional layers of snow. Yet, as a general rule it can be said that a snowfall of less than 6 to 10 inches (15 to 25 cm) seldom leads to avalanche releases.

The building of a *snowpack* in many ways resembles the process of *sedimentation.* This is the deposition of unconsolidated materials by water, ice, or wind resulting in layering. Such *stratification* is easily revealed in any cut made into a snowpack.

The physical conditions inside the pack are strongly influenced by the temperature distribution within the mass and by the pressure exerted by the weight of the overlying snow. Studies conducted at the Swiss Federal Institute for Snow and Avalanche Research emphasize that it is especially the contrasting temperatures within the snowpack that lead to the formation of an unstable grainy type of snow. The coarser grains tend to develop closer to the ground than near the cold surface air. While the air may be extremely cold at high elevations, and especially during the long winter nights, the ground engages in a considerable *heat flow* into the bottom of the snowpack. In response to this heat the snow crystals engage in a process called **sublimation.** In sublimation a solid—in this case the ice crystals—passes directly into the *vapor form* without melting first. We all have observed how piles of snow evaporate at below-freezing temperatures. This process also works in reverse, that is, water vapor may pass directly into the solid ice form.

Water vapor exerts pressure. **Vapor pressure** is the pressure contribution made by water molecules to the total gas pressure produced by the atmosphere. The amount of water vapor, and therefore its pressure, is greater at the higher temperatures in the lower portions of the snowpack than in the colder layers above. Thus the vapor will diffuse from high-pressure to low-pressure zones moving along the so-called pressure gradient (Miller; 1971, pp. 21 and 55). This vapor diffusion, and the subsequent subli-mation on the surface of the colder ice particles in the upper snow layers, results in the formation of a very unstable type of ice particle, *hoar frost.* These ice crystals do not form bonds with each other owing to their angular shape and fragile nature (Fraser; 1966, p. 64). This material has very low shear strength and is very similar to loose sand. Thus, hoar frost may act as a "ball bearing" to the mass of snow resting on it and facilitates its sliding downhill.

Whether or not such slide will result, and at what speed, depends to a large extent on the *slope angle.* However, there is no general agreement as to how steep a slope must be because of the great variability of snow characteristics. Masses of wet snow may release at angles below 15 degrees while stable snow, frequently encrusted with ice, may remain on a slope of 50 degrees. It seems that avalanching takes place most frequently on slopes of 30 to 50 degrees (Armstrong; Williams, 1986, p. 74).

Avalanche Types

Over the years the classification of avalanches has become more sophisticated and includes a consider-able number of control factors and categories. This classification system is in response to a better under-standing of cause-and-effect relationships. One of the most useful and widely accepted classifications is that offered by Professor R. Haefli and Dr. M. de Quervain (Swiss Federal Institute for Snow and Avalanche Research; 1955), which is discussed in detail in the book by Colin Fraser entitled *The Avalanche Enigma.* For the scope of this discussion let it suffice to describe the two basic avalanche types.

The first type, mostly composed of dry loose snow, is called a *powder avalanche* [Plate 23]. It usually embodies a great amount of air and may move at speeds up to 125 miles per hour (200 km/h). Such speed can be ascribed to the fact that this type of avalanche may ride on a cushion of air trapped underneath. A similar behavior was described in connection with the glowing ash clouds of volca-noes. The damage inflicted by such powder avalanches, in addition to that caused by the direct impact of the snow, can be traced to the severe *air blast* and the **turbulence** within the mass. This topic is well presented in the book by Betsy Armstrong and Knox Williams, *The Avalanche Book:*

Plate 23. This photograph shows a powder avalanche hurtling down a steep valley slope. Note the violent billowing of the loose snow as the entrapped air explodes outward. (Photo-graph used by permission of the Swiss Federal Institute for Snow and Avalanche Research).

Because avalanches can travel at very high speeds, the resulting impact pressures can be sig-nificant. Smaller- and medium-sized events have potential to heavily damage wood frame struc-tures (impact pressures of 1 to 15 pounds per square inch). But on the other end of the scale, extremely large snow avalanches possess the force to uproot mature forests and even destroy structures built of concrete (impact pressures of more than 150 pounds per square inch). Actual destruction would, of course, occur only when such an avalanche came in contact with some type of structure—which represents just a very small fraction of the total avalanche events.

Some reports of avalanche damage describe circumstances which cannot be easily explained simply by the impact of large amounts of fast-moving snow. Some observers note that as the avalanche passes, some buildings actually explode, perhaps resulting from some form of vacuum created by the fast-moving snow. Other reports indicate that a structure was destroyed by the 'air blast' preceding the avalanche.

(Armstrong; 1986, p. 70)

The *wet-snow avalanche,* the second major type, is also called a **ground avalanche.** This category involves large masses of wet, loose snow that fre-quently carries along boulders, soil, and trees and is, depending on size and speed, the most powerful avalanche. Yet, the fact that ground avalanches form

Plate 24. Thick, compacted snowpacks can detach from underlying snow layers and form what are called slab avalanches. (Photograph used by permission of the Swiss Federal Institute for Snow and Avalanche Research).

Plate 25. Snow, under stress of its own weight, is capable of slow plastic deformation; but eventually part of the mass may break off and may trigger an avalanche. (Photograph used by permission of the Swiss Federal Institute for Snow and Avalanche Research).

regularly in the spring, and tend to follow known tracks, explains that they do not do as much harm to life and property as do the more unpredictable powder avalanches (Fraser; 1966, p. 88). Compacted snowpack may also detach from the underlying unstable snow layer in the form of a large cohesive slab and is called a *slab avalanche.* This type of avalanche usually involves a thick surface layer with a thickness ranging from 5 to 12 feet (1.5 to 3.5 m) [Plate 24].

Trigger Mechanisms for Avalanches

Similar to the trigger mechanism of landslides, snow avalanches start mainly because of some mechanical disturbance. A frequent cause is the fall of a snow *cornice,* an elongated shelf-like slab of snow that forms on the *downwind* side of a mountain peak (fig. 3.1). Under the pull of its ever-increasing weight it will first bend downward because snow, like solid ice, is capable of slow *plastic deformation* [Plate 25]; but eventually it may break off and trigger an avalanche *on the slope below.*

The fact that *cornices* develop on leeward sides of mountain ridges allows observers to determine the prevailing winds in a given region. This correlation assists in assessing which side of a mountain represents the greatest avalanche threat because of *orography* which refers to the uplift of air masses along wind-facing slopes. This mechanical uplift is responsible for the heavy snowfall on such slopes.

Several factors may start an avalanche: rock- and icefalls, unstable cornices, sonic booms, a tree falling over in a storm, or a snowpack fracturing [Plate 26]. A number of avalanches, especially the slab avalanches, are initiated by skiers. It is estimated that, in the European Alps, about 80 percent of all avalanche victims are skiers. This percentage does not include casualties caused by unusual avalanche disasters that may destroy entire villages.

One of the worst avalanche seasons on record is the winter of 1950–51 when abnormally heavy snowfalls afflicted the central Alps of Europe for more than three weeks. Several series of avalanches occurred in Switzerland, Austria, and Italy killing

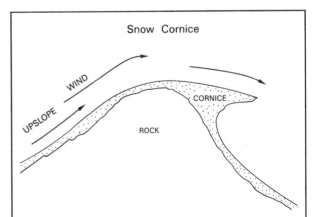

Figure 3.1. A cornice is an elongated wind-built protrusion of dense snow that forms on the downwind side of mountain peaks. Eventually it may break off and thus trigger avalanches.

Plate 26. Large icefalls can detach from glaciers and thus trigger avalanches below. This photograph shows severely fractured glacial ice below the Jungfrau Mountain in the Bernese Alps of Switzerland. (Photo by C.H.V. Ebert).

Plate 27. Wooden and steel barriers are placed on the upper slopes of avalanche-prone sections of mountains. Such barriers arrest and stabilize dangerous masses of snow. (Photo by C.H.V. Ebert).

about 700 people and destroying more than *2,000 houses.*

Recent studies of avalanche disasters show a marked increase in avalanche hazards because of more intensive human activities in avalanche-prone areas. A noticeable rise in the number of accidents and fatalities in the United States began in the 1970s; this increase parallels the growing popularity of skiing, mountaineering, and snowmobiling in this country. Similar trends were observed in other countries such as Canada, Switzerland, Austria, France, and Japan (McClung, Schaerer; 1993, p. 13).

The danger of avalanches can be reduced by educating people so that they can understand the causes and the nature of this mountain phenomenon. This is especially important for novice skiers and climbers. Also, many technical measures can be taken to reduce avalanche risks: construction of retaining and deflecting walls; placing of protective gallery tunnels over hazardous highways or railroad sections; leaving protective stands of forest above villages; monitoring known avalanche tracks by paying special attention to changes in the snow conditions [Plates 27 and 28]. In addition to these tech-

niques it is also a constructive practice to trigger avalanches before their potential for harm becomes too great, or to clear a particular track before it is opened to the public. This triggering involves "shooting down" threatening snow mass with explosives and even mortars.

Plate 28. Avalanche barriers (see the mountain in the background) and tunnels are in wide use in Switzerland to protect settlements and roads. (Photo by C.H.V. Ebert).

DISASTERS INVOLVING THE HYDROSPHERE

Earth has been called the *"water planet."* Even the most cursory glance at a globe readily reveals that blue, the color representing water, is dominant. More precisely, water covers about 71 percent of the earth's surface. Yet, the oceans, which represent approximately 93 percent of the water of the *hydrosphere,* constitute only a small fraction (0.023 percent) of the total mass of planet earth. The remainder of the hydrosphere can be found in the atmosphere, in lakes and rivers, and as water contained in the pores and fissures of rock formations.

Water is the only natural substance that can exist in solid, liquid, and gaseous form within the temperature ranges of our planet. Furthermore, it is the essence of our life, of the biosphere, as we know it on earth. *Protoplasm,* the biophysical basis of life, contains 80 percent water. Thus we may truly call water the great supporter of life. Yet, one flood in China in 1887, of the Yellow River, killed more than 2 million victims, and the 1970 Bay of Bengal storm surge drowned about 400,000 people.

It was pointed out before that we have to view our planet as a complex system with many interacting components. We must understand that the hydrosphere is functionally interrelated with earth materials, with the atmosphere, and with the biosphere. Disasters involving water may be the result of excessive amounts of precipitation, rapidly melting snowpacks, landslides blocking rivers, earthquakes triggering seismic sea waves, powerful storms piling up water against the shores, or of uncontrollable runoff from land robbed of its protective vegetation cover. Massive runoff may also occur from paved-over land that no longer can absorb water to feed it gradually into natural drainage systems. Although the ultimate causes of disasters involving the hydrosphere may be traced elsewhere, the disasters discussed in the following chapters either occur in water, involve water, or deal with the behavior of water in response to energy input.

TSUNAMI WAVES AND STORM SURGES

Key Terms

storm surge	wave periods	standing wave
wave trough	GOES	delta
wave crest	hurricanes	monsoon
continental shelf	midlatitude cyclones	polder
submarine	fetch	astrobleme
canyons	tropical cyclone	extinction
wavelength	spring tide	

The expression *tidal wave* is deeply entrenched in the vocabulary of many people and in that of popular publications. It is frequently, and incorrectly, applied to two large wave types that have nothing to do with tides: the *tsunami*, or seismic sea wave, and the *storm surge*. Though both can wreak havoc with coastal regions, or tidal zones, they are natural phenomena that are associated with earth disturbances and with atmospheric storm systems, respectively.

Formation of Tsunamis

The name tsunami, deriving from the Japanese language, literally means *"harbor wave."* Japan is earthquake country, and many shores and harbors have been devastated because of these gigantic waves, which may reach heights of 200 feet (60 m), although such a height is rare.

Tsunamis are encountered in ocean regions that are affected by plate movements, especially in plate subduction zones, and where seismic and submarine volcanic disturbances are frequent. It is not surprising, therefore, that throughout history these waves have brought death and destruction to many islands and coastal communities. This is especially so within the realms of the Pacific and the Mediterranean Sea. It is also easy to understand that such waves, often following earthquakes, are called seismic sea waves even though not all of them are caused by such disturbances.

It is more correct to say that tsunamis result from a variety of geologic processes, which include earthquakes, sudden crustal subsidences beneath the ocean, massive marine sediment slides, and submarine volcanic eruptions. Recent studies reveal that there is a direct relationship between plate tectonics and the triggering of such waves. As the plates are subducted into *trenches,* as discussed in chapter 1, they tend to be fractured by minor faults (Menard; 1977, p. 102). This can result in a sudden drop of a large area of ocean floor near the trench margin. On the other hand, such faulting can also produce sudden uplift.

In the first situation there is an abrupt drop of the water mass, while in the second situation there is a massive uplift of the water (fig. 4.1). Therefore, the initial leading edge of a tsunami wave system can be either a *wave trough* or a *wave crest,* the elevated part of a wave. If the crustal motion is downward a trough is formed; if the movement is upward a crest results. This helps to explain that whenever a trough arrives at a coast, a marked drop in sea level occurs and the waters recede from the shore (Duxbury; 1984, p. 270).

A smaller number of tsunamis are generated by sudden, massive detachments of sediments. Such slides may occur on the *continental shelf* margins where immense masses of water-saturated sediments accumulate. They become unstable when disturbed by seismic tremors or possibly by large internal ocean waves. These large waves travel within the ocean at interfaces of waters of different densities.

As the large volumes of sediments detach and avalanche down the shelf sides, or move rapidly into *submarine canyons,* they may exert a *plunger effect,* which suddenly displaces extensive amounts of water (fig. 4.2). This water mass then behaves in a similar fashion as water affected by faulting on the ocean floor.

Volcanic disturbances, unless unusually violent and possibly involving the collapse of a volcanic island, do not regularly generate sufficient energy to produce tsunamis. Examples where they did, however, will be discussed later in this chapter.

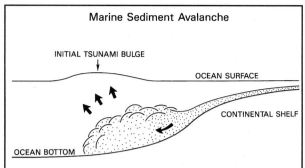

Figure 4.2. A massive sediment avalanche, descending along the slope of a continental shelf, can produce a "plunger effect" by displacing large volumes of water. This action can trigger a tsunami.

Tsunami Characteristics

After the energy of a disturbance is transmitted to the water, wave energy begins to travel just as in any wave system. However, it is important to realize that tsunami waves, even though they move through deep oceans, behave like *shallow-water waves.* The reason for this is their great length, which may exceed 100 miles (160 km) as compared to the average ocean depth of about 12,200 feet (3,720 m). Whether an ocean wave is classified as a *deep-water* or *shallow-water wave* depends on the relationship between its length and the depth of the water through which it passes. If the water depth is greater than one-half of the *wavelength* (here defined as the distance between two crests) the wave energy is not affected by the ocean bottom. On the other hand, if the depth of the water is less than L/20 (1/20 of the wavelength) the wave is strongly influenced by friction with the ocean bed. It is easy to visualize that the ratio between the length of a tsunami and ocean depth makes it behave like a shallow-water wave.

The wave energy in a tsunami thus is distributed from the ocean surface to the ocean floor and over the entire length of the wave (Duxbury; 1984, p. 270). This wave energy moves at speeds of more than 400 miles per hour (643 km/h); yet, a ship in its path would hardly notice such a wave because its *height* in the open ocean may only be 1 or 2 feet (31 to 62 cm) over a distance of 100 miles (160 km)! Why, then, is such a wave so disastrous in its impact on the shore?

Once the tsunami wave moves into the shallower waters of the *continental shelf,* where ocean depth quickly changes to a few hundred feet (normally

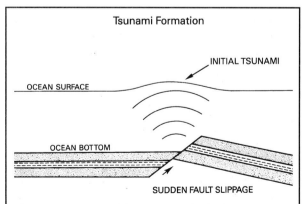

Figure 4.1. A sudden, strong fault slippage on the ocean floor can supply the energy to generate seismic sea waves (tsunamis).

about 400 feet or 122 meters), the wave energy is abruptly concentrated into a much smaller mass of water. This means that, similar to the way a breaker forms as a wave approaches a beach, the tsunami's wavelength is drastically shortened, its speed decreases, and its height grows dramatically.

The sharp reduction in speed may begin far off-shore depending on the width of the continental shelf and on the depth of the water. In the open ocean, at average ocean depth, a tsunami moves at about 656 feet per second (200 m/s). When the wave encounters a depth of about 330 feet (100 m) its speed is reduced to about 99 feet per second (30 m/s), and at 165 feet (50 m) its forward motion declines to about 72 feet per second (22 m/s) (Knauss; 1978, p. 221).

At this stage the tsunami becomes the destructive phenomenon for which it is so well known. While the lower part of the wave is held back by the friction with the bottom, as is the case with a breaking wave, the top of the mass of water crashes forward with unstoppable force. The resulting height of the tsunami may be from that of a normal large breaker to a towering height of 200 feet (60 m). How large and how destructive it may become depends on a variety of factors: the distance it has to travel from its point of origin, obstacles in its path, variable ocean depth, and, as stated above, the configuration of the continental shelf. Since wide shelves use up more wave energy, it seems logical that the wave under such conditions has less power and does less damage.

Obviously it is very difficult to get accurate measurements of tsunami heights. Early reports on the tsunami that struck the area of Merak, on Java, in connection with the cataclysmic eruption of Krakatoa (1883), referred to a wave height of 180 feet (55 m) (Simkin; 1983, p. 76). Other events, as the one that destroyed the ancient island of Stronghyle (1450 B.C.?), now called Santorini, are said to have been much larger than those associated with the Krakatoa disaster (Ebert; 1986, p. 435).

The tsunami is frequently visualized as a single wave, but more correctly the tsunami is a *wave system* that spreads out from a point of energy release. A number of waves may originate from such a source with **wave periods** of 10 to 20 minutes, or longer. This means that a given coast may be exposed to a series of tsunamis arriving at considerable time intervals. The first wave does not have to be the highest one. It is possible that subsequent waves are much more dangerous. This situation could be the result of *constructive wave interference,* the merging of an oncoming wave with a preceding wave that was reflected back from the shore. It may also occur that subsequent seismic shocks release additional wave energy into the ocean. Wave activity may continue for many hours, and each wave may have different energy levels. It has been observed that the third to the eighth waves tend to be the largest (Bernstein; 1971, p. 58).

Tsunami Disasters

It was pointed out that the occurrence of tsunamis correlates well with maritime zones of active plate subduction. For this reason many of the better-known tsunami disasters take place in the rimlands of the Pacific and in the Mediterranean Sea. On a worldwide basis tsunamis happen, on the average, at least once a year. In the Pacific alone about 270 damage-causing seismic sea waves were generated between 279 B.C. and 1946 (Ward; 1978, p. 52).

Possibly one of the most dramatic events occurred about 3,400 years ago in connection with the final destruction of the volcanic island of Stronghyle, now Santorini, in the Aegean Sea. Whether the obliteration of the island was brought about by the collapse of the volcano or by a cataclysmic eruption is still debated. However, the result was a huge *caldera* surrounded by three arc-shaped islands that resemble the setting of Krakatoa. Whatever the specific circumstances, the destruction of the volcano undoubtedly triggered gigantic seismic sea waves that spread across the Aegean and into the eastern Mediterranean. Islands and the variable depths of the sea deflected and modified the tsunamis. It seems unlikely that the waves, by the time they reached the shores of the Nile Delta and other eastern coastal sections, exceeded 20 feet (6 m) (Ebert; 1986, p. 438). On the other hand there is no question that the tsunamis must have been much higher closer to their source. It is not surprising that speculations arose in many quarters whether this event was directly, or indirectly, involved with the abrupt decline of the Minoan culture on Crete and with happenings referred to in biblical history (Cazeau, Scott; 1979, p. 197).

Another disaster on July 21, A.D. 365, bears out the fact that the Mediterranean Sea is subject to the formation of severe seismic sea waves. A series of strong earthquakes, originating along submarine fault lines, struck in the eastern basin of the Mediterranean. It was said that the waters around Sicily, Greece, and Egypt receded before the tsunami waves rolled over the low-lying islands and shores. Alexandria, in the outer Nile Delta, was the city affected most severely. Here, about 50,000 people died, and fishing boats were carried far inland with some ending on the rooftops of houses. It was during this disaster that the famous 600-foot (183 m) lighthouse of this flourishing harbor city was destroyed.

Other well-known disasters involving seismic sea waves include the devastation of Port Royal, Jamaica, in June 1692. Three earth tremors leveled most of the houses and subsequent tsunamis, ranging in height up to 24 feet (7.3 m) drowned hundreds of persons, bringing the total casualties to more than 1,600. Another earthquake, with large seismic sea waves following, destroyed Lisbon, Portugal, on November 1, 1755. In consequence of the impact of the quake, the ensuing great fire, and the tsunamis a total of 50,000 people lost their lives (Nash; 1976, p. 336).

Without doubt, the most famous of all tsunami disasters was the one that resulted from the final eruption and subsequent collapse of the volcanic island of Krakatoa, which took place on August 27, 1883. The island, located in the Sunda Strait between the islands of Java and Sumatra, was the home of three volcanoes that had not been active since an eruption in 1680–81. Explosive eruptions began in May 1883 when the smallest volcano, Perboewatan, became active. The climax followed almost exactly three months later when the largest volcano, Rakata, after almost 19 days of violent eruptions, destroyed itself in one last cataclysmic event. The final collapse allowed the sea to enter the partially emptied volcanic chamber, which, upon contact with the cold water, exploded. A cloud of pulverized debris and steam rose to an estimated height of 83,000 feet (25,300 m) into the sky.

The resulting tsunami waves rolled over the coastal lands of Sumatra and Java, killing more than 36,000 people and destroying more than 200 towns and villages. At Merak, where the largest tsunami reached the top of a hill 130 feet (40 m) in height, nearly 3,000 victims drowned (Simkin; 1983, p. 77). From other locations it was reported that the height of the seismic sea waves ranged from 20 to 120 feet (6 to 36 m).

The tsunami that struck Hilo, Hawaii, in 1946, visiting that port city, oddly enough, on April Fools' Day, was one of the most carefully observed events; it also had a far-reaching consequence as far as the development of the Pacific Tsunami Warning System is concerned.

A group of scientists, including marine geologist Francis Shepard, happened to be in Hawaii while stopping over on their journey to Bikini Island in connection with an atomic bomb test. A series of tsunami waves, originating from a submarine earthquake near Unimak Island, Alaska, reached Hawaii in 1 hour and 30 minutes. Dr. Shepard described the onrushing waves which were accompanied by hissing and roaring sounds. He also attested to the fact that the third and fourth waves appeared to be the highest (Bernstein; 1971, p. 58).

The damage done by the great waves was not uniform because some of the locations were more exposed than others. Heights of 36 feet (11 m) were reported at Oahu, and a 56-foot height (17 m) occurred at Hawaii at Pololu Valley. As is the case with tsunamis the waters lowered substantially before the onslaught of the massive waves. Several persons who could not contain their curiosity to view the exposed beach, with several ships lying on their sides, were caught by the first towering wave from which there was no escape [Plate 29].

The waves appeared at about 15-minute intervals, smashing ships, piers, and warehouses. Coastal highways were washed away, and entire rows of houses were demolished and swept inland. Fortunately, only 160 persons were killed, but the property damage exceeded 25 million dollars.

As early as 1923 it was possible to predict successfully the arrival time of seismic sea waves, thanks to the research work done by seismologists of the Hawaiian Volcano Observatory (Menard; 1974, pp. 185–86), but the then available technology in the field of communication was not adequate. The first working warning system, in part inspired by the

Plate 29. This scene shows the powerful tsunami that struck Hilo Harbor, Hawaii, on April 1, 1946. Note the doomed man in the lower left of the picture. (Photo credit: NOAA; National Geophysical Date Center).

events of April 1, 1946, was established in Honolulu in 1948.

Human indifference, and possibly short memories, were demonstrated fourteen years later when a series of severe earthquakes rocked the coastal lands of Chile starting on May 21, 1960. The earthquakes lasted until May 30, demolishing more than 100,000 buildings and about one-fifth of the country's industrial installations. Powerful seismic sea waves invaded the coastal lands and added to the destruction. Despite the fact that the citizens of Hawaii received a tsunami warning six hours before the first wave arrived, few people paid attention to it. This negligence caused the death of more than 60 people in Hilo.

The same wave system traveled 9,000 miles (14,480 km) to Honshu, Japan, where it drowned almost 200 people. By that time the wave height was reduced to an estimated 15 feet (4.5 m), less than half of its original height that was reported at 35 feet (10.7 m). Another series of powerful tsunamis resulted from the earthquake that struck Alaska in the Prince William Sound area on March 27, 1964. The seismic sea waves did considerable damage in Resurrection Bay at Seward [Plate 30].

To predict the arrival time of tsunami waves is of critical importance. The precise determination of the epicenter of a given earthquake makes it possible to calculate the arrival time to an accuracy of a few minutes. This can be done because of the sufficient knowledge that exists about the depth distribution of the Pacific Ocean (Pond; 1978, p. 188). However,

Plate 30. On March 27, 1964, a tsunami wave —triggered by the Alaska Earthquake— carried this ship inland in Resurrection Bay at Seward, Alaska. (Photograph credit: NOAA; National Geophysical Data Center).

just as is the case with the prediction of earthquakes and volcanic eruptions, it is very difficult to foretell whether a given seismic event will actually result in a tsunami and, second, whether the tsunami will do any specific amount of damage when it arrives.

With the development of the *Geostationary Operational Environmental Satellite* (**GOES**) it is now possible to issue tsunami alerts within five minutes. The GOES receives transmissions from sensitive *tidal sensors* which are installed along the coasts of North and South America.

As abnormal changes in ocean level occur in response to seismic disturbances, the sensors emit signals which are picked up by the satellite. GOES immediately transmits the signal to the base station at Wallops Island, Virginia, from where it is sent to the Hawaii Tsunami Warning Center in Honolulu. The Center then issues tsunami watches or warnings to all areas that are potentially affected. (Barnes-Svarney, 1983, p. 256).

Threats from Space

Directly related to tsunamis, yet of rare occurrence, is the effect of meteorite or asteroid impacts on our planet. There is no doubt that the most cataclysmic tsunami-type waves were created by such impacts. This assumption seems to be quite plausible when one recalls that 71 percent of Earth's surface is covered by oceans. Whether such impacts refer to a meteorite or to an asteroid depends on the size of the celestial fragment. Asteroids, such as Ceres, are much larger and may attain a diameter of 600 miles (1,000 km); fifteen asteroids are known to have a diameter of 155 miles (250 km).

How many asteroids exist is not known but about 4,000 have been classified and their orbits are reasonably well identified. The so-called "Apollo group" is potentially the most dangerous to us because their members move across Earth's orbit. The other category belongs to the "Aten group" and falls directly within Earth's orbital path (Murck et al., 1997, p. 256).

The scars left behind by meteorite strikes are referred to as *astroblemes* or impact craters. Many astroblemes have disappeared over time because of erosional forces on our planet. Smaller fragments burn up in the atmosphere to become "shooting stars." More than 80 astroblemes are known and new ones are being detected. The gigantic craters, which were blasted out of the earth's surface, survived longest in hard-rock areas such as the Canadian,

Scandinavian, and Siberian shields. Two large impact locations are at Sudbury, Ontario, and the Lake Manicougan site in Quebec. A smaller and more recent event left behind the well-known Barringer Crater in Arizona which has a diameter of 3,940 feet (1,200 m) and is well preserved (Plate 31). The youth of the crater, estimated at about 50,000 years, and the dry climate of the region protected its geologic features.

The effects of a meteorite strike depends on several factors: (1) the mass of the meteorite, (2) the collision velocity, (3) the angle with which the earth's surface is struck, and (4) the geology of the impact location. The higher the collision velocity, and the steeper the entry angle, the greater the impact pressure will be. Impact velocities may be as high as 33 mi./s (54 km/s) to as low as 1.5 mi/s (2.4 km/s) (Wood; 1979, p. 38).

One of the great questions dealing with Earth's history is the sudden disappearance of the dinosaurs and more than half of marine animals species at the end of the *Cretaceous era* about 65 million years ago. Some light was shed on this event with the discovery of an unusual concentration of the element *iridium* in a clay layer which is a transitional feature between the Cretaceous and Tertiary eras. Iridium is not common on Earth but was found in larger concentrations in meteorite material (Melosh; 1989, p. 222).

Seismic research at the northern end of Mexico's *Yucatan Peninsula* detected a huge buried impact crater with a diameter of probably more than 150 miles (240 km) (fig 4.3). An investigation into the nature of the local sediments discovered normal oceanic deposits at a depth of 1,640 feet (500 m) overlain by a layer of rock material which had melted under the impact of this killer meteorite (Segar; 1989, p. 166). The crystallized rock-melt is called *tektite* and is typically associated with impact areas.

The strike of a major meteorite, let alone that of an asteroid, is a stupendous event with almost unimaginable consequences: (1) the potential extinction of life forms, (2) massive geologic changes on

Plate 31. Aerial view of the Barringer Meteorite Crater in Coconino Co., Arizona, which was formed about 50,000 years ago. It has a diameter of 3,940 feet (1,200 m). (Photo credit: United States Geological Survey).

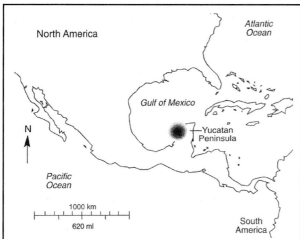

Figure 4.3. The impact of a large meteorite, which struck in the northwestern coastal area of the Yucatan Peninsula, may have caused the extinction of the dinosaurs about 65 million years ago, Seismic research detected the so-called Chicxulub Crater.

the earth's surface caused by impact compression, excavation, and by the deposition of debris, (3) shock compression waves in the atmosphere with subsequent reaction between molecular nitrogen and oxygen which leads to the formation of nitrogen oxide. Nitrogen oxide can degrade upper atmosphere ozone in a process which is similar in connection with atmospheric nuclear explosions (see Chapter 15), (4) creation of the so-called "nuclear winter" effect as dust and smoke block solar radiation, and (5) triggering the largest possible tsunami waves, potentially hundreds of feet in height, smashing into the low-lying parts of the continents and obliterating islands. While we do not know the frequency of such occurrences, we do know that they have taken place and will occur in the future. Could it be a blessing that we do not know when and where?

Origin of Storm Surges

Surges, in general, could be called an *abnormal rise* in the water level of lakes, estuarine inlets, and in ocean bodies. Some surges are almost unnoticeable and may occur at regular intervals as the result of tide-generating forces. Others may reach heights of 13 feet (4 m), or more, when associated with storm systems such as *hurricanes* or very severe *midlatitude cyclones.* If surges are the result of storms, with an accompanying marked drop in atmospheric pressure, and if the rise of the waters causes death and

destruction, they are usually called storm surges or "*surge tides.*" The latter expression, like the term *tidal wave,* is not correct unless, as it has happened in some instances, a storm-generated surge coincides with a high tide. It is the nature and effects of storm surges that are of particular interest to us here.

As most natural phenomena, storm surges are not caused simply by one factor but by the interaction of several. Yet the two main causes resulting in an abnormal rise of sea level are wind friction with the surface of the water and the upward-bulging effect of the water surface if atmospheric pressure is reduced sharply and rather suddenly.

Wind energy is transferred to a water surface by friction. This has basically two effects: The water is actually pushed forward in a process called *laminar* (layered) *flow,* and second, wind-generated waves are created. In the open ocean the water can freely move away from the wind action, but if such wind-generated surface flow moves into a semi-enclosed bay, or an arm of the sea, the water has no place to go and will pile up against the shore. It is easy to see that small islands, *deltaic* plains, reclaimed tidal lands, and generally unprotected coastlines are especially vulnerable to such surges.

High wind velocities alone are not instrumental in producing storm-surge conditions. The high winds must also be *consistent* in both direction as well as in speed for a considerable length of time. For this reason a stationary, or very slow-moving, storm is much more dangerous than a powerful but rapidly moving system, the latter usually resulting more in wave and wind damage. Furthermore, the wind must blow over an extensive stretch of water, called a **fetch,** to build up the necessary surge flow. It appears that an effective fetch should not be less than about 400 to 500 miles (645 to 805 km).

The second factor involved in producing a marked rise in sea level is a significant reduction in atmospheric pressure over a large and well-defined area. Such a sharp decline in pressure is associated with the centers, or *eyes,* of hurricanes and with some strongly developed midlatitude cyclones. Both storm types will be treated later in more detail in chapter 6.

These storms, especially the hurricane or *tropical cyclone,* develop very low atmospheric pressure, which may be as much as 2.5 inches of mercury (100 millibars) lower than the pressure level outside the storm system (Knauss; 1978, p. 226). This low pressure can lead to a 3-foot (0.90 m) rise in sea level. In itself this may not appear to be a great factor; however, if combined with the surge flow, a high tide, towering waves, and in a confined area, the potential danger becomes quite evident.

Nature of Surges

We have seen that both wind friction and the lowering of atmospheric pressure conspire to produce storm surges. How severe these surges may become is influenced by many other factors, so each situation has its unique parameters. In some instances, as will be shown in one of the case studies dealing with the 1953 North Sea Surge, the coincidence of a storm-generated surge with a high tide can be especially devastating. This is particularly so if it happens to be the above-average *spring tide,* which occurs twice a month at full moon and at new moon phases. In addition to the tidal influence there is also a less-known phenomenon, the *standing wave,* or *seiche.*

The standing wave is a periodic fluctuation of the surface of lakes or semi-enclosed bodies of water. This type of wave can be compared to water sloshing back-and-forth in a tub after it has been raised at one end and then placed down. In nature such a wave is the effect of several forces, which include minor earth tremors beneath a body of water, momentary strong winds from a passing storm, or a sharp change in atmospheric pressure as a result of an advancing *weather front.* In all cases the surface of the water becomes more or less tilted, as seen from the side, and subsequently engages in a seesaw motion, which is usually in the direction of the longest dimension of the body of water. The maximum vertical displacement of the water surface occurs along the shores, corresponding to the rim of a container, while the central point, the *node,* stays at the same height.

The wave moves back and forth within a given *time period,* which is primarily determined by the *length* and *depth* of the basin. Assuming a simple, rectangular model the period (T), in seconds, can be determined by the formula

$$T = \frac{2L}{\sqrt{gh}} = 0.64\frac{L}{\sqrt{h}}$$

where L and h are the length and depth of the basin, respectively, both given in meters, while g (9.8 m/sec^2) is the *gravity acceleration constant* (Thurman; 1985, p. 257).

Standing waves of this sort have very different characteristics in both period and height. Lake Erie (9,940 square miles; 25,745 square km) has such a periodic change of level of 3.2 inches (8 cm) with a period of 14.3 hours (Gross; 1971, p. 110); but in other bodies of water a seiche may be several feet in height. Tides and storm surges not only may reinforce each other but, if coinciding with a standing wave, also can produce exceptionally high water levels, particularly in semi-enclosed seas such as the North Sea (Ingmanson; 1985, p. 159).

The height of a storm surge is also influenced by the configuration of the continental shelf. It is interesting to note that a wide and gently sloping shelf helps to build up the storm surge whereas the opposite is true for the tsunami. Furthermore, the severity of a storm surge is also determined (1) by which side of the storm is approaching the coast, and (2) by the angle between the coastline and the storm track. A right angle seems to result in the highest rise of the water. Also, in the Northern Hemisphere, maximum storm surges tend to develop about 9 to 18 miles (15 to 30 km) to the right of the storm track (Costa; 1981, p. 429).

Storm Surge Disasters

The preceding section discussed the forming of storm surges and explained that this natural hazard, caused by intense low-pressure systems, in an ever-present threat to low-lying coastal zones and islands. Some of such areas, especially the northern shores of the Bay of Bengal, repeatedly suffered severe surges that killed hundreds of thousands of people; yet, the high-risk *delta* lands of the Ganges remain populated by millions.

The Ganges Delta, at the head of the Bay of Bengal, is shared by India and Bangladesh; the latter was formerly called East Pakistan. This region has been devastated many times by massive storm surges that were brought on by violent tropical cyclones (hurricanes); these storms typically move from the open waters of the Indian Ocean into the more confined waters of the Bay of Bengal (fig. 4.4). Bangladesh, in particular, is threatened by such flood surges. Since the tropical cyclone season overlaps with the drenching rains of the summer monsoon, a double-barreled threat from floods repeats itself annually. Historically, *monsoon* floods that cover 20 percent of Bangladesh occur one year out of two, and 37 percent of the territory is stricken by extensive flooding one out of ten years.

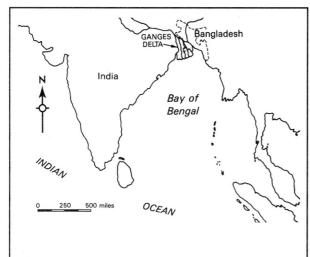

Figure 4.4. The Ganges Delta, at the head of the Bay of Bengal, has seen repeated devastation brought on by typhoon-generated storm surges. ──────────

In addition to climate, other factors contribute to the chronic flood hazard. For example, the delta lands are subsiding under the weight of sediments added by the Ganges and its tributaries. Severe soil erosion in the upstream regions, caused by extensive deforestation and subsequent cultivation, has greatly increased the sediment load of the river; the addition of such sediments also reduces the water-carrying capacity of the streams by filling the riverbeds. This riverbed aggradation is further enhanced by the construction of dams. In some sections the riverbed of the Ganges has been raised by as much as 16 to 33 feet (5 to 7 m) since the building of the Farakka Barrage in India. Furthermore, intensive urbanization, by creating vast areas of paved-over land surfaces, also aggravates the flood problems of the Ganges river basin (Khalequzzaman; 1992, p. 17).

One of the worst storm surges struck the delta lands in October of 1737 and killed an estimated 300,000 people. Another surge visited the area in 1864, and in 1876 a major surge drowned 250,000 victims; however, the flood event of 1970 proved to be the worst on record that ever occurred in that part of the world.

In October of that year, about 21 days before the disastrous surge of November 12 struck the same area, the population of the Ganges Delta was alerted to the potential dangers of an approaching storm. The first storm did not do a great deal of harm. Thus, when the second and much more powerful cyclone headed for the upper Bay of Bengal, the people did not become greatly alarmed. The warnings issued apparently failed to stress that this second storm was much more dangerous than the first one.

The delta lands of the Ganges, with a population then estimated at about six million, consist of a maze of low-lying tracts of land and islands, some of which formed only recently from the silt carried by the Ganges and its tributaries. In many instances, unless immediately swept away by the rapidly rising water, the people became trapped on these small islands and promontories extending into the Bay of Bengal. The waters rose to a height of 10 to 30 feet (3 to 9 m), as estimated by survivors in the area. Official readings by the Bangladesh Water and Power Development Authority were given at 15 feet (4.5 m) (Islam; 1974, p. 19).

The damage of this surge was staggering: 300,000 to 500,000 people drowned, half a million cattle destroyed, and over 70 percent of the rice crop wiped out. Yet, the survivors chose to stay, and many thousands of persons have since moved into this area. Why would people, despite such evident dangers, cling to this location? The majority who had lived through several storm surges stated that this had always been their home and that in good years the land would give their families the food they needed (Islam; 1974, p. 22). More recent events in 1987 and 1988 flooded 40 percent and 60 percent of Bangladesh, respectively. In 1991, a devastating tropical cyclone produced a 20-foot (6 m) wall of water that obliterated entire islands in the delta region and killed 140,000 people (Parker; 1992, pp. 3–4).

In response to these repeated disasters, the Bangladesh Flood Action Plan (FAP) was drawn up in 1989 by the Bangladesh government in cooperation with the World Bank. This step was the first out of four planning phases aimed at improving the conditions under which the population of the delta lands lives.

An interesting and controversial question is whether the people who live in these chronically endangered areas would be inclined to remain there if no outside help were given through government-financed flood control plans and through other assistance and relief programs. On the other hand, there is no answer either to the question as to what other places these poverty-stricken people could go and start a new life.

Another storm surge disaster, spawned by a very strong midlatitude cyclone, moving from the North Atlantic into the North Sea, devastated low-lying *polder* (an area reclaimed from the sea) lands of the

Netherlands and reclaimed tidal marshland, or *fens,* in England, in January 1953.

The geographical setting of the North Sea, open toward the Atlantic in the north and narrowing toward the coastlands of England, the Netherlands, and West Germany in the south, can be compared, although in mirror image, to the situation of the Bay of Bengal in its relationship to the Indian Ocean. This resemblance is emphasized by the fact that more than 12 devastating surges plagued this area between 1825 and 1976; yet, the January 31, 1953, disaster pushed the sea higher than generations could remember (Vereenigin ter Bevordering van de Belangen des Boekhandels; 1953, p. 1). This flood inundated about 300 sq.mi (780 km^2) of fenlands in England, drowning more than 300 persons (Strahler, Strahler; 1973, p. 406), killed about 2,000 victims and forced thousands from their homes in the Netherlands. Here, an estimated 9,653 sq.mi (25,000 km^2) were ravaged by the waters of this surge.

The unique violence of this storm surge was the result of several conspiring factors. A severe midlatitude cyclone had moved very slowly from southwest of Iceland toward the mouth of the North Sea where it became almost stationary (fig. 4.5). Fierce northwest-erly winds, estimated to be near 93 miles per hour (150 km/h), blew consistently into the North Sea with a fetch of about 580 miles (933 km). It was estimated by a number of authorities that possibly more than 5 billion cubic yards (4 billion m^3) of Atlantic water were pushed into the North Sea. This resulted in a surge measuring about 6 feet (2 m) along the east coast of England while, further south, the waters rose by an average of nearly 10 feet (3 m) along the shores of the Netherlands (Robinson; 1953, pp. 134–141). To this must be added a spring tide that, coinciding with the storm surge, was predicted to be about 8 feet (2.5 m) above the normal tidal range.

The statistics can never describe all the damage and all the suffering imposed by this rampage of nature. Throughout history the Dutch have fought against the encroaching sea by building dikes and storm protection structures. The largest was the Delta Plan, which was initiated in 1957 to be completed in the 1980s. Again and again the precious soils of the reclaimed polder lands have been exposed to the salt of the sea. This included the human-induced flooding in the fall of 1944 when, in connection with military operations during World War II, many dikes were bombed or blown up.

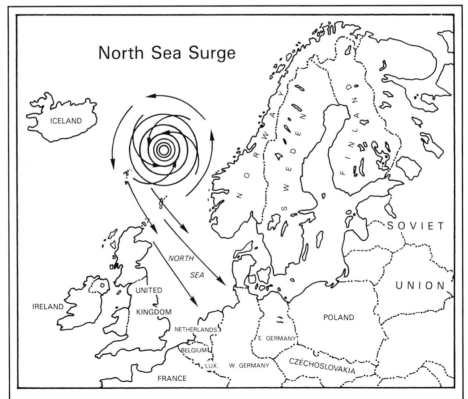

Figure 4.5. The devastating North Sea Surge of 1953 was generated by a powerful midlatitude cyclone that moved very slowly from the North Atlantic into the North Sea.

The last storm-surge disaster to be mentioned here is the well-known destruction of Galveston, Texas, in 1900. After Labor Day weekend of that year a hurricane forced a surge against the low-lying sandbar island on which part of this city is located. The highest section was just about 8 feet (2.5 m) above sea level. This was high enough to protect the community against the normal tide of about 2.5 feet (0.7 m) and ordinary storms; however, this "elevation" was no match against a surge that reputedly reached a height of 16 feet (4.8 m) topped by waves of 8 feet (2.5 m). More than 6,000 persons out of a population of 38,000 died in this disastrous event. Most of the victims drowned, but a number were killed by flying debris, including roof tiles, and by collapsing buildings as the fury of the storm roared across the doomed island.

Galveston was rebuilt and protected by a large *seawall,* which proved itself when subsequent hurricanes struck the area. Although there can be no perfect guarantees against nature's power, carefully planned preventive measures can sometimes help.

RIVER AND URBAN FLOODS

Key Terms	
stream terraces	cyclonic depressions
hydraulic action	El Niño
flood crest	hydrophobic
flash flood	urbanization
drainage basin	hydrology
jet stream	

In the introductory presentation of disasters—involving the hydrosphere—it was stressed that the hydrosphere itself, as well as the events associated with it, is functionally interrelated with the other components of the earth system. This fact is particularly important to remember when we probe into the causes of either natural or human-induced flooding.

Life functions on Planet Earth are so intimately involved with water that most people hardly ever stop to think about this truly miraculous substance. They take it for granted unless the lack of it threatens their existence, or the fury of floods overwhelms entire communities. Floods caused by tsunamis and storm surges were covered in the preceding chapter. The potentially destructive nature of rivers, often unwisely modified by man, is examined at this time.

Floods, similar to most disasters, are quite diverse in their origins, patterns, types, and also in their impact. For example, the sudden failure of a dam is quite different from the slow and ominous rising of a stream.

Contrary to some perceptions, floods may occur in just about *all* types of climates and in all kinds of terrain; however, their nature is decisively shaped by the parameters of the environment in which these floods form. The common denominator of all floods is that an *abnormal* and *uncontrollable* amount of water is encountered that damages—or completely destroys—the land and structures and may drown thousands of people.

Floods, just as other natural disasters, have accompanied human history throughout most parts of the world. Among all rivers of known record, China's Huang Ho—the Yellow River—stands out with its record of large-scale flood disasters. This river, so appropriately referred to as "China's Sorrow," has flooded 1,500 times with one flood going back to 2297 B.C. (Olson; 1970, p. 67). This record includes two incidents where the dikes were destroyed to stop enemy armies (Edwards; 1982, p. 129). Millions of people have been drowned by this river; but for thousands of years this region, ravaged by both floods and famines, has been and still is one of the most populous areas of that country.

The Yellow River is a classic example which illustrates the powerful attraction of people to rivers despite the inherent dangers of floods. The Yellow River, as symbolized by its name, has carried billions of tons of *silt* throughout the turbulent years. It offered water, transportation, food, and soil to the people living on its banks. How much silt has been carried by this river is illustrated by the fact that its water surface now flows 25 feet (7.6 m) above the surrounding floodplain; the river bed was raised by the settling sediments and is constrained by natural and human-made *levees.*

Floods, like earthquakes and volcanic eruptions, appear to be on the increase. However, we have to be careful to distinguish between floods that occurred, mostly unnoticed, throughout eons of time and those now being reported. More people, all over the world, are encroaching upon the floodplains of rivers, and more people, and more communities than ever before, are threatened by floods. The generally level topography of floodplains lends itself well to the building of homes, factories, and highways. Many cities in the United States, New Orleans and Cincinnati being good examples, are built almost entirely on floodplains. It is estimated that about 15 percent of this nation's urban land is located in such hazard zones (Moran et al.; 1980, p. 450). Thus it can be stated that floods, as a form of disaster, are becoming more frequent because (1) more people venture into flood-prone zones, and (2) many areas become more flood-prone because humans make this type of environment more vulnerable to floods.

While floodplains are attractive to many people, and for different reasons, another factor induces people to move into these potentially dangerous places. Flood control structures, such as levees and diversion channels, tend to instill a false sense of security. Yet, many such facilities are designed to handle only average flood conditions (Moran et al.; 1980, p. 452). Moreover, it appears that most individuals have poor memories and tend to forget unpleasant experiences. Flood awareness quickly dissipates over time. Maintaining the alertness of people and inducing them to engage in continuous planning for flood emergencies requires a constant hazard-awareness campaign, particularly in places where flood potentials are high but the probability of flooding is infrequent (Perry et al.; 1981, pp. 123–124).

Causes of Floods

Left alone, a river system will seek equilibrium with the physical parameters of its environment. This nat-ural law means that a river, as it passes through its various stages of *fluvial-geomorphic* development, demands space within which it can function. This space includes the main channel and its tributaries, a floodplain, or the overflow lowland bordering the river, and the **stream terraces** that normally define the lateral dimensions of the river's domain (fig. 5.1). How much water, and over what amount of time, will find its way into the stream system depends on a number of factors. No universal hierarchy of importance can be assigned to such factors; their individual import varies according to the specific physical setting. The ultimate factor, of course, is the amount of water that a given stream system is expected to accommodate. If input of water, for one reason or another, exceeds a river's capacity the channel will overflow.

Such *overflow floods* may be a regular event to which civilizations have adapted for thousands of years. Until shackled by the construction of several *barrages,* including the huge Aswan High Dam, the Nile regularly flooded the lower sections of its valley and thus supplied water for irrigation and silt to restore soil fertility. Other examples of *seasonal flooding* can be found in the monsoon lands of Asia. Here, the runoff from the torrential *monsoon* rains swell rivers such as the Ganges, Irrawaddy, Brahmaputra, Mekong, and other rivers in that part of the world. People living in such an environment have learned to accommodate their life-styles to such recurring events by (1) engaging in limited seasonal migration to higher grounds, (2) constructing their homes on the higher stream terraces above the most threatened sections of the floodplains, and (3) building houses on stilts sufficiently high to outlast all but the worst floods.

Seasonal flooding may also be produced by spring snowmelt in areas of abundant snow accumu-

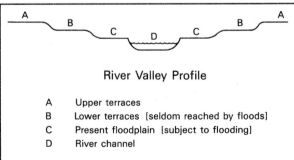

Figure 5.1. Low-lying floodplains (C) are subject to frequent inundations and should not be used for construction and development unless adequately protected. The upper stream-cut terraces (A,B) are much safer from flooding.

lation and marked seasonal temperature changes. How rapidly a snowpack will melt depends primarily upon the energy input in the form of *solar and terrestrial radiation,* and upon the air temperature. If rain falls on the snow, the melting rate may accelerate, especially if the heat content of the rain is high. Moreover, the severity of the runoff is greatly affected by the snowpack's ability, even if only temporary, to absorb and hold the rainwater. This absorption factor varies with the amount of rainfall, the thickness of the snow layer, and the physical conditions of the snow itself. For these reasons identical rainstorms on snow may produce quite different *flood potentials* (Ward; 1978, p. 33). In addition to the precipitation factor one must also consider the influence of topography, soils, and vegetation conditions on flood potentials.

The role of *topography* is a composite of *slope steepness, slope length,* and also *slope orientation.* A short slope obviously reduces the time for water to enter the surface material, especially if it is composed of bare rock. In a similar way a steep slope will diminish the absorption capacity of a hillside.

Slope orientation, that is, the position relative to prevailing winds and to the sun, also plays an important role in producing flood potentials. In the Northern Hemisphere, for example, a heavy accumulation of snow is more exposed to solar radiation on the southern side than on the north-facing slopes. This factor accelerates melting and quick runoffs, leading to more flooding in the valleys below the southern slopes. Precipitation, in the form of either snow or rain, is much heavier on the *wind-facing* slopes; this is called the *orographic effect* (fig. 5.2). As air is forced upslope it will cool rapidly by expansion and soon may reach its *condensation temperature.* This happens because air at lower temperatures cannot

hold as much moisture as at higher temperatures. Depending on this condensation temperature either water droplets or ice particles develop and form heavy cloud formations. The considerable amounts of precipitation that may fall from these clouds can lead to heavy runoff. This effect illustrates why windward sides of mountains tend to create greater flood potentials than the downwind slopes.

Ground conditions, and especially the types of soils present, also constitute a major factor in assessing flood potentials. *Impervious* rock surfaces, as well as soils with a high *clay* content, resist effective water *infiltration.* This behavior is especially noticeable when clayey soils are already saturated with water from earlier rainfall or from melting snow.

Several soil factors determine how quickly water is absorbed by soil and subsequently allowed to pass through it. Movement into the soil is mostly controlled by *texture,* or particle size, by the *structure* of the soil, and also by the amount of organic matter contained in it. The greater the quantity of organic material, and the coarser it is, the more water can enter the soil (Donahue et al.; 1977, p. 85). Infiltration of water, and *percolation* within the soil, is also very much affected by frozen soil material. In the spring some of the subsoils may still be frozen. This condition makes a soil impervious or will markedly slow down *hydraulic conductivity.* All these soil factors lead to heavy runoffs instead of feeding the water slowly into the stream system.

The absence or presence of vegetation also affects flood potentials in several ways. Plants are capable of removing large quantities of water from the soil to support their life functions, including *photosynthesis.* Plants, therefore, assist in preventing soils from becoming saturated. A dense vegetation cover also slows down both soil erosion and runoff, and the soil-plant system stores water like a sponge. Unwise forestry practices, culminating in the reckless destruction of the tropical rainforest, lead to the denudation of sensitive areas, thereby increasing flood potentials of many regions.

Nature of Floods

The preceding discussion has shown how a variety of factors contribute to the threat of flooding. It was also pointed out that the rise and fall of rivers is in response to natural conditions, and that humans need to understand such processes and should adapt to them. It is important to keep in mind that each flood is different in intensity and behavior; therefore, each must be appraised within the context of its total setting.

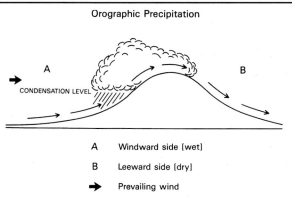

Figure 5.2. Orographic precipitation may occur when moist warm air is forced to rise up the windward slope of a mountain barrier.

Minor flooding occurs each year, but it is the unusually big ones that get our attention. These large floods are the result of relatively infrequent combinations of weather and hydrologic conditions. We hear at times expressions such as "a 10-year flood," "a 25-year flood," or even "a 100-year flood." These terms refer to the magnitude of a flood that tends to occur once in 10 years, 25 years, and 100 years, respectively. This, by no means, is to imply a certainty but is merely a probability or projection based on past flooding history. In most cases, unfortunately, detailed flood histories—hydraulic and atmospheric data—are either not available or are too inadequate for accurate flood potential assessments. Theoretically it is possible to have two, or more, 25-year floods in one season; yet, they may not occur at all within 25 years, or longer. These expressions, therefore, are more descriptive of their magnitude than precise in their meanings.

Floods can be categorized according to their *causes* (snowmelt, heavy rainfalls, ice-jam floods, rain-on-snow floods, and many other types) and *magnitude.* They can also be described in terms of *single* or *multiple* peaks, speed of development, and their behavior within their flood paths.

Probably the most common river flood is the *single-event* flood. This flood is produced by intensive precipitation over a number of hours, or even days, over a given drainage basin of a river (Ward; 1978, p. 19). One has to realize that only 1 inch of rainfall (2.54 cm) on 1 acre (0.4 hectare) of land represents about 26,000 gallons (98,384 liters) of water. This explains why one vigorous storm system, especially when moving slowly, can easily cause severe flooding if other factors are also present.

Maximum amounts of precipitation deriving from midlatitude cyclones are associated with the *frontal systems* of such storms. Along such weather fronts there is an uplift of warm and moist air over colder and drier air. In the Northern Hemisphere these fronts usually develop in the south-facing half of the storm system. It is for this reason that the location of the storm's path relative to a river's drainage basin is a major factor in how much precipitation can be expected. If the storm path is north of the basin, the weather fronts will pass over it, and heavier precipitation amounts can be expected. If the storm track is to the south, the frontal action is avoided and much less precipitation is the rule. A more detailed explanation of weather fronts can be found in connection with disasters involving the atmosphere in chapter 6.

In the midlatitudes, where series of storm systems follow each other in the spring season, *multiple flooding* may result in successive flood peaks. Such type of flooding can become very damaging because it extends over prolonged time periods. This sequence means that the soils become saturated as their water storage capacity is exceeded, flood reservoirs are full, and many structures could be weakened by persistent water pressure and exposure to moisture.

Minor flooding usually can be handled by most river channels. But when channels are filled to capacity, without spilling over into the adjoining floodplain, they are said to be in their *bankful stage.* During this stage the river engages in **hydraulic action,** dislodging materials from the bottom of the channel and from the river banks, while *abrasion* grinds down the larger particles into finer sediments (Greenland, DeBlij; 1977, p. 356). Both processes are important contributors to the deepening of the channel. Once the water flows over into the floodplain the river is said to be in flood stage. Engineers consider a river to be "in flood" when its water has risen to an elevation (flood stage) at which damage can occur in the absence of protective work (Coates; 1981, pp. 365–366).

As large quantities of water from the upper reaches of the drainage basin find their way into the main stream channel, a **flood crest** builds up and will work itself downstream. Such a flood crest, also called a *flood wave,* increases the water level of the stream; thus the flow velocity will increase because the flood wave will move faster than the stream. This kind of crest usually is very high in the narrow-channel area upstream, where it moves at high velocity, but tends to be of relatively short duration. As the highwater bulge moves down the river channel it loses height and increases in length. This is accompanied by a decline in the flow velocity so that the flood crest stays longer in the lower stream sections (fig. 5.3). The total time needed for such a flood wave to be discharged may take a few hours, or days, for smaller streams to several weeks in major river systems (Hoyt; 1955, p. 7).

Flood conditions may take several hours, or days, to develop completely. This means that threatening weather and hydrological factors can be monitored and warnings may be issued in time for people to take precautions. However, one type of flood, the **flash flood,** may unleash its force practically without warning and can result in severe damage and death.

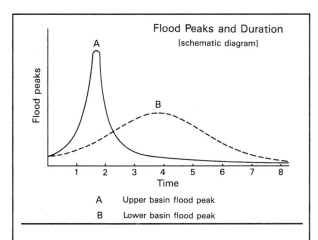

Figure 5.3. Flood peaks (A) tend to be higher and of shorter duration in the upper sections of a river basin. Floods in the lower basin are lower but persist over longer time periods (B).

◎ The Great Mississippi Basin Flood of 1993

The **drainage basin** of the Mississippi embraces a total of about 1.25 million square miles (3.3 million km2); it is surpassed only by those of the Amazon and Congo rivers. The Mississippi has never been called "America's Sorrow"—in contrast to the Yellow River which is often referred to as "China's Sorrow"—but its vast basin, which receives the waters of some 100,000 streams, has produced many devastating flood disasters. The catastrophe of the summer of 1993 exceeded in volume and in damage all floods that have occurred in the United States in modern times.

Construction of flood control levees started before the turn of the century, and by 1927 the Corps of Engineers was confident that the levees were able to constrain the mighty Mississippi; however, in April 1927 a major flood shattered this optimistic assessment: more than 16 million acres (6.5 million hectares) were flooded and over 300 people perished.

In subsequent years, in response to the 1928 Congressional Flood Control Act, the battle against the Mississippi was resumed by building more levees and large diversion channels. Major floods in 1937 and 1973 defeated the struggle against the river and inundated 8 million acres (3.3 million hectares) and 16.5 million acres (6.7 million hectares), respectively. Over 270 persons lost their lives in the two events (Clark; 1982, pp. 80–82), but the impact of these floods paled against the havoc

wreaked by the event of 1993.

The casualties associated with the 1993 flood were relatively few. About 50 people drowned, but 54,000 had to be evacuated. Fifty thousand homes were destroyed or severely damaged, more than 20 million acres (8 million hectares) were under water [Plate 32], the harvests in most of the flooded areas were lost, more than 600 billion tons of topsoil washed away, and 18 percent of the Federal levees, as well as 78 percent of the non-Federal levees, failed or were overtopped (NOAA; Natural Disaster Survey Report, 1994, pp. 1–4 to 1–5).

What forces generated this monstrous flood? It appears that several unusual conditions conspired in producing this disaster: (1) persistent heavy rains occurred over a period of 9 to 10 months preceding the major flooding. In most areas, the amount of precipitation averaged 150 percent above normal, (2) the soils became saturated and were unable to absorb any additional moisture. This situation resulted in very rapid runoff and raised many streams to bankfull or flood levels, (3) abnormally low atmospheric sea-level pressure prevailed over the central and western North Pacific Ocean, and this low pressure also reached into western United States. The **jet stream** shifted accordingly and attained a distinct southwest-to-northeast orientation. This position allowed a clockwise circulation over the eastern United States to push moist air masses from the Gulf of Mexico into the Mississippi basin, and (4) the almost stationary jet stream was associated with a fixed surface cool front which forced moisture-laden air from the south to overrun the colder air to the north. This uplift, in combination with several **cyclonic depressions,** produced heavy precipitation (NOAA; Natural Disaster Survey Report, 1994, pp. 3–15 to 3–17).

Several other causes have been mentioned as possible triggers for the abnormally high precipitation and the then prevailing atmospheric circulation patterns. Such factors included the eruption of Mount Pinatubo (see Chapter 2), the greenhouse effect (see Chapter 14), and ocean surface temperature anomalies brought on by an oceanic phenomenon called "**El Niño**" (see Chapter 10). El Niño allowed warm equatorial waters to move unusually far northward to become a source for large amounts of atmospheric moisture and for disturbances in atmospheric circulation.

Major floods which occur in very large areas, such as the Mississippi basin, are not "one blow" events but extend over considerable time periods—

Plate 32. Members of the National Guard and local residents labored day and night to build barriers against the rising waters of the Mississippi during the 1993 flood disaster. (Photograph credit: National Weather Service; Office of Hydrology; Paul Griffin, Department of Defense).

days, weeks, or even months—and show marked differences from place to place within the total area affected by the flood. The Great Flood of 1993 began in early June and lasted through August. Moreover, the flood was a composite event that involved a number of tributary river basins. Some of these basins developed rapid rises in stream levels which led to very intense and relatively short-duration flood events that could qualify for being referred to as "flash floods." This type of flood may unleash its force practically without warning and is a very dangerous flood phenomenon.

Flash Floods

Flash floods may form anywhere in the world, but they are especially unpredictable in some arid environments where rainfall is scarce and quite erratic. Dried-out soil surfaces in such areas contribute to flood potentials. These soils may be initially *hydrophobic*, that is, they have a very low affinity for water, and repel water infiltration until they are sufficiently wetted. Subsequent infiltration rates may slow down as small soil particles, washed along the surface, may clog soil pores. In many instances flash flooding is associated with powerful, local thunderstorms that produce sudden torrential downpours turning dry ravines and desert washes into raging torrents [Plate 33].

A flash flood of this kind occurred on August 17, 1954, in northern Iran where total annual precipitation varies a great deal but hardly ever exceeds about 8 inches (20 cm). A severe thunderstorm near the town of Farahzad soaked the south-facing slopes of the Elburz Mountains. A deluge of water rushed into the normally dry, steep-sided valleys and converged on the Moslem shrine of Imamzadeh Davoud. A wall of water, estimated at 90 feet (27.4 m) in height, overwhelmed a large number of pilgrims, killing about 1,000 people (Clark; 1982, p. 98).

Another well-known flash flood disaster killed 140 people in the Big Thompson Canyon disaster in

Plate 33. Sudden downpours from mountain thunderstorms can turn dry ravines and desert washes into raging torrents. These flash floods are unpredictable and therefore very hazardous to unwary campers and hikers. (Photograph credit: R. C. Holmgren).

Colorado on July 31, 1976. The steep topography, in combination with surface materials of very low water-absorption capacity, could not handle the massive runoff from a sudden series of torrential downpours that reached a maximum intensity of more than 12 inches (30 cm) between 6:30 P.M. and 11:00 P.M. (Coates; 1981, p. 376).

Dam Failures

In addition to floods that result from various natural causes there are also disastrous events linked to humans as a modifier of our environment. Failures of dams, caused either by selecting unsuitable construction materials, by the lack of maintenance, or by reason of locating such structures in geologically unstable settings, have destroyed immense amounts of property, washed away farmland, obliterated entire communities, and killed thousands of people in all parts of the world. For example, excessive rainfall brought on the collapse of an earthen dam in India in 1979, killing about 5,000 persons.

One of the worst flood disasters, involving a dam failure in the United States, took place in Pennsylvania on May 31, 1889, when an old earthen dam collapsed. The resulting flood killed 2,200 people of Johnstown. This disaster was the result of human neglect, combined with poor judgment, and a two-month period of unusually heavy rainfall.

The earthen dam, originally constructed to impound water for the Pennsylvania Canal, had been completed by 1853. The dam, 70 feet (21 m) high and nearly 900 feet (275 m) in length, held back a reservoir that varied in depth from 40 to 100 feet (12 to 31 m) and had a maximum capacity of 17,747,022 cubic yards (13,568,060 m³).

The facility was sold in 1857 and suffered many years of neglect which was manifested by several leakages mainly around the main drain pipes. In the late 1870s the reservoir, then called Lake Conemaugh, and the dam were sold to private interest, and the area became known as the "South Fork Fishing and Hunting Club." A road was built across the length of the dam, which lowered it by about 2 feet (61 cm). A grating was placed over the spillway to prevent the game fish from escaping from the lake. But this action trapped a large amount of floating debris and unfortunately clogged the outlet.

For two months heavy precipitation had fallen on the region, including a 14-inch (35.6 cm) snowfall in April. This was followed by almost incessant rainfall, which accelerated the melting of the snow. Abundant rain fell on May 30 and continued into Friday, May 31. By noon of that day the water of the reservoir started to spill over the dam and the structure deteriorated rapidly. About 3:15 P.M. the central portion of the dam gave way, and a flood wave estimated at 30 to 40 feet (9 to 12 m) roared toward Johnstown located about 15 miles (24 km) down the valley (A.S.C.E., Transactions; 1891, pp. 343–344).

Had the people of Johnstown been aware of the impending danger? Why did 2,200 of the inhabitants drown? Different sources give different answers. It was said that a warning came by telegraph as late as 2:25 P.M. saying that "the dam is getting worse and may possibly go" (Butler; 1976, p. 160). On the other hand it is also possible that the population of Johnstown had been alerted, year after year, to the potential failure of the dam, and the people began to ignore these warnings, paying a terrible price for their complacency.

Another type of flooding is brought on almost entirely by rendering the land's surface impervious through paving, building airports, and highways, that is, through general urbanization. The negative impact of urbanization on flood potentials is our next topic.

Urban Flooding

In many ways humans have drastically changed the surface of the earth through deforestation, agriculture, introduction of different vegetation types, through extensive construction involving the movement of large amounts of earth materials, and by enlarging the areas covered by towns and cities: **urbanization.** In each case, both on a small or on a large scale, the **hydrology** of an area is affected. This change becomes quite evident when one studies *hydrographs* of urbanized areas. Hydrographs are graphs of water levels and rates of flow as a function of time.

Urban hydrographs usually show sharp peaks of maximum water levels and a marked reduction in time between the occurrence of maximum precipitation and the resulting runoff. This analysis is being referred to as the *lag time* (Gregory, Walling; 1979, p. 70).

Two major environmental modifications are primarily responsible for drastically altering the rainfall-runoff relationship: (1) making the land surface impervious by covering it with pavement and construction work, and (2) installing storm sewer systems that collect urban runoff and rapidly discharge large volumes of water into stream systems. The impact of the first factor is illustrated by the following: If a 10-acre (4 hectares) shopping plaza is covered by a 10-inch (25.4 cm) snowfall, corresponding to 1 inch (2.54 cm) of rainfall, it would generate a runoff of about 260,000 gallons (984,000 liters, or 984 m^3) of water. This amount of runoff pours into the storm sewers and quickly finds its way into a discharge area, usually a stream. In a very fitting way this process can be described as a *flashy discharge,* which is responsible for the abnormally rapid changes in water levels in urban streams [Plate 34].

This easily traceable effect of urbanized regions on flood potentials becomes obscured whenever

Plate 34. When heavy rainfall exceeds the capacity of storm sewers, widespread flooding may occur quite independently from the overflow of nearby rivers. (Photo by C.H.V. Ebert).

unusually heavy rainfall occurs that may lead to general flooding; then the hydrographs of urban and rural areas tend to become more alike. In a study on the effects of urbanization on floods it was found that the impact of urbanization is felt most in floods of relatively low magnitude such as 10-year floods (Hollis; 1975, pp. 431–435).

There is no easy solution to this problem. On the one hand it is necessary to remove excess water from streets, parking lots, and recreational facilities. On the other side the resulting floods may cause severe stream channel erosion and floodplain inundations. In some favorable locations it may be possible to conduct the excess water into natural or artificial *catchment areas* that allow the water to slowly percolate back into the groundwater. This technique was applied on Long Island, New York, where infiltration rates are high because of coarse sandy ground materials (Strahler, Strahler; 1977, p. 199). Another way to make a stream less prone to flooding is to modify its channel. This effort may include the deepening of the channel, strengthening the banks to prevent erosion, and constructing protective levees and diversion channels.

These two factors, making the land surface impervious and the use of storm sewers, definitely are the main culprits in urban stream flooding, but other typical suburban and urban activities also influence the local hydrology. These activities include the washing of streets, waste treatment, the watering of lawns and shrubs, and the dumping of snow into low-lying tracts or even directly into the streams; all these actions affect the levels of streams.

More significant, of course, than the last-mentioned factors is the basic folly of placing homes, at times entire subdivisions, and industrial facilities in vulnerable floodplains. What attracts both developers as well as home buyers to floodplains? Floodplains tend to be level tracts of land. This means that fewer terrain modifications are necessary for the construction of street networks and buildings. Moreover, floodplains are made up of *alluvial soils,* which often contain considerable amounts of sandy and gravelly materials, a fact that makes excavations less difficult and therefore less costly. Another consideration is that many floodplain locations, especially when left in their natural conditions, are aesthetically attractive: They offer a home for a variety of wildlife and in many instances allow for access to a stream for boating and other recreational activities.

The problem of urban flooding will not go away easily because *urban sprawl* appears to outpace flood control efforts. Such efforts are very costly and require a great amount of careful planning and time. After World War II, urbanization in the United States, including service facilities and transportation, consumed nearly 1 million acres (405,000 hectares) a year. It is estimated that the amount of urbanized land in the continental United States will reach 75 million acres (31 million hectares) by the year 2010, compared to 18 million acres (6.5 million hectares) in 1950 (Wagner; 1971, p. 374). In the western region of New York State urban sprawl and a variety of flooding problems, associated with a local stream, present a difficult situation for the Town of Amherst, and for adjoining communities. The following discussion illustrates the interaction between *physical* and *human-made* factors causing flood problems in a rapidly developing urban setting.

⊚ Ellicott Creek: A Complex Flood Problem

Ellicott Creek, with a total length of 47 miles (76 km), drains an area of about 110 sq. miles (285 sq.km) in Erie, Genesee, and Wyoming Counties in the western part of New York. This creek is the major tributary to Tonawanda Creek, which discharges into the Niagara River. Ellicott Creek begins about 22 miles (35 km) to the east of Buffalo. It collects runoff from the hilly country in its headwaters and then follows a general northwesterly direction through gently rolling to flat topography. The widest portion of the creek's floodplain is in the Town of Amherst which includes the Village of Williamsville. After having plunged over the 60-foot (18 m) Onandaga Escarpment in Williamsville the creek engages in a relatively quiet flow with an average gradient of 2 feet per mile (38 cm/km).

The Town of Amherst is a fast-growing suburban community that is located on the northern city limit of Buffalo. In 1930, Amherst was a small town of 13,181 inhabitants in a predominantly rural setting. By 1950, the population had risen to 33,744; and in 1985 it exceeded 112,000. In 1972 agricultural and vacant land still accounted for 50.2 percent of the total land area of Amherst, but it declined to 39.8 percent by 1985. This change was accompanied by a corresponding increase in developed land (Town of Amherst; Master Plan, 1985).

In response to this increase in population the demand grew for more roads and highways, more homes, more schools, and for an ever-increasing number of commercial establishments. All this could be accomplished only by rendering an increasing percentage of the land area impervious through paving and construction. The ambitious Master Plan of the Town of Amherst envisions that 89.7 percent of the still vacant land will be zoned residential, 4.5 percent commercial, 3.2 percent industrial, and 2.6 percent community facilities. There is no question that this overall trend, leaving only 7.3 percent of the town's total area undeveloped, already had a major impact on the hydrology of the region. Recalling the discussion of runoff from impervious surfaces, let us look at just one example that underscores the potential severity of the developing problem.

In the northern section of Amherst is a large 67-acre (27 hectares) shopping plaza. About 58 acres (23 hectares), or 88 percent of this plaza, is impervious. This surface generates a runoff of 1.5 million gallons (5.7 million liters) of water in response to 1 inch (2.54 cm) of rainfall or about 10 inches (25.4 cm) of snow.

Floods repeatedly have done severe damage to Amherst. Serious flooding occurred in 1929, 1936, 1960, 1977, and in 1985. Damaging floods may occur at any time of the year if conditions are right; however, almost all major floods in this area took place in the late winter months or in early spring during periods when melting snow and rainfall coincide. From flood to flood more homes were damaged and more property was lost because an increasing amount of urban development had moved into the floodplain sections. Such development also includes the largest portion of the new 1,200-acre (486 hectares) Amherst Campus of State University of New York at Buffalo.

One of the worst floods occurred one month after the "Blizzard of January, 1985," which left a 3-foot (91 cm) snowpack in the area. During the critical days of February 21 to 24, when mild temperatures melted the snow, an additional rainfall of 1.5 inches (3.8 cm) fell, which accelerated the snowmelt. This precipitation increased the total runoff from the snowpack that, by itself, had an estimated water content from 3 to 5 inches (7.6 to 12.7 cm) [Plates 35 and 36].

Ellicott Creek crested at 11.2 feet (3.4 m) about 2:00 A.M. on Monday, February 25 (Wuerch; 1985), widely overflowing into the low-lying sections of Amherst and causing millions of dollars in damage. A considerable amount of flooding in areas outside the floodplain apparently was caused

Plate 35. In February 1985, a rapid snowmelt of a 3-foot (91 cm) snowpack, and a rainfall of 1.5 inches (3.8 cm), caused extensive flooding in the Town of Amherst, New York. (Photo by C.H.V. Ebert).

Plate 36. In February 1985, a rapid snowmelt, ice-clogged storm sewers, and hydrostatic pressure from high-running Ellicott Creek turned some streets in Amherst, New York, into streams. (Photo by C.H.V. Ebert).

by the inability of the storm sewers to handle the massive runoff from the town. Similar flooding of this nature, although much less severe, was observed after periods of heavy downpours from severe thunderstorms, and also from prolonged frontal precipitation, which did not cause any significant rise in the creek itself. This is a good example of typical *urban flooding* caused exclusively from the runoff from the impervious areas and the inadequacy of drainage facilities.

A large flood control project, planned for decades, is now completed. Channel corrections, construction of extensive diversion channels, and strengthening of the stream banks are the main components of the flood control and erosion control project. Whether the expected beneficial effects can match the area's plans for further urbanization only the future can tell.

DISASTERS INVOLVING THE ATMOSPHERE

3

**P
A
R
T**

Key Terms	
weather	troposphere
climate	homosphere
air masses	carbon dioxide
meteorology	stratosphere
barometer	jet stream
radar	ozone layer
RAWIN	tornado
remote sensing	

Pliny the Elder (A.D. 23–79), the famous Roman naturalist, stated his belief that air is the principle of life, penetrating all of the universe. There is no question that the atmosphere is a critical component to all living things on the surface of the earth and, in an indirect way, to most life-forms in the ocean.

We cannot perceive the *atmosphere,* that in its normal condition is without odor, taste, or color. However, we can observe many processes that take place in this immense envelope of gases: clouds, lightning, rain, snow, wind, and many other atmospheric phenomena. These features, some physical and some chemical in nature, are the manifestations of the dynamic aspects of the atmosphere and its interactions with other substances (dust, ice particles, smoke) that are being introduced into the atmosphere by both nature and humans.

The atmosphere is one of the most powerful influences on human life (Hidore; 1969, p. vii). Although one immediately tends to think of **weather,** the momentary physical state of the atmosphere at any given place, or of **climate,** the sum total of weather elements that reflects long-term averages and extreme states of the atmosphere, many other atmospheric functions and effects are often overlooked. Without the atmosphere, space particles, such as *meteorites,* would slam into the earth instead of burning up within the atmosphere. Harmful types of *radiation,* including most of the dangerous *ultraviolet* rays, would destroy life on earth if they were not absorbed by our protective mantle of gases. Moreover, the atmosphere, as a result of *differential heating,* develops areas of contrasting pressure that induce the exchange of **air masses** of different properties. This heat exchange, similar to that performed by ocean currents, balances our *latitudinal heat budget* by transporting heat and moisture from the equatorial regions to the mid- and higher latitudes. Without the atmosphere the temperatures of the earth would rise to over 200° F (93° C) during the day and drop to near −300° F (−185° C) at night (Critchfield;

1960, p. 8). Similarly, moist air is moved from the oceans into the continental interiors to supply needed moisture. The annual amount of water evaporated from the world's oceans is estimated to be near 0.024 percent of the total volume contained by the oceans, or 80,160 cubic miles (33,400 km^3) of the total 328 million cubic miles (1.3 billion km^3) (Schaefer; 1987, p. 57).

Wind, the exchange of air between high pressure and low pressure centers, also transports dust, pollen, pollution ingredients, and even insects over considerable distances. The same wind may erode soil, generate major dust storms and blizzards, drive towering waves across the oceans, accompany violent hurricanes and tornadoes, and form dangerous squalls and shear lines. These aspects will be discussed in the following chapters.

From these basic concepts mentioned, we can readily see that the role of the atmosphere is, indeed, extremely complex. It is difficult to visualize the depth of the ocean, or the height of Mount Everest, but it is even more perplexing to get a clear perception of the makeup and the spatial dimensions of the earth's atmosphere.

The serious study of the atmosphere probably began when early peoples developed systematic farming, realizing the basic dependence on favorable weather and seasons. A similar awareness must have occurred to sailors as they waited for advantageous winds to carry their ships across the seas. Yet, scientific studies of the atmosphere, leading to the field of **meteorology,** had to await, as was true also for *oceanography,* the development of suitable instrumentation. Important steps were the invention of the *thermometer* by Galileo in 1593, and the **barometer** by Torricelli around 1643. About 250 years later kites carried thermometers into the air, followed by balloons in 1909. By 1919, airplanes carried diverse meteorological instruments to collect weather data.

More accurate wind data was obtained shortly after the development of *radar.* This led to the **RAWIN** (*radar wind*) system based on the reflection of radar waves from metallic surfaces carried aloft by weather balloons to elevations of 100,000 feet (30,500 m). With the subsequent appearance of rockets and weather satellites humans created better methods for exploring the far reaches of our atmosphere. On April 1, 1960, the first weather satellite, TIROS I (Television and Infrared Observation Satellite), was launched and opened the era of global weather observation (Hubert; 1967, p. 11).

Satellites, space probes, and *remote sensing* techniques have given us a much better understanding of the nature and dimensions of the atmosphere. We now know that the space about us is not "empty," but that there is a certain continuity of matter. Thus one can say that the atmosphere of Earth, in a strict sense, blends with that of the sun; however, that portion which supports life on our planet, and in which weather processes take place, is compressed into a layer called the *troposphere.* The troposphere varies in thickness from about 5 miles (8 km) near the poles to 9 to 10 miles (14 to 16 km) along the equator. Stated in different terms: On the average about one half of the earth's atmosphere is found below 18,000 feet (5,490 m). This means that many of the world's highest mountains extend considerably beyond half the mass of the atmosphere.

Beyond the troposphere we encounter several other layers of much lower density and somewhat different composition. Below 50 miles (80 km) the atmosphere is relatively well mixed and is referred to as the **homosphere.** The gases, which are quite constant in composition, include nitrogen (78%), oxygen (21%), and argon (0.93%). These make up the bulk of the atmosphere. Two other gases, both variable and quite important in their influence on atmospheric conditions in general, and on the environment in particular, are **carbon dioxide** (average 0.03%) and *water vapor.* Of the two, water vapor varies most ranging from almost zero to about 4 percent. The great variability of water can be partially explained by the fact that it can exist in the atmosphere as gas, as a liquid, and as ice. In all its states water is a critical ingredient in practically all weather processes. Yet, the water contained in the atmosphere accounts for only 0.001 percent of the hydrosphere compared to 98 percent of the water contained in the ocean (Gross; 1971, p. 3).

Above the homosphere we encounter a more stratified arrangement of gases in a similar way as it occurs with liquids of different density (Miller; 1971, pp. 5, 6). Immediately above the troposphere ties the **stratosphere,** which extends to about 31 miles (50 km). Very few weather phenomena are associated with the stratosphere, but these are of particular interest to us because they influence our weather systems and thereby affect life on earth. Two of these phenomena are the **jet stream,** a high-velocity upper-air current, and the **ozone layer.**

The jet stream, which exists in several bands, is a relatively narrow wind stream encountered at elevations ranging from 20,000 to 40,000 feet (6,000 to 12,000 m). It moves at speeds of about 50 mph (80 km) in the summer to above 100 mph (160 km) in the winter. These speeds pertain to its core, which makes up part of the jet stream's total width of about 300 miles (482 km). The jet stream was first noted in World War II, when B-29 bombers flew from the Marianas to Japan. The pilots consistently reported strong westerly winds in excess of those normally expected (Lehr et al.; 1957, p. 43).

The ozone layer, with its greatest concentration at about 15 miles (25 km) elevation, is of prime importance to our existence because it absorbs large quantities of ultraviolet radiation. This type of radiation would prove to be exceedingly harmful, if not lethal, if it were to strike the earth's surface in full force. The role of ozone, as an important part of so-called *photochemical smog,* will be discussed later. At this point we are ready to look at some types of atmospheric events that may lead to hazardous conditions and at times to overwhelming disasters. The first two are the hurricane, or tropical cyclone, and the most violent of all storms, the *tornado.*

HURRICANES AND TORNADOES

Key Terms	
vorticity	cloud seeding
willy willy	Operation Stormfury Americas
baguio	twister
easterly waves	cold fronts
intertropical convergence	latent energy
northeasterly trade winds	wall cloud
eye	cumulonimbus
Coriolis effect	SKYWARN
convection	

The earth's atmosphere, because of pressure differences within it, is constantly in motion. Some of these motions are in the form of gentle breezes, others involve fierce winds in connection with storm systems. Basically, there are three major low-pressure weather disturbances: the *hurricane,* the *tornado,* and the *midlatitude cyclone.* The latter, showing as a *Low* on our weather maps, is classified as a nonviolent storm; yet, many severe weather phenomena are associated with it. They may be triggered along frontal systems as will be discussed in chapters 7, 8, and 9. All of the above storm types display various degrees of **vorticity,** which is, by general definition, a rotational circulation of air around a low-pressure center.

Before we concentrate on the violent nature of these weather disturbances it is important to point out that these storms are essential in many ways: They equalize heat energy buildup in the atmosphere, they supply needed precipitation, and they produce winds that can perform many useful tasks. Winds drive windmills and turbines, create wave energy on the oceans, cool tropical shores, and foster many forms of recreation, such as sailing and soaring. In contrast to nature's impact in the form of hurricanes and tornadoes, these ecological and peaceful functions of the atmosphere are simply taken for granted and do not make the headlines.

Development of Hurricanes

Hurricanes, large-scale tropical storms, develop over just about all tropical oceans. The warm waters supply the enormous amount of *thermal energy* required to generate and to sustain these storms. The energy flow in a hurricane in one day can amount to the equivalent energy released by four hundred 20-megaton hydrogen bombs (NOAA/PA 76008; 1977, p. 11).

The worldwide occurrence of hurricanes is reflected by the different names given to these violent storms: "hurricane" in the West Indies and North America, "*willy willy*" in Australia, "cyclone" in India and Bangladesh, and "*baguio*" in the South China Sea and the Philippines. As a general rule these tropical storms affect mainly the east coasts of the continents, or wherever low-latitude warm ocean currents predominate.

Hurricanes tend to form in the low latitudes (5 to 20 degrees) as weak *tropical depressions.* These may be indicated by cloudiness, by moderate wind velocities, and by precipitation. Such depressions, also referred to as **easterly waves,** appear most frequently in the late summer when the **Intertropical Convergence,** or I.T.C., is at its greatest distance from the equator. This low-pressure zone attracts the **northeasterly trade winds** from the Northern Hemisphere and the *southeasterly trades* from the Southern Hemisphere. The low-pressure zone shifts in response to the annual migration of the sun, as do all planetary wind and pressure belts.

Not all tropical depressions evolve into hurricanes. If they do, they usually go through a stage called *tropical storm* with winds up to 75 mph (120 km/h). If the winds exceed this velocity the storm is classified as a hurricane.

Hurricane Characteristics

A fully developed hurricane may have a diameter of 100 to 300 miles (160 to 480 km). From above, it is almost circular in appearance with winds blowing counterclockwise (Northern Hemisphere) around a relatively calm center, the so-called "*eye*" of the storm. At sea level the atmospheric pressure may drop below 28 inches of mercury (Hg) (948.2 millibars). One of the lowest pressure readings was associated with a hurricane near the Philippines (1927) when the barometer dropped to 26.185 inches Hg (886.7 millibars). Another record low of 25.9 inches Hg (877 millibars) occurred in 1958 in the same region (Eagleman; 1983, p. 172).

The eye is the location of the lowest atmospheric pressure; however, it is not reached by the high-velocity winds (up to 200 mph, or 320 km/h), because *centrifugal* forces and the **Coriolis effect** of the earth's rotation keep them from reaching the center of the storm (fig. 6.1). Thus, while the winds and towering clouds roar around the eye, the weather

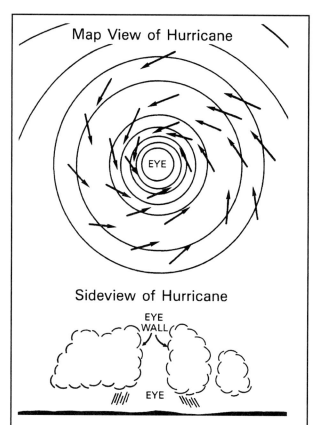

Figure 6.1. A map view of a hurricane shows its closely spaced near-circular isobars and a counterclockwise (Northern Hemisphere) wind system. A hurricane's typical diameter is about 300 miles (480 km). The area of lowest pressure (the eye) measures about 25 miles across (40 km) and is bounded by the eye wall.

conditions in this calm zone are typified by relatively clear skies and wind velocities as low as 10 mph (16 km/h).

The mature hurricane can be likened to a huge whirlpool [Plate 37]. The winds on the outer edge blow around the storm and slightly toward the center. As they approach the central *vortex,* the velocity increases until the highest speed is obtained in the inner zone around the eye. Here, as marked by the towering clouds of the *eye wall,* the moist warm air is chimneyed upward, releasing the heat of condensation needed to sustain the storm system. As upward **convection** and *condensation* take place, the thermal energy is changed into kinetic energy of motion, drawing fresh supplies of moist warm air from the outside into the storm (Taylor; 1954, p. 252).

As long as the hurricane remains in contact with warm ocean surfaces, fresh amounts of water vapor are drawn into the vortex. This not only leads to the

Plate 37. This satellite photograph of Hurricane ALLEN (August 9, 1980) shows the near-circular appearance of such tropical storms. Note the spiraling cloud pattern around the eye of the storm. (Photograph credit: NOAA; National Hurricane Center).

release of huge quantities of energy during the process of condensation, but also produces torrential rains. Once the hurricane moves inland, the supply of oceanic water vapor is cut off, and the storm begins to lose strength. Yet, even though the system weakens, and may no longer qualify to be called a hurricane, its remnants generate precipitation in the range of 6 to 12 inches (15 to 30 cm) in the areas over which the disturbance passes. Another reason for the storm losing strength upon entering a land area is the friction with the land surface; because the friction with land is much greater than with the ocean, the storm usually loses speed.

The overall movement of hurricanes is guided by a number of factors, which include upper atmospheric winds, other weather systems, and warm ocean currents. These factors can be monitored, and the path of a hurricane can be reasonably well predicted. This seems to be true for about 70 percent of these storms (Eagleman; 1983, p. 206). Some hurricanes, however, may move quite erratically and may even become stationary. On the average they travel at about 10 to 20 mph (16 to 32 km/h).

Which factor is dominant in influence on a hurricane's path depends entirely on the specific setting of each storm. One of the worst hurricanes (HAZEL, October 5–18, 1954) originated in the vicinity of Trinidad. Subsequently it followed the warm waters of the Caribbean Sea, which it left between Haiti and Cuba to set a more northerly course. The subsequent route adhered to the warm Gulf Stream until it moved inland near the border between South and North Carolina. Then the storm veered toward an almost stationary midlatitude cyclone north of Lake Ontario in Canada. The merging with this low-pressure cyclone resulted in torrential rainfalls accompanied by flash-flooding in the Greater Toronto area, causing about 90 fatalities.

Hazards Associated with Hurricanes

The *main hazards* encountered with hurricanes include: (1) *high winds,* (2) *torrential rains,* (3) *storm surges,* and (4) *large ocean waves.* Damage thus is done both on land and at sea.

Any person who has lived through a hurricane will immediately recall the roar of the winds, the debris hurtling through the air, the splintering of glass panes, the ominous groaning and cracking of the rafters under the force of the storm. Yet, wind may be the least destructive aspect as far as damage and loss of life are concerned. Of course, there are exceptions as demonstrated by Hurricane CELIA (1970), which battered Corpus Christi, Texas, with gusts up to 161 mph (260 km/h). It was the costliest storm ever to strike that state (NOAA/PA 77019; 1977, p. 11).

However, the violence of CELIA was overshadowed by hurricane HUGO (September 16–22, 1989) which roared through the Caribbean and southeastern United States, and by the most destructive hurricane in American history, ANDREW.

Hurricane ANDREW struck Florida's east coast at 3:00 A.M on August 24, 1992, after spreading chaos and death in the Bahamas. Subsequently the huge storm exited the Florida peninsula, crossed the Gulf of Mexico, and moved inland west of New Orleans, Louisiana. ANDREW, with peak winds clocked at 165 mph (265 km/hr), caused an estimated combined material damage of about $22 billion in Florida and Louisiana. Luckily the number of human casualties, in all areas affected, remained close to fifty; this is remarkably low in view of the intense physical damage that resulted from this hurricane.

How much damage results from wind pressure depends not only on the wind velocity but, above all, on the type of structure that must withstand the pressure [Plate 38]. Large external surfaces enclosing mostly empty space, such as warehouses and barns, are especially vulnerable to wind pressure. Pressure from winds increases with the square of the wind

Plate 38. This beach house on the Gulf of Mexico was ripped apart by the fierce winds and waves of a hurricane. Placing such structures along hurricane-prone coasts entails a high risk. (Photograph credit: NOAA; National Hurricane Center).

velocity. This law of nature means that a tenfold rise in wind speed increases pressure 100 times.

Severe flooding, caused by the deluge of rain associated with hurricanes, is also a frequent hazard. Both the excessive amount of rain and the short time period of occurrence conspire to make hurricane-generated rainfall so dangerous. A good example of such disastrous rain is that brought on by Hurricane AGNES (June 14–23, 1972). This storm formed near the Yucatan Peninsula of Mexico, moved across the Gulf of Mexico, and then up the eastern seaboard from Florida to New York. The belt of heavy precipitation was 250 miles (400 km) wide and caused destructive flooding in many areas, resulting in about $2 billion damage and 118 deaths (NOAA/PA 76008; 1977, p. 15).

Many rivers reached record flood stages; for example, the James River, Virginia, crested at 36 feet (9.5 m) matched by widespread inundations in the valleys of the Potomac, Susquehanna, Genesee, and Ohio. The worst flood damage occurred in the hill country of Pennsylvania and in the southern regions of bordering New York. Hardest hit were Corning, Kingston, Harrisburg, and Wilkes-Barre.

The third hurricane hazard, storm surges, was covered in chapter 5; however, it is appropriate to point out that storm surges are considered to be the most dangerous aspect of these storms. About nine out of ten hurricane deaths result from drowning in surge flooding. The case studies of Galveston,

Texas (1900), with 6,000 fatalities, and that of the Ganges Delta, Bangladesh (1970), with more than 300,000 deaths, illustrate this point. The bulge of water, corresponding in size to the eye of the hurricane, may be from 40 to 100 miles (64 to 160 km) wide. One of the highest storm surges in the United States occurred in connection with Hurricane CAMILLE (August 14–22, 1969). It measured 24.2 feet (6.3 m) in the Pass Christian-Long Beach area of Mississippi and resulted in more than 140 deaths in that region.

Hurricanes, as well as some midlatitude cyclones, may produce mountainous ocean waves. Most narratives of hurricane disasters deal with their impact on land, on communities, and on land-based installations. Yet, powerful hurricanes have wreaked havoc with ships all over the world. On September 26, 1954, the Japanese ship *Toyo Maru* capsized in a typhoon with a loss of 794 people, and on September 22, 1957, the West German merchant marine training ship *Pamir* sank in a violent hurricane in the North Atlantic with only 5 out of 86 crew members surviving (Nash; 1976, pp. 704, 705).

Both *wave action* and *wind pressure* can doom ships. Variable wind direction and wind speed usually restrict the height of ocean waves to about 7 to 8 feet (about 2 m), although some storm-generated waves may attain heights of 33 feet (10 m). Hurricane winds of 73 to 82 mph (117 to 132 km/h) may build waves in excess of 60 feet (18.2 m). The largest ocean wave, with a height of 112 feet (34 m), measured authentically, was seen in February 1935 by the crew of the U.S. Navy tanker *Ramapo* (Thurman; 1985, p. 216).

One of the worst marine disasters in connection with a hurricane happened to the U.S. Third Fleet in the Philippine Sea on December 18, 1944. During that storm three U.S. destroyers of the Farragut Class (USS *Hull* and USS *Monaghan*) and one of the Fletcher Class (USS *Spence*) sank. The storm caused a total loss of 790 men and destroyed well over 100 aircraft (Calhoun; 1981, p. 216). The waves, as estimated by the crew on the destroyer USS *Dewey*, reached a height of 60 feet (18 m) so that another destroyer, the USS *Maddox*, which was 1,500 yards away from the USS *Dewey*, completely vanished from sight (Calhoun; 1981, p. 52). Later investigations into the stability of ships indicated that one of the aircraft carriers, the USS *Langley*, rolled consistently to 35 degrees on both sides. The USS *Dewey* rolled once to 75 degrees, but survived!

Modification of Hurricanes

With the exception of bringing needed rain to some areas, hurricanes have few redeeming features and spell human tragedy and high costs in damage. According to a tabulation by NOAA, 54 of the more damaging hurricanes, which occurred between 1900 and 1976, exacted a death toll of 12,098 in the United States alone. Three of these storms were in the $500-million to $5-billion range of damage category, and 21 caused from $50 million to $500 million damage.

Much loss of life and some property damage could be avoided by taking timely precautions and by learning ahead of time about the nature of these storms. However, the immense power of hurricanes can totally overwhelm communities and make the precautions appear to be futile. This realization has motivated scientists to search for ways to either reduce the strength of hurricanes or, as is hoped, to control or even eliminate them.

Since 1956 research and **cloud seeding** experiments have been conducted by what is now NOAA's National Hurricane and Experimental Meteorology Laboratory in Miami, Florida. These efforts were combined under the name "*Operation Stormfury.*" The basic thrust of the experiments was to reduce the high wind velocities in hurricanes by seeding the clouds with dry ice. The dry ice particles, with a temperature of $-108.4°$ F ($-78°$ C), change large quantities of water vapor directly into ice crystals. The ice crystals will grow at the expense of already existing water droplets because water molecules escape more easily from liquid water than from the more rigid ice crystal structure. This fact accounts for the water vapor pressure being higher over the liquid than over the ice surfaces, and the water molecules move from higher to lower vapor pressure areas. The initial seeding was done just outside the ring of highest wind velocities in the outer margins of the eye wall. It was hoped that the massive release of heat, generated by the freezing process, would widen the eye. This would flatten the *pressure gradient* and cause a reduction of wind velocities. A modified series of experiments, dubbed "*Operation Stormfury Americas,*" was introduced later. The newer method proceeded with seeding outer cloud belts further away from the eye to create a secondary eye wall about halfway out from the center of the hurricane. This procedure led to a marked flattening of the pressure gradient.

Between 1961 and 1971 only four storms were seeded. Three showed clear signs of beneficial modification, but the results of the last seeding, Hurricane GINGER in 1971, did not produce tangible results. Public concerns, both in the United States and in Mexico, as well as the mixed results, led to a suspension of the seeding operations, but research in hurricane modification continues (NOAA/PA 76008; 1977, p. 19).

Formation of Tornadoes

The *tornado,* also referred to as a "*twister*" in the United States, is a short-lived, violent local storm with a rotating column of air. It may, or may not, be in contact with the earth's surface. Its roaring sound, and the fact that tornadoes frequently are associated with severe thunderstorms, probably is responsible for its name. The word *tornado* stems from the Spanish word *tronada,* or thunderstorm.

Tornadoes occur in just about any part of the world, but they are most numerous in North America, especially in the central and southeastern regions of the United States. The total number varies from year to year, but it is estimated that about 90 percent of the world's tornadoes occur in the United States. The explanation for this fact can be found in the *physical geographical setting* of North America.

The forming of tornadoes is linked to a variety of atmospheric conditions and with severe weather systems which include strong thunderstorms. Thunderstorms may develop either as powerful, single convectional storms, or *supercells,* or they form in connection with well-developed cold fronts typical of fast-moving midlatitude cyclones. In either case the air masses that spawn tornadoes must contain a great amount of moisture with large amounts of latent energy.

Latent energy produces changes in the physical state of a substance, such as water, without increasing the temperature of such a substance. For example, to evaporate 1 gram of liquid water, about 600 calories are needed to change this water into water vapor. As this water is uplifted and cooled, the heat used for evaporation is released back into the air. This release of latent energy is a critical process in generating thunderstorms as is discussed in the next chapter.

Most significant tornadoes, rating F3 or higher on the so-called Fujita scale, occur in connection with supercells and are usually located within the right rear quadrant of the storm. Weaker tornadoes, at least as a general rule, tend to develop along the leading edge of a single, convectional thunderstorm or in those that spring up as a line of storms along a weather front (NOAA/PA 82001; 1982, p.2).

In North America, in contrast to most other continents, the major mountain chains tend to run in a north-to-south direction. Examples of this are the Rocky Mountains and the Appalachians. Their alignment allows tropical moist air masses, from the Gulf of Mexico and the Caribbean, to clash with cold dry air from the Canadian interior in the central open plains. The great amount of energy contained in the tropical air may generate severe thunderstorms. When uplifted by colder dense air, as along cold fronts, the warm moist air can spawn heavy frontal thunderstorms. Thus it is not surprising that nine states (Kansas, Iowa, Texas, Arkansas, Oklahoma, Missouri, Alabama, Mississippi, and Nebraska) average more than five tornadoes each year (Hidore; 1969, p. 79). A second world area known for a considerable number of tornadoes, although fewer than in the United States, is Australia. This continent also lacks any major east-to-west mountain ranges.

The life span of tornadoes varies from less than a minute to tens of minutes. For any given place tornadoes may not last more than an average of half a minute (Blair, Fite; 1957, p. 233). Considering the violence of these storms and the suddenness of their formation, it is understandable that many characteristics of tornadoes are very difficult to assess. More recently developed technologies, such as *Doppler radar* and satellite imagery, have made great inroads in exploring the nature of tornadoes. The application of Doppler radar revealed for the first time the internal wind velocities and structural characteristics of tornado vortices.

The Doppler effect basically represents a change in the observed frequency of light, sound, and of other wave energy, caused by movement of the source or that of the observer. When a source approaches an observer, or sensor, the effective length of a given wave becomes shorter than it would be if no motion were involved. If the source recedes from the observer, or sensor, the wavelength increases.

In studies of vortex-wind velocities in tornadoes, the Doppler radar is aimed at the movement of precipitation particles as they move either toward or away from the sensor. *Electromagnetic* waves are sent toward the particles and the frequency shift of the radar "echo" is then evaluated. Key research on vortex development, but using Doppler radar data, was carried out by Professor Joe R. Eagleman cited later. Several conditions seem to be consistently present before tornadoes form:

1. An ample supply of moist warm and unstable air on the surface
2. A mechanism that allows cool drier air aloft to overrun the warm surface layer
3. Wind shear between the warm surface layer and a cold air layer which moves rapidly across it.
4. A strong upper-air wind flow (usually at an elevation between 10,000 to 20,000 feet or about 3,050 to 6,100m) passing around the flanks of a cumulonimbus supercell.

The first aspect, i.e. that of moisture and latent heat was discussed earlier. The second condition occurs when a rapidly moving cold front aloft advances faster than the surface cold front and thus entraps temporarily the warm unstable air (fig. 6.2). This position could induce an upward spiraling motion of the warm air into the cold air aloft analogous to an "upward drain" that leads to vortex development.

The third condition evolves from a horizontal wind shear between a strong above-surface wind and the unstable surface layer. The source of the cold air could be a part of a so-called cold-air avalanche (see Chapter 7) or strong westerly winds aloft. The initial shear produces a horizontal rotating "tube" (fig 6.3 A) which subsequently is drawn into the convec-

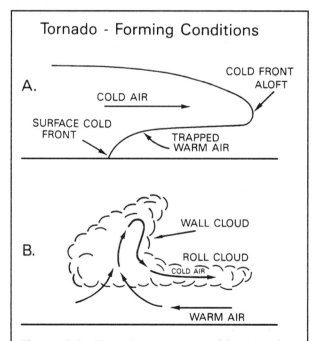

Figure 6.2. Trapping warm unstable air under cold air aloft assists in forming tornadoes. This situation may develop with a rapidly advancing cold front aloft (A) or when cold air descends from a cumulonimbus cloud and spreads on top of warm surface air (B).

tional updraft of the thunderstorm. The vortex then assumes a more vertical position in the area of maximum updraft (fig. 6.3 B). The release of latent heat from the uplifted moist air, as well as the friction between the rising warm air and compensating descending color air, further intensifies the vortex.

The fourth scenario deals with the development of a tornadic vortex within a thunderstorm supercell which is subjected to a strong upper-air wind flow. As the wind blows against the trailing side (usually the western side in midlatitude U.S.) of the cumulonimbus, the wind flow tends to separate and move around the southern and nothern flank of the supercell. The friction between the westerly wind and the cloud sides initiates a vortex-like circulation in the southern and northern sections of the towering cloud. This "*double-vortex*" system was detected by the use of dual Doppler radar (Eagleman; 1983, p. 212). If a fully-grown tornado vortex will evolve it develops from the southern one of the two vortices which has a counterclockwise cyclonic rotation.

Each tornado has its own characteristics; yet, most of them form behind the so-called **wall cloud,** that is, behind the usually heavy rain that pours out of the leading sector of the thundercloud or **cumulonimbus.** This veil of rain explains why most tornadoes are obscured during their initial phase of development. They usually start as a short funnel-shaped extrusion beneath the thunderstorm several thousand feet (about 600 to 900 m) above the earth's surface. The early vortex may appear as a rotating whitish cloud. As it works itself downward, it may become

temporarily invisible as condensation decreases within the somewhat warmer air. Subsequently, upon touching the earth's surface, the vortex sucks up dust and debris, and attains its well-known ominous black funnel appearance [Plate 39].

Tornado Characteristics

The two most fear-inspiring features of a tornado are its roaring noise, likened to hundreds of freight trains, and the threatening black, twisting funnel cloud, or vortex, extending from the parent cloud. Sometimes a series of two or more tornadoes form from the same cloud. They may dissipate, lift off the ground, and touch down again in quick succession. This explains why tornado warnings must be issued on an area basis, rather than for a specific location.

The total destruction often caused by tornadoes is the result of (1) their extremely *high rotary wind velocities,* and (2) their *partial vacuum* that exists within the vortex [Plate 40]. The short life span, the unpredictable location within a storm system, and the violence within the tornado itself usually prevent the direct use of instruments to measure accurately wind speeds and pressure conditions. Estimates, based on engineering studies of tornado damage, indicate that horizontal wind speeds in the vortex may be higher than 300 mph (480 km/h).

The low pressure inside the vortex can only be estimated. It is possible that the pressure may drop to as low as 17.7 inches Hg (600 millibars), or about 60 percent of the normal atmospheric pressure (Eagleman; 1983, p. 99). If the outside atmospheric pres-

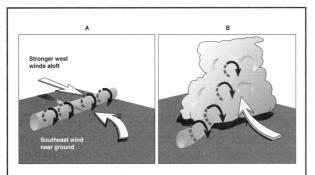

Figure 6.3 (A and B). Horizontal wind shear between an above-surface wind and opposing surface currents can produce a spinning "tube" (A). As the thunderstorm matures the tube is uplifted by strong convectional updrafts (B). The rotating funnel is further strengthened by convection and may form a tornado (After NOAA "Tornadoes—nature's most violent storms." PA 92052, p.4).

Plate 39. When a tornado touches the ground the vortex sucks up soil and debris. At this point in its development the funnel cloud attains its ominous black appearance. (Photograph credit: NOAA; National Severe Storms Laboratory).

Plate 40. The awesome power of a tornado is the result of its extremely high rotary wind velocities that may be higher than 300 mph (480 km/h) in combination with the low-pressure suction effect exerted by the vortex. (Photograph credit: NOAA; National Severe Storms Laboratory).

sure, caused by the direct passage of a tornado, were to drop anywhere from 1 to 5 inches Hg (34 to 169 millibars), the net excess pressure inside a tight building could rise from 70 to 400 pounds/sq. ft (340 to 1270 kg/m^2) (Blair, Fite; 1957, p. 234). This pressure build-up, in combination with the force exerted by the high winds, can lead to a total disintegration of a structure [Plate 41]. The power of a tornado is illustrated by the facts that, according to the U.S. Weather Service records, 2 × 4 wooden boards have been driven into and through brick walls, and a railroad car, weighing 83 tons, was lifted off its tracks. Another tornado (Mississippi, 1975) carried a home freezer for a distance of over a mile (about 2 km).

Conforming to the overall tracks of weather systems in North America, tornadoes tend to approach from a west-southwesterly direction and subsequently move toward the northeast. This is particularly the case if they are associated with advancing cold fronts. The length of their paths is highly variable and averages about 4 miles (6.5 km); yet, one tornado (May 26, 1917) was reported to have traveled a distance of 293 miles (470 km) across Illinois and Indiana lasting 7 hours and 20 minutes. The forward motion falls into the range from 25 to 40 mph (40 to 64 km/h); some storms become almost stationary, others speed along at almost 70 mph (112 km/h).

Plate 41. Most structures, as shown in the case of this totally demolished farm building, disintegrate under the force of a tornado. (Photograph credit: NOAA; National Severe Storms Laboratory).

Their path of destruction averages about 300 to 400 yards wide (330 to 440 m). In a few cases the damage spread over a width of a mile (1.6 km), or more.

Occurrence of Tornadoes

For some regions in the United States the threat of tornadoes is an ever-present fear. On the average the largest number of tornadoes occur in the southern plains of Texas, Oklahoma, and Kansas. Between 1953 and 1980 Texas averaged 119 tornadoes and suffered an average annual death toll of 11 persons (NOAA/PA 82001; 1982, p. 5). The numbers for Oklahoma are 53 tornadoes and 8 deaths per average year. There is an annual migration in the occurrence of tornadoes. The spatial concentration of these storms clearly correlates with the geographical location of moist warm air masses coming in conflict with colder northern air.

In February this conflict zone centers on the central Gulf States, then migrates into the southeastern Atlantic States by March, dominates the southern Great Plains in May, and shifts into the northern Great Plains and the Midwest in June. Most of the tornadoes in this country occur between March and July, with the month of May being by far the most "*tornado-prone*" time (fig. 6.4).

The frequency of tornadoes in the United States falls within a range of 500 to 600 per year; however, some years had a much greater number. Some of the worst years were 1967 with 912 tornadoes and 1973 when more than 1,100 were reported. One of the most intense outbreaks of tornadoes took place on April 3–4, 1974, when 148 were spawned along a strong cold front that stretched from southern Canada to Alabama. They caused 315 deaths and more than $600 million in damage (NOAA/PA 77027; 1978, p. 2). Because of better instrumentation, and with a more efficient reporting network, the average number of tornadoes has seemed to increase over the past decades. This may not be an actual increase but the result of more effective detecting devices. A similar phenomenon was discussed in connection with the "*increasing*" occurrence of earthquakes (chapter 1).

Even though the total number of annual tornadoes in the United States is large, the statistical probability that a particular place will be visited by this deadly storm is quite small. In an area most frequently subjected to tornadoes the chance of a tornado strike, at a given location, is about once in 250 years (NOAA/PA 77027; 1978, p. 4). However,

mathematical predictions for most natural disasters are quite unreliable and show many deviations. For example, between 1892 and 1977, Oklahoma City was struck 32 times by tornadoes. Baldwin, Mississippi, was struck twice by tornadoes during a 25-minute period on March 16, 1942, and Codell, Kansas, suffered tornadoes three times, all on May 20, in 1916, 1917, and 1918! The infrequency and unpredictability tend to lead to an uninformed population ignoring prevention and survival measures, but anyone who has gone through a tornado event will never forget that experience!

Tornado Warning Systems

In view of the almost total destruction brought on by tornadoes, it is necessary to develop a good understanding of this dangerous storm type, to prepare communities for such events and their aftermaths, and to issue warnings if warranted.

NOAA's National Severe Storms Forecast Center is located in Kansas City, Missouri. Here, meteorologists monitor the weather systems of North America, and they look especially for atmospheric conditions that pose potentials for tornado develop-

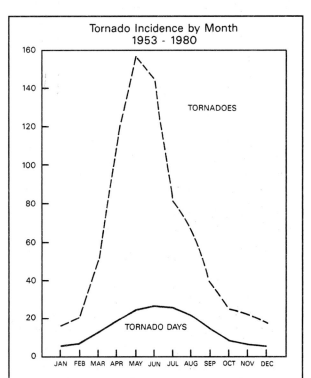

Figure 6.4. Tornadoes may occur in any month, but most of these violent storms in the United States appear between April and July with a peak concentration in May. (Modified from: NOAA, Tornado, PA 77027, 1978).

ment. It is not possible to predict either the exact time, or the place, where a tornado may form. However, general areas (25,000 sq. miles or 64,750 sq. km) can be identified where the occurrence of severe thunderstorms, and potentially tornadoes, is probable. In that case *Tornado Watches* are issued and communicated to the appropriate weather stations (NOAA/PA; 1982, p. 5).

Subsequently, local offices of the National Weather Service, working in close cooperation with trained tornado spotters (***SKYWARN*** System), prepare to detect tornadoes by the use of radar or by direct observation. Once a tornado shows on radar, or is confirmed by a qualified tornado spotter, an official *Tornado Warning* is issued via radio and television stations, and the appropriate community authorities are alerted. To reach some people, who may not be able to listen to radio or television warnings, a steady-tone signal (3 to 5 minutes duration) by Civil Defense sirens may be used. Standard National Weather Service radio frequencies (162.40; 162.475; 162.55 MHz), and amateur radios further aid in spreading the warning.

Much remains to be done to comprehend fully the formation and nature of both hurricanes and tornadoes. Sophisticated electronic equipment, new types of weather satellites, and further advances in the atmospheric sciences will assist in reaching that goal.

THUNDERSTORMS AND LIGHTNING

Key Terms

squall line	intracloud	triggered lightning
cumulus	sheet lightning	stepped leader
static electricity	intercloud	ball lightning
supercooled droplets	cloud-to-ground	plasma
return stroke	ground-to-cloud	St. Elmo's fire
compression waves		

Ancient people trembled with fear, and even modern humans stand in awe whenever sky meets earth, when thunder roars and blinding lightning bolts strike. Mythology ascribed these phenomena to the gods' anger, or to their titanic battles. Modern science has found many answers pertaining to thunderstorms and lightning, but numerous questions remain to be answered.

Many people are deeply afraid of these storms. Such fear, in many cases, is brought on by incorrect notions often rooted in superstition. Moreover, the suddenness of a close lightning strike, accompanied by the terrifying crashing sound of the thunderclap, can be a nerve-shattering experience.

Most people see in thunderstorms a threat to life and property. This is not without cause since, on a yearly basis, 20,000 homes are destroyed by lightning, and more than 100 people are killed in the United States alone; yet, by comparison to other disasters, or measuring this death toll against the number of human beings killed annually by automobiles (more than 40,000 a year), the average annual fatalities resulting from lightning are quite low. Statistics indicate that, between 1959 and 1984, a total of 2,574 people lost their lives to lightning strikes. During the same 25-year period approximately 1 million persons were killed in car accidents.

Thunderstorms, numbering about 16 million per year for the world, develop in just about any place on earth with the exception of the polar latitudes; yet, on very rare occasions thunder was heard even in high latitudes because of weather fronts that reached beyond their customary regions. However, most thunderstorms form in the moist tropical zones well supplied with solar energy and high levels of moisture.

Causes of Thunderstorms

Thunderstorms can be triggered by a variety of meteorological conditions. At least two fundamental *prerequisites* must be present before any type of thunderstorm can develop: (1) *a large amount of warm, unstable air with a high water vapor content (latent energy),* and (2) *a mechanism that uplifts this air through convection (thermal uplift)* or by mechanical means. Regardless in what way this occurs, it will lead to an increase in air mass instability. This situation means that the uplifted air mass will be warmer, and more unstable, than the air that surrounds it. In many instances several of the factors, discussed in the following passages, will combine in this process.

All thunderstorms have some common features, but how they form has a marked influence on their intensity and associated weather phenomena, such as the amount of electrical discharges, internal turbulence, types and quantities of precipitation and, as we have already observed in chapter 6, the development

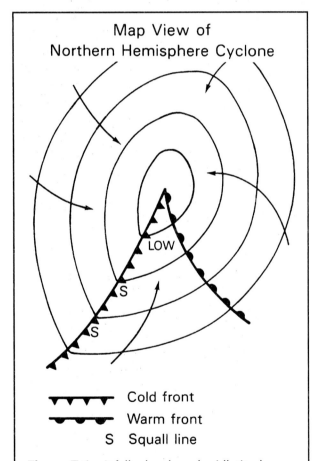

Cold front
Warm front
S Squall line

Figure 7.1. A fully developed midlatitude cyclone may cover one-third of the United States. The storm has a counterclockwise (Northern Hemisphere) wind system and displays distinct cold and warm fronts that separate cold and warm air masses. The cold front is preceded by the squall line (S), which is marked by turbulence, wind shear, and thunderstorms.

of tornadoes. What, then, are the different conditions that bring about the needed uplift of an air mass and render it unstable?

The first major category develops in connection with the frontal systems of midlatitude cyclones and therefore is classified as *frontal thunderstorms.* While other subclassifications exist, we can divide these types into (1) *cold-front thunderstorms,* and (2) *warm-front thunderstorms.*

A mature midlatitude cyclone is composed of contrasting air masses that are drawn into its low-pressure center. When these air masses meet they form *fronts.* These fronts represent imaginary surfaces that separate the air masses. Where the air mass surfaces intersect the earth's surface they are called *surface fronts.* The fronts can be shown quite readily on weather maps and in the model cyclone shown in fig. 7.1. The segment of the fronts that is above the earth's surface is referred to as *front aloft.*

Cold-front thunderstorms develop in the warm air, which is uplifted by the sloping surface of the advancing cold air mass. Distinguishing features of such storms are: (1) their *linear* distribution along the front, (2) their *clear visibility* as they approach, and (3) their relatively *short duration.* The latter, of course, depends on the forward speed of the frontal system [Plate 42].

Warm-front thunderstorms form when the warm air mass, usually drawn into the cyclonic storm from the southern quadrant (Northern Hemisphere), overruns the retreating cold air (fig. 7.2). The slope of warm fronts is much gentler than that of cold fronts,

Plate 42. Cold-front thunderstorms develop along the sloping surface of the advancing cold air mass. Here, the uplift of warm moist air produces towering thunderclouds. (Photograph credit: NOAA; National Severe Storms Laboratory).

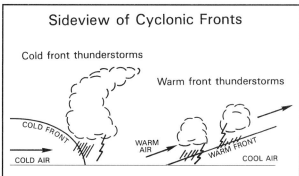

Cold front thunderstorms

Warm front thunderstorms

COLD FRONT

COLD AIR

WARM AIR

WARM FRONT

COOL AIR

Figure 7.2. A sideview of a midlatitude cyclone shows the vigorous uplift of moist warm air along the cold front. This action can produce cold-front thunderstorms. Warm-front thunderstorms form when warm air overrides colder surface air along a warm front.

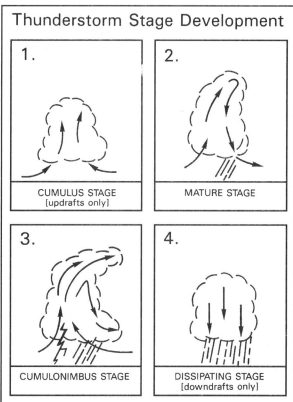

1. CUMULUS STAGE [updrafts only]

2. MATURE STAGE

3. CUMULONIMBUS STAGE

4. DISSIPATING STAGE [downdrafts only]

Figure 7.3. The development of convective thunderstorms usually passes through four cloud stages: (1) the cumulus stage (updrafts only), (2) the mature stage (both updrafts and downdrafts), (3) the cumulonimbus stage (violent updrafts and downdrafts, possibly tornadoes, and wind shear), and (4) the dissipating stage (downdrafts only).

the ratio being 1:80, or even 1:200, in contrast to about 1:40, respectively. This means that the sloping surface rises 1 unit of height per 40 units of distance for the average cold-front slope.

Warm-front thunderstorms are much more numerous and markedly scattered, compared to cold-front storms. They tend to be less intense and produce showery but generally nonviolent weather (AF Manual 105–5; 1962, pp. 12.2–3). Additional thunderstorm development may occur in the general zone of convergence, in the center of the midlatitude cyclone, as well as in a bandlike region, or *squall line,* preceding the cold front.

The second major category of thunderstorms is not associated with frontal systems but is the result of *convection,* or *thermal uplift,* produced by intense surface heating. As the air is heated its capacity to hold water vapor increases, and large amounts of vapor are picked up from various surface sources. As the air warms it expands, becomes buoyant, and begins to rise. This rise does not involve the entire air mass but proceeds in individual *thermal cells.* This process is the result of the variable heat absorption capacities of different earth surfaces, such as soils, vegetation, or water. As the water vapor in these thermal cells condenses, additional heat *(latent heat of condensation)* is released. This heat release makes the warm-air cell progressively warmer relative to the surrounding air and forces continued ascent.

The growth of such convectional cells is more systematic, and slower, than the formation of thunderclouds along fronts. There are *four* distinct stages in the development of thermal thunderstorms (fig. 7.3).

1. The *cumulus* stage. This is the initial *fair weather* cumulus cloud so typically seen during the afternoons of sultry summer days. This cloud, which may or may not evolve into a thunder storm, may rise to about 22,000 feet (6,700 m) [Plate 43].

2. *The mature stage.* This could be called "an overgrown cumulus cloud." This cloud has both *updrafts* and *downdrafts.* The downdrafts usually occur in connection with precipitation falling from the upper portion of the cloud, dragging cooler air downward. This type of cloud stage may reach beyond 30,000 feet (9,150 m).

3. *The cumulonimbus stage.* This huge cloud extends to elevations of 40,000 to 60,000 feet (12,200 to 18,300 m) with updrafts of 40 to a maximum of 90 feet/sec (12 to 27 m/sec) (Taylor; 1954, p. 227). Massive downdrafts, mostly in the leading portion of the cloud, reach about

Plate 43. This photograph shows a convective cumulus cloud that is so typically observed during the warmest part of a summer day. Such clouds may rise to about 22,000 feet (6,700 m) but may not always produce thunderstorms. (Photo by C.H.V. Ebert).

Plate 44. This massive cumulonimbus cloud, photographed over the east coast of Nicaragua, displayed incessant lightning and was fed by the indraft of moist unstable air from the Caribbean Sea. (Photo by C.H.V. Ebert).

the same velocities as the updrafts. This type of cloud is the most violent stage of the storm with torrential rain or hail, incessant lightning, and possibly tornadoes [Plate 44]. These cumulonimbus clouds may prevail for at least 1 hour in contrast to 10 to 15 minutes for the smaller cumuli types. Furthermore, some thunderstorm clouds, or cloud conglomerations, with diameters of 30 miles (48 km), have been tracked for several hours (Riehl; 1965, p. 103).

4. *The dissipating stage.* This phase is reached when the storm has expended its energy and only downdrafts occur. Lightning becomes less frequent, and precipitation may fall as steady rain.

In addition to frontal and convectional situations, other conditions can produce these storms. One is the so-called *orographic thunderstorm,* which develops in air masses uplifted along the wind-facing side of mountain ranges. They frequently remain stationary for many hours, may form single or multiple cells, and are commonly accompanied by hail. This type is very typical along high mountains, such as the Rocky Mountains. The principle of the orographic process was first discussed in chapter 5, in connection with flood potentials on the windward side of *orographic barriers.* Other types of thunderstorms may also be generated by massive volcanic eruptions, and some can be triggered over large area fires as will be mentioned later in connection with forest fires and firestorms.

Occurrence of Thunderstorms

Because a large supply of moist warm air is needed for the development of thunderstorms, it is not surprising that the bulk of the world's annual 16 million storms is associated with tropical regions rich in both moisture and heat. Some of these locations, such as Panama, Java, and the equatorial zones of Africa and South America, may have as many as 200 days with thunderstorms a year (Blair, Fite; 1957, p. 225).

In the United States the areas with maximum concentrations of thunderstorms correlate strongly with the availability of moist unstable air from the Gulf of Mexico and the Caribbean. Another concentration shows the effect of our north-to-south mountain ranges, as mentioned in chapter 6, in connection with the development of tornadoes.

While you are reading this line about 2,000 thunderstorms are in progress somewhere in the world, and lightning strikes at a rate of about 100 per second. In the United States alone about 100,000 thunderstorms are recorded annually, with property losses in the hundreds of millions in dollars (NOAA/PA 83001; 1985, p. 1). During the 25-year period from 1959 to 1984 all states, with the exception of Alaska and Hawaii, reported deaths caused by lightning. The Gulf coast of Florida, according to a tabulation in the U.S.D.A. Yearbook of 1941 (Climate and Man), shows the greatest number of days with thunderstorms in the Tampa area, averaging about 80 days. Florida, between 1959 and 1984, had a total of 255 fatalities caused by lightning, or nearly

10 percent of the U.S. total (NOAA/PA 83001; 1985, p. 4). Ninety-six percent of these deaths occurred from May through September when thunderstorms are most numerous (fig. 7.4).

The Nature of Lightning

The nature and effects of electricity must have fascinated people since ancient times. The Greeks gave the name *"elektron"* to amber, or fossilized tree resin. They observed the phenomenon of static electricity, which would build up in this golden substance when rubbed against fur or wool.

The generation of **static electricity,** and its release in the form of sparks, became a recognized process with the invention of the Leyden jar in 1746. Benjamin Franklin constructed similar condensing devices and experimented with sparks that could kill small animals. Subsequently, looking for sparks on a grand scale, he carried out his well-known experiments using a kite to attract and conduct lightning. Since his era, more than 200 years ago, scientists have searched for the answers surrounding the generation and characteristics of lightning bolts. Although it is relatively easy to produce fairly large electrical sparks in the laboratory, and to study their properties, this becomes more difficult in nature. The violence that rages inside the towering cumulonimbus clouds presents almost unsurmountable obstacles to carry out controlled, planned observations and measurements.

All scientists agree that strong *opposing charge fields* must exist before a lightning bolt can form.

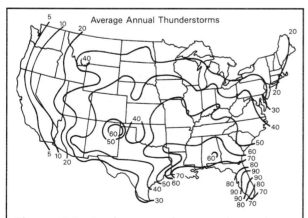

Figure 7.4. As shown on the map above, the greatest annual number of thunderstorms in the United States occur near the Gulf coast of Florida near Tampa. Florida also leads in the highest number of deaths caused by lightning. (Based on NOAA/PA 83001; 1985).

Furthermore, the various types of bolts are well known, as are their parameters of length, shape, speed, temperature, and their electrical potentials. What is less clear, and still quite controversial, is the way in which the opposing charge fields form and locate within the cloud. How many theories about *charge separation* exist is difficult to say; however, since 1885 more than seventeen ways have been proposed (Hill; 1986, p. 473).

Principles leading to the buildup of opposite charges involve: (1) friction between falling raindrops, or ice particles, and air, (2) temperature gradients within ice particles causing charge separation between layers (thermoelectric effect), (3) raindrops, having double layers composed of a negative outer shell and a positive inner core, atomizing as a result of wind friction, (4) charge separation emanating from freezing and melting processes, and (5) induction of charges when particles fall, or rise, through already existing electrical fields.

Another theory, based on *convectional updrafts,* advances the idea that lightning precedes heavy precipitation, instead of falling precipitation particles leading to the buildup of electrical charge fields. This *"convective"* theory was first proposed by Dr. Bernard Vonnegut in 1955. When he trained radar on thunderclouds over Grand Bahama Island, he noted that very little precipitation existed in the storms before lightning took place. After lightning occurred, torrential rains fell (Fless; 1982, p. 8). This theory maintains that falling precipitation, and the collision of particles, could not generate the electric field potentials needed to initiate lightning. The electrification is seen as the result of powerful updrafts and downdrafts moving minute charged particles within the cloud.

Dr. Vonnegut's theory was strengthened by geophysicists Charles Holmes and Dan Holden. They used a specially designed radar unit that could detect both cloud particles and the ionized channel of the lightning stroke. One result of their experimentation confirmed that the particles of precipitation appear to grow thousands of times faster after a lightning stroke than they would have through the generally accepted *"collision"* theory (Garelik; 1984, p. 66). Regardless which theory is correct, and they all contain correct as well as unworkable aspects, separate regions of opposite electrical charges must build before lightning can take place.

It has been established that particular electrical fields locate within a fully grown cumulonimbus cloud as well as on the earth's surface and its protrusions. *Positive* areas build in the upper portion of the cloud which extends far into the freezing zone. Strong updrafts are capable of carrying **supercooled droplets** and ice crystals to these elevations. A large, elongated core of the cloud tends to be dominated by *negative* charges, which usually occupy the largest part of the cloud base.

The earth's surface, normally carrying a negative charge, develops a positive charge through induction by the strong negative potential of the cloud base. This positive field on the earth's surface tends to follow the cloud like a shadow (Ahrens; 1982, p. 405). The *induced* positive charge concentrates on high points, such as towers, telephone poles, tall buildings, and on isolated trees. The fact that this concentration of positive charges takes place on high and exposed objects clearly demonstrates the functions of a *lightning rod*. The positive charge at the end of the metal rod is to attract the lightning bolt and to conduct the electrical discharge to the ground. A properly installed lightning rod is an effective protection against lightning damage to buildings. It is estimated that more than 90 percent of buildings destroyed by lightning were not protected in this way (Eagleman; 1983, p. 133).

Research, conducted by meteorologists of the Department of Atmospheric Science at State University of New York at Albany, indicates that about 5 percent of all lightning strikes lower a positive charge to the ground. These very destructive bolts, taking an unhampered route through the dry atmosphere, issue from the positive fields of the cloud and connect with negative charge fields on the ground. This relatively newly discovered lightning type, studied in the United States, Japan, and Sweden, seems to be associated with weaker convective storms or with the periphery of more intense ones. The lightning occurs outside the regions of strong updrafts and apparently is not followed by a return stroke (Cejka; 1987).

Dry air is a poor conductor of electricity, so enormous electric potentials must be generated before the air breaks down and permits a discharge of electricity to travel through it. If raindrops fill the air, it seems that this breakdown occurs when there is a voltage gradient of 100,000 volts/m over a distance of about 160 feet (50 m) (Gedzelman; 1980, p. 355).

The human eye is poorly equipped to see the true nature of a lightning flash. When the luminous image is still "visible" to us, similar to the aftereffects of a photographic flash, the lightning strike has already disappeared. Moreover, what we perceive as one

powerful streak of light in reality is a composite of several phases through which a lightning bolt passes.

As a lightning discharge emerges from a cloud it moves in a *steplike* fashion, each step being about 55 yards (50 m) long and requiring about one microsecond (one millionth of a second). After an incredibly short pause, about 50 millionths of a second, the bolt continues in similar steps until it is about 50 to 60 yards (45 to 55 m) from the earth's surface. Then a separate upward stroke, as in a short-circuit, connects with the approaching bolt, and the **return stroke** takes place. This upward stroke, following the same path as the original lightning flash, is much larger and brighter (Ahrens; 1982, p. 407). Yet, this return stroke is not the end of the lightning event. In a typical lightning sequence, which we perceive as one bright image, there are usually about four leaders and return strokes. The whole process usually terminates within less than two-tenths of a second.

The *temperature* inside the lightning channel cannot be measured directly but is estimated to be about 54,000° F (30,000° C). The intense heating of the air surrounding the lightning bolt results in an explosive expansion of the air. The **compression waves** that accompany this violent expansion produce the thunder. When lightning occurs in the immediate vicinity, the thunderclap sounds like a sharp crack and is of relatively short duration. More distant thunder is known for its rumbling noise and reverberations. This is the result of the sound waves being reflected, both from air layers of different density as well as from hills and large buildings. The sound of thunder, depending on atmospheric conditions and topography, can be heard as far as 20 miles (32 km).

Lightning Types

The strength, and the location, of contrasting electrical charge fields determines what type of lightning will take place. In many thunderstorms the individual lightning bolts are obscured by falling rain, mist, or by other clouds. For this reason the number of lightning strokes is generally underestimated by the observer. However, sensitive electronic recorders, often used by power companies, and even a plain radio, reveal the almost unceasing barrage of electrical discharges generated by a thunderstorm [Plate 45].

In line with this statement it is interesting to note that more than 60 percent of all lightning takes place inside a cloud (Eagleman; 1983, p. 114). This **intra-**

Plate 45. A barrage of lightning bolts is released from this thunderstorm as it crosses an urban area. What appears to be a single streak of lightning actually is a composite of many individual steps. (Photograph credit: NOAA; National Severe Storms Laboratory).

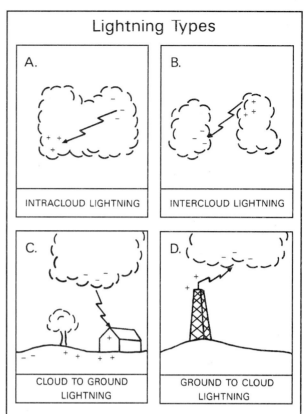

Figure 7.5. Four of the more common types of lightning include: (A) intracloud (60 percent of all lightning), (B) intercloud, (C) cloud-to-ground, and (D) ground-to-cloud (also called triggered lightning).

cloud discharge is veiled by the cloud mass and may lead to what is sometimes called *sheet lightning.* If such lightning occurs at very high elevations, and at some distance away, the sound is refracted aloft and may not be heard from the ground. This type of lightning can be impressive, but it is harmless as far as the earth's surface is concerned (fig. 7.5A).

Another form is called *intercloud* lightning because it strikes from cloud to cloud, usually at higher altitudes. These spectacular zigzag bolts can be seen quite readily as they breach at times considerable distances between clouds (fig. 7.5B). Although of no danger to observers on the ground, these discharges may become hazardous to aircraft as will be discussed later in this chapter.

The best-known lightning, and the most feared one, is the *cloud-to-ground* type, which hurtles from the dark skies toward the earth (fig. 7.5C). It may be reassuring to know that by the time we hear the crashing thunder the immediate danger has already passed. The intense heat generated by such lightning bolts, and by its return stroke, can split trees and may turn sandy surfaces into a glassy crust.

Closely resembling the cloud-to-ground type is the *ground-to-cloud* lightning, which moves from ground features toward a cloud (fig. 7.5D). Lightning behavior of this kind was observed in the 1930s in connection with electrical discharges that apparently emanated from the tip of the Empire State Building in New York. It was difficult to demonstrate clearly that the lightning bolts really originated in that way.

It was not until much later that studies, made by the Swiss Mt. San Salvatore Lightning Observatory, revealed that such lightning is quite frequent under certain conditions. Two metal towers had been erected for this study. Of the more than 100 lightning strokes recorded, about 80 percent originated from the tips of these towers.

As this research indicates, favorable places for triggering this type of lightning are high mountain peaks, tall television towers, radio towers, and skyscrapers. *Triggered lightning,* as this phenomenon is also called, apparently starts with a *stepped leader,* as in cloud-to-ground lightning, but no return stroke from the cloud takes place.

It was mentioned before that dry air is a poor conductor of electricity, and that moisture particles act as stepping stones for *lightning discharges.* When surface air is very dry and is overlain by a layer of somewhat moister air, lightning tends to travel sideways after leaving the cloud base. If a moist surface layer of air is encountered at some distance away from the cloud, the bold abandons its horizontal path and turns downward and thus becomes a lightning strike *"out-of-the-blue."* Obviously, this kind of lightning constitutes a great danger to unsuspecting persons who feel safe because they are miles away from the threatening storm cloud.

Another somewhat strange type of lightning is the so-called *"ball lightning."* Its name derives from the fact that it occurs in the form of a luminous ball-shaped sphere. So far no definitive answer to this phenomenon has been found, but it has been observed by many people.

The apparent size and the duration of ball lightning vary considerably. The differences in the descriptions of ball lightning are understandable when one realizes the elements of surprise and fright which accompany such an event. The size of the fiery ball seems to range from a few inches to over six feet (2m); it may just float in the air, or it may travel along wires and fences to end its performance in a noisy explosion. In one reported incident the fireball—estimated at about 19 inches (50cm) in diameter—moved down a hallway toward a person who had to step aside to let it pass (Eagleman; 1983, p. 121).

No one exhaustive explanation for ball lightning exists in the scientific literature; however, it is generally believed that this type of lightning is made up of a spherical mass of *plasma,* a highly ionized gas in which free electrons balance the charges of positive ions. Such plasma has many unusual characteristics that normally are not found in common gases.

Strong electrical charges, especially noticed in arid mountainous environments, can build up on the earth's surface. This author, while climbing a mountain range outside Kabul, Afghanistan, observed the effects of such an electrical potential, feeling a strong tingling sensation on the top of his head and at the fingertips. In some situations, such as on dry rock surfaces of high mountains, on ship masts, and also on the tips of airplanes, this buildup of electrical charges can produce an eerie bluish halo effect, called *St. Elmo's fire.* In itself this phenomenon does not present any danger.

Lightning Hazards

In the beginning of this chapter it was stated that in the United States between 1959 and 1984, more than 2,500 persons were killed by lightning. Almost 7,000 individuals were injured during the same time period. In many instances both injury and death could have been avoided with a better understanding of the nature of lightning.

Many of the more common injuries caused by lightning are severe external and internal burns, as the electric current passes through the body. Yet, a surprising number of victims survive this horrible experience, especially if first aid is applied immediately. In some instances the electric jolt paralyzes the respiratory system, produces cardiac arrests, or results in irregular heartbeats. Only the quick application of artificial respiration may save the victim.

Persons are especially endangered when they stand taller than the surroundings, as standing on an open golf course, or on a hilltop. Taking refuge under an isolated tree, or standing near metal farm equipment, or wire fences, crouching in a wet depression, or even lying flat on the ground is asking for disaster. The latter situation, lying down, should be avoided because electrical currents travel along the ground surface. Sitting on some dry clothes, which serve as insulator, offers some protection. When in a group of people, ask them to stay at least several yards apart (NOAA/PA 83001; 1985, p.5). If time permits, go into any building, and stay in your car if caught in a thunderstorm. If lightning should strike the car, remain inside for at least half an hour to allow the electrical charge to dissipate, then jump out of the car without touching the car and the ground at the same time.

There are few places that are truly safe from lightning strikes; however more than half of all deaths from lightning probably occur outdoors. Unfortunately, the human body is an excellent conductor of electricity because of its high moisture and salt content. Dry skin, on the other hand, has a considerable resistance to electric conductivity. This resistance leads to the greatest intensity of burning and charring where the electric current enters and leaves the body. The total conductivity of a human body is, of course, increased manifold if a person is connected to a good conductor: wet ground, water, or any metal base.

The electric shock almost always results in unconsciousness and in disturbance of the heartbeat (arrhythmia). If the currents flows through the brain stem it may cause the heart to stop beating altogether and breathing to cease (Clayman; 1989, p.395).

Additional and often fatal damage develops from the intense internal heating of the body. The resistance of body tissues to the movement of electrons causes heat to rise similar to the heating of a wire which conducts electricity. The physiological impact of lightning can thus be summarized as: (1) causing unconsciousness, (2) stopping heart and breathing functions, (3) raising internal temperatures to damaging or fatal levels, and (4) producing severe skin burns. Recovery from the effects of lightning to the brain and nervous system is similar to recovering from a stroke.

In addition to the direct burn and electric shock effects, there is also a mechanical impact caused by a *pressure wave* that surrounds a lightning stroke. Peak overpressures of at least 20 atm, or 293 pounds, per square inch can be produced by the extreme heating in the lightning channel. This shock wave is able to cause damage to a distance of over 3 feet (1 m) (Krider; 1985, p. 216).

In general, lightning strikes involving aircraft are less dangerous than generally assumed, as far as all-metal planes are concerned. The metal skin is an excellent conductor of electricity and the airplane, for a short moment, acts like a wire permitting the electrical surge to pass through it. Lightning seems to travel along the greatest distances, such as from wingtip to wingtip, and from nose to tail (Eagleman; 1983 p. 121). The physical damage to the airplane is usually restricted to small punctures in the hull. The greatest damages, and often with disastrous results, occur to sensitive electronic equipment. After being subjected to strong electrical surges such equipment may become unusable, making the plane uncontrollable (AF Manual 105-5; 1962, pp. 12–18).

There are few incidents where lightning could be cited as the direct cause of aircraft accidents resulting in high fatalities. Two disasters, both involving *dirigibles,* took place on December 21, 1923, and on May 6, 1937. The first happened to the French dirigible "Dixmude" (the former German Z-114), which apparently was struck by lightning over the Mediterranean Sea on a voyage to North Africa. It ended in a fiery explosion that killed all 52 persons aboard. The second disaster befell the German dirigible "Hindenburg" on May 6, 1937, as it was being winched down for a landing at Lakehurst, New Jersey. Thunderstorms had passed over the field, and it is possible that the dirigible could have accumulated a large electrical charge. Whether the subsequent explosion was caused by triggered lightning from the landing mast, or whether an electrical discharge issued from the airship's hull, or whether sabotage played a role in the fiery finale of the huge dirigible will never be known for sure. Field Marshall Goering, head of the German Luftwaffe, feared any rumor of sabotage and called the disaster "an act of God" (Mooney; 1975, p. 260). It is noteworthy to state that both the "Dixmude" and the "Hindenburg" were filled with highly inflammable hydrogen. Had they been buoyed by nonflammable helium, these disasters probably would not have taken place.

One of the worst aircraft disasters caused by a lightning strike occurred on December 8, 1963, when a Boeing-707, after taking off from Baltimore, was struck by lightning. The plane crashed near Elkton, Maryland, killing 81 persons; luckily, more than 70 people had disembarked in Baltimore.

DUST STORMS AND BLIZZARDS

Key Terms		
khamsin	prefrontal squall	haboob
blizzard	humus	wind-chill factor
dust storms	loess	hypothermia
surface creep	black dusters	anticyclone
saltation	sharav	carbon monoxide
dust devils	relative humidity	

The sky changes from a dark gray to an ominous black. For a few moments the sun is still visible as a pale disk, but then it disappears as if frightened by the howling winds driving thick clouds of particles through the air. Visibility is reduced to near zero, and human and beast take shelter wherever possible. Lightning bolts pierce the darkness, but the thunder sounds muffled and diffused against the roar of the storm.

This description passage contains some of the common storm characteristics one could encounter in Egypt, in the Sudan, in Russia, in the southwest of the United States, and also in New York, South Dakota, or even in England. Yet, the storms in question can be very different in origin and characteristics. In North Africa the skies may turn dark with enormous clouds of dust when visited by the dreaded *khamsin;* but in New York the sky looks gray when thick clouds and swirling masses of snow herald the arrival of a *blizzard.*

Rapidly moving cyclonic fronts can create two types of storms: blizzards and dust storms. Blizzards are restricted to the higher latitudes and are associated with vigorous midlatitude winter cyclones. Blizzards cover extensive areas, and usually they can be well predicted because their movement coincides with the advance of strong, well-defined cold fronts. *Dust storms* can form under any strong wind system, in any part of the world, as long as loose particles can detach from the earth's surface and enter the turbulence of the atmosphere. Dust storms may be quite local, especially when they are produced by thermal thunderstorms or even by tornadoes. Other dust storms, such as the monster storms of the American southwestern regions, immortalized by the destructive *Dust Bowl* of the 1930s, are also tied to frontal weather systems, and in many ways resemble snow blizzards. Large-scale dust storms may also occur in the dry higher latitudes of the world, such as western China, large arid regions of the former Soviet Union, and in the northern Great Plains of North America.

Sven Hedin, the famous Swedish explorer, encountered a severe dust storm on April 28, 1895, when he traversed the interior desert of Takla-Makan in what is today the northwestern province of Sinkiang in China. In his words: "There was little actual daybreak to speak of. Even at noon the darkness was more pronounced than at dusk. It was like marching at night. The air was filled with opaque clouds of drift-sand. Only the nearest camel was dimly visible, like a shadow in the otherwise impervious mist" (Hedin; 1925, p. 154).

Causes of Dust Storms

Two basic conditions must be present before a dust storm develops, namely (1) an ample supply of detachable surface material, and (2) a wind system capable of lifting and transporting the particles in the atmosphere.

The term "*dust*" is not a scientific classification. An analysis of dust particles, in a typical dust storm, reveals that we are dealing with medium-sized silt (about 0.03 mm) and clay (below 0.002 mm). Even though high wind velocities may move larger particles, such as sand and even gravel, this movement takes place on the surface and is called *surface creep.* Larger particles also move in leaps and bounds near the surface, a process referred to as *saltation.*

The finer components are picked up by wind velocities of about 15 to 30 mph (24 to 40 km/h), and the small particles are lifted into the air by strong updrafts or by turbulence. The updrafts are most frequently encountered under strong convectional thunderstorms and along quickly moving cold fronts, as was discussed in chapter 7. *Turbulence,* a more irregular air motion, occurs when the advance outburst of colder air from a fully developed thunderstorm touches the ground. Severe ground turbulence is also produced by vortex storms, such as tornadoes and *dust devils,* as well as along the *prefrontal squall* lines of midlatitude cyclones.

Any vegetation growth protects a soil surface against wind erosion. Dust storms, therefore, are generated in areas that either have been denuded, by overgrazing or radical deforestation, or where the dry climate does not support a coherent plant cover. Even in regions of bunch grass or sage brush, which are types of discontinuous vegetation, dust storms occur.

Whether soil particles are picked up by wind also depends on (1) the moisture content of the soil, and (2) whether the particles are cemented by either mineral or organic matter. The erosiveness of soil is just about inversely proportional to its *humus* content (Whitaker, Ackerman; 1951, p. 74). Dust storms, therefore, develop in arid regions as well as in sub-humid areas where soils are naturally low in humus or have lost their humus.

Whether dust is lifted high up into the air or drifts closer to the ground depends not only on the initial vigor of the uplift but also on the internal temperature stratification of the air itself, which affects its *stability* or *instability.* When the internal temperature decrease with altitude is small, as in stable air, the dust will not rise very high, and clear skies prevail aloft. But, when there is marked instability, or large temperature contrasts with increasing elevations, the dust-laden heated surface air will rise to great heights and may totally obscure the sky. Under these conditions, frequently encountered in low-latitude desert environments, dust may be lifted to 15,000 feet (4570 m) and spread over wide areas (Federal Aviation Agency AC-006; 1965, p. 136). While flying within arid land regions, such as Iraq and Iran, it is common to see a clearly marked upper limit to a dust layer at about 5,000 feet (1525 m), which may prevail for several days (Cressey; 1960, pp. 101–102).

The conditions leading to dust storms often have been blamed on humans' activities, primarily the cultivation of erosion-sensitive tracts in subhumid areas. Although it is true that reckless land use in many parts of the world accelerates wind erosion, dust storms are a natural phenomenon in all low-rainfall regions (Smith; 1958, p. 56). Dust storms have occurred throughout the earth's history. Wind-carried *loess,* deposits of fine siltlike material, can be found in many regions of the world. Some of these deposits were formed during periods of dry climates that, by now, have been replaced by more humid conditions.

A study by J. C. Malin, investigating dust storms between 1850 and 1900, shows that they occurred in the then uncultivated southwestern regions of the United States. The research revealed that dust storms were encountered scattered throughout the 50-year period. The wind erosion problems, however, certainly became worse after the land was put to the plow (Whitaker, Ackerman; 1951, p. 183). The terrible "*black dusters*" of the 1930s, partly induced by the unwise cultivation of land in the Oklahoma-Texas Panhandle, will be treated later in this chapter and in chapter 10.

It is interesting to note that naturally occurring soil dust in the atmosphere, deriving mostly from desert surfaces, represents about 15 percent of all particulate material in the atmosphere, but that human-induced agricultural dust accounts for less than 1 percent (Allen; 1983, p. 133).

Dust Storm Types

What type of dust storm develops depends on how it forms, on the amount of soil particles available, and on the general climatic conditions in the area involved. Although each dust storm differs in intensity and in its life cycle, one can distinguish three basic types:

1. *the midlatitude frontal*
2. *the low-latitude frontal*
3. *the thunderstorm-generated type*

The midlatitude frontal dust storm is typically associated with strongly developed cold fronts of cyclonic depressions. As these fronts pass erosion-prone lands, the vigorous uplift of surface air caused by the advancing wedge of the cold air raises up large amounts of soil particles. North America, with its sharply contrasting air masses (see chapter 6), is known for its strong weather systems and storm activities. It is not surprising that dust storms of the frontal type are a well-known phenomenon in the United States. These storms have occurred a hundred years before the drought of the 1930s (Smith; 1958, p. 63), and probably also in prehistoric times.

The strong winds preceding cold fronts, with gusts at times reaching 30 to 60 mph (48 to 96 km/h), are capable of lifting and transporting incredible quantities of soil particles. Though it is impossible to accurately measure the total amount removed and transported, estimates state that a moderately sized storm, covering about 5,000 square miles (12,950 km^2), may move 7 million tons of soil. By comparison, some of the monster storms, or "black dusters," of the 1930s covered areas of more than 175,000 square miles (454,000 km^2), and transported more than 200 million tons of soil dust (Bennett; 1939, p. 121). Some of this material was carried aloft for thousands of miles.

In some instances the dust mixes aloft with various forms of precipitation. Thus, snow and hail may become strongly colored according to the origin of the dust. On March 9, 1918, snow and sleet fell over Madison, Wisconsin, depositing 13.5 tons of dust per square mile (2.6 km^2) in that area. In a similar way, although not associated with precipitation, a reddish-brown dustfall deposited about 30 tons on the same

amount of area at Lincoln, Nebraska, during a storm in 1933 (Blair, Fite, 1957, p. 241).

The second type of dust storm, also associated with cyclonic weather systems—although in the lower latitudes—is the *khamsin*. This storm, throughout history, brought stress and devastation to northern Egypt and to other parts of the Mediterranean area.

The name khamsin (also kamseen) comes from the Arabic language, meaning "fifty." It is typically encountered in Egypt, beginning in March, and tends to occur throughout a period of about fifty days, a fact that possibly accounts for its name. But this name could be misleading because this hot, southeasterly wind does not blow uninterruptedly; it springs up suddenly and at irregular intervals. The mechanism for this dust storm, emanating from the reaches of the Sahara to the south, is the passage of late-spring *cyclonic depressions.* These travel eastward from the North Atlantic across the Mediterranean Sea. As the southern quadrant of these cyclones face Egypt, the hot, dust-laden desert air streams into the low-pressure trough. As wind velocities increase, large quantities of desert dust are picked up. The skies may darken for several days until the cold front passes; then, the northwesterly flow of cooler air from the sea displaces the desert air and clears the sky.

As this type of cyclonic depression migrates toward the Middle East, the hot southerly wind blows across southern Israel, the Red Sea, and Saudi Arabia. In Israel this wind is called the **sharav,** and it may produce severe heat waves. If the wind passes over the open waters of the sea, the humidity may rise markedly. This results in oppressively hot and humid conditions, especially along the coastal regions (Israel Geography; 1973, pp. 106–108).

If the desert wind is encountered directly, without being modified by the Mediterranean, it is possible for the **relative humidity** to drop to less than 10 percent while the air temperature rises to 122° F (50° C) (Ahrens; 1982, p. 298). This combination of extremely low humidity and excessively high temperatures dessicates crops and can do great damage. An interesting question arises when we look at a passage in the Old Testament: "And Moses stretched forth his hand toward heaven; and there was a thick darkness in all the land of Egypt three days" (The Holy Scriptures; 1955, pp. 148–149). The southerly winds and darkened skies, associated with the passing of the khamsin, tend to prevail for about three days at a time.

A smaller type of dust storm, yet more violent, is the ***haboob.*** Its name also derives from the Arabic and means "strong wind." This storm is most typically encountered in the Sudan, especially around Khartoum; it is most numerous during the hottest part of the year, from May to about September. However, since it is triggered by well-developed thunderstorms, it may occur at any time. Haboobs have also been reported throughout the Middle East, in Iran, India, western Africa, Australia, and at times in southern Arizona.

This storm varies considerably in size depending on whether it forms from a single convectional thunderstorm, or from a series of cells that may develop along air mass frontal systems. The latter can be observed when the summer monsoon moves northward toward the drier regions in India. A similar situation occurs in Africa when the more humid equatorial air masses move north into the Sudan and into southern Egypt.

The basic mechanics of this dust storm can be found in the powerful cold-air downdrafts issuing from the leading edge of a thundercloud. The violent downdraft pours out of the cloud base and spreads out ahead of it. The gusty turbulent wind, the haboob, lifts up the surface dust and produces a rolling dark cloud. This much cooler air stands in sharp contrast to the very high ground and air temperatures that precede such storms. Bare-ground temperatures in low-latitude deserts can easily reach 170° F (77° C), heating the air to 120° to 130° F (49° to 54° C). The sudden drop in temperature, accompanied by a quick change in wind direction, characterizes the haboob. Momentary wind speeds of over 60 mph (100 km/h) have been experienced in these outbursts. If rain falls, it is found behind the advancing haboob (Eagleman; 1983, p. 235).

Tornadolike winds and the sharply reduced visibility constitute the main hazards. Tents and lightly built houses can be destroyed by the violent wind gusts. The 1980 American attempt to rescue the hostages held in the American Embassy in Teheran, Iran, failed in part because of the interference by haboobs. The unpredictable dust storms originated in connection with thunderstorms in the desert interior. Fine talcumlike dust penetrated into the cockpits of the helicopters, raising the inside temperature from 88° to 93° F (31° to 34° C), and reduced visibility to zero. Dust clouds had not been forecast in the pilots' weather briefing. The masses of dust in the flight path of the helicopters were generated by thunderstorms that developed suddenly about 50 miles (80 km) to the west. The dust forced one helicopter to return to the carrier USS *Nimitz,* while another had to be abandoned in the desert for mechanical reasons. This reduced the total number of helicopters to six; this number was considered the absolute minimum to continue the planned but unsuccessful mission (Ryan; 1985, pp. 70–75).

Haboobs have been likened to overgrown *dust devils* because, from a distance, they may look like a column of swirling smoke. The color of these storms may vary from yellowish to reddish, depending on the color of the surface soils. Compared to the khamsin, the haboob is relatively short-lived, lasting from one to two hours. True dust devils are much smaller and generally harmless convective whirls, or eddies. They usually last only a few minutes. Dust devils can be observed in all parts of the world, especially when soils are dried out, and when ground temperatures are very high. They may reach about 30 feet (10 m) in diameter and hardly ever extend beyond 6,000 feet (2,000 m) in altitude (Moran, Morgan; 1986, p. 279).

Blizzards

In the United States alone, annual snow-related deaths account for an average of about 90 lives. One of the worst years was 1958 with 345 deaths; this was exceeded in 1960 when 354 persons died of snow-related causes. About one-third of all victims are killed in car accidents, less than one-third die of exhaustion and overexertion, while about 13 percent fall victim to exposure and fatal freezing (NOAA/PA70018; 1975, p. 2). In most cases, here and abroad, it is the *blizzard* that exacts such a number of deaths. This storm type is by far the most violent and the most dangerous winter storm.

The origin of the name blizzard is not quite clear. Terms such as *blysian* (Old English), *blizzer,* and *blaze* seem to be related in their meaning of "burning." Whatever the meaning, a blizzard blinds with its blowing snow and "burns" (bites or blisters) the skin with its low temperatures, rendered even more dangerous by the ***wind-chill factor.***

Although blizzards do occur in many parts of the world, they are most common in the midlatitude interiors of North America and Eurasia. They are characterized by high winds filled with powdery masses of snow. The snow is either partially or entirely picked up from the ground. Visibility in many of these storms is reduced to zero. While the total amount of snowfall may be less than that of a regular winter snowstorm, the blizzard accounts for huge, dunelike snowdrifts that paralyze traffic and may trap people in their cars.

At what point a snowstorm is called a blizzard is somewhat arbitrary; blizzardlike conditions may momentarily occur with almost any severe winter storm. However, for the sake of consistency, the National Weather Service issues blizzard warnings when wind speeds of at least 35 mph (56 km/h) combine with considerable amounts of falling or blowing snow, and visibility is dangerously restricted. Should the wind velocities exceed 45 mph (73 km/h) and are accompanied by a high density of falling or blowing snow, with temperatures of 10° F (−12° C) or lower, a severe blizzard warning is issued.

It is the combination of *high wind velocities* and *low temperatures* that poses a deadly threat to people by causing a condition called **hypothermia,** or subnormal body temperatures. High winds result in exceedingly rapid heat loss from the body. This is not just the effect of the low temperatures of the air but is brought on by the wind-chill factor. For example, when the outside air temperature is 10° F (−12° C), and the wind blows at 45 mph (73 km/h), the rate of body heat loss is equivalent to −38° F (−39° C) under calm conditions (NOAA/PA 70018; 1975, p. 3).

Hypothermia sets in when the body temperature falls more than 4° F (about 2° C) below the normal level of 98.6° F (37° C). If this state lasts for several hours, death could occur. Anyone whose body temperature falls below 90° F (about 32° C) has anywhere from a 17 to 33 percent chance of dying (American Medical Association; 1982, p. 723). Older people are especially endangered because their bodies cannot regenerate heat as quickly as those of younger individuals; moreover, the sensory mechanism, which detects a drop in body temperatures, slowly loses its efficiency in elderly persons. Unaware of this danger older people may die without warning. Furthermore, heat loss can be made worse when wearing wet clothes because water conducts heat about 25 times more efficiently than air. This means that heat loss from both skin and clothes is greatly accelerated. A person caught in wet clothes, as may happen in a snowstorm, can develop potentially fatal hypothermia at temperatures considerably above the freezing mark, especially if there is high wind velocity.

Blizzard Disasters

The hazards associated with blizzards (*low temperatures; windchill; low visibility; unsurmountable snowdrifts*) can occur in all terrains. Consequently,

blizzard conditions have not only endangered people in populated areas, in cities, and in the plains of the western regions of North America, but they have also trapped skiers, mountain climbers, explorers, and groups of people who attempted to cross mountain passes.

A mountain blizzard disaster struck a group of more than 100 persons who migrated from Springfield, Illinois, to California. This was the ill-fated Donner Party, organized by George and Jacob Donner in 1846. On October 28 the expedition reached the Sierra Nevada when, suddenly, a violent blinding snowstorm hit. The blizzard continued for another eight days and was followed by additional storms (Ed. Bantam/Brittanica; 1978, p. 147).

Trapped by waist-high snow and bitter cold, the people were caught in their make-shift camp. Food supplies ran out quickly so that the people had to slaughter first their cattle, then the horses, and eventually even dogs to stay alive. By mid-December the first victims died, and a small party of sixteen, eleven men and five women, left the camp to get across the mountains. Lack of food and more blizzard conditions eventually reduced the group to two men and the five women. Without food, starving, and nearly out of their minds, some of them were said to have eaten the flesh of those who died. Of the original Donner Party only forty-eight survived (Nash; 1976, p. 154).

Another renowned storm disaster occurred also in the 19th century when a powerful blizzard struck the northeastern United States in March 1888. That storm killed 400 people, including 200 victims in New York City alone.

The meteorological setting of this blizzard partially resembled the collision between a cold front over the Great Lakes and Hurricane HAZEL (1954) as it moved up the American east coast (see chapter 6). In the 1888 blizzard, a low-pressure cyclone moved up the east coast and pushed warm moist air up the slope of a rapidly moving cold front. Sunday morning of March 11, 1888, began as a cloudy rainy day in New York. Although the Weather Bureau had predicted somewhat cooler weather and light snow moving in, the subsequent event caught New York's population in complete surprise and utter unpreparedness.

The cold front struck before noon. The temperatures plunged, and a violent storm blanketed the area with more than 2 feet of snow by Monday morning. Winds rose to hurricane force, and in a short time all the streets became impassable. Drifts were reported to have reached up to 18 feet (5.5 m), and the temperature stood at −4° F (−20° C). There was no electricity, no surface transportation, and the city was paralyzed (Butler; 1976, p. 196). New York City received most of the publicity, but many other cities, from Boston to Philadelphia, were also hard hit by this blizzard.

Many severe winter storms have struck the northeastern United States since this event, causing death and hardship; however, the massive blizzard of January 1977, which roared into Buffalo, New York, stands out as a prototype of such a disaster.

◎ The Buffalo Blizzard of 1977

On Friday, January 28, 1977, at around 11:00 A.M., the worst blizzard in its history hit the Buffalo area of western New York. During the storm the temperatures remained near 10° F (−12° C), winds gusted up to 70 mph (113 km), and heavy blowing snow reduced visibility to zero.

The overall preceding weather conditions contributed to the incredible impact of this blizzard which raged at full force, without abatement, for more than 17 hours. Heavy snowstorms in early December, and at intervals throughout the month, coupled with record-breaking low temperatures, brought about 93 inches (234 cm) of snow, which is 13 inches (33 cm) above the normal average for the entire year. Lake Erie, with a total surface area of approximately 10,000 square miles (25,900 km²), was entirely frozen at the end of December, and a large amount of snow, estimated at 30 inches (76 cm) in thickness, covered the ice. When the blizzard struck, the fierce winds picked up a lot of this snow, in addition to the new snow associated with the storm, and carried it into the metropolitan area of Buffalo (Rossi; 1978, p. 286).

The meteorological situation could not have been more "favorable" for producing this monster storm. A deep low-pressure system had moved eastward just to the north of the Great Lakes. It was followed by an almost stationary high-pressure **anticyclone** that built up over the continental interior. The counterclockwise winds around the Low, and the clockwise circulation around the High, led to a persistent west-to-northwest flow of extremely cold air into the western New York region and southern Ontario, Canada. The considerable pressure difference between the two weather centers produced very high wind velocities.

At 5:00 A.M. the Buffalo Weather Service issued a forecast indicating that near blizzard conditions would develop late in the afternoon of that fateful January 28. Six hours later, at 11:00 A.M., the forecast was changed to a blizzard warning (U.S. Army Corps of Engineers; 1977, p. 3). While the warning was broadcast, the cold front hit downtown Buffalo; it approached like a white wall. People on the sixteenth floor of a large bank building could feel the floor shudder and heard the plate glass windows creak (Bahr; 1980, p. 46). Within half an hour most businesses and industries sent their employees home early.

Within a short time people realized that they were trapped in the city. Transportation systems came to a stop, and cars became stuck in the rapidly accumulating massive drifts. Moreover, many people became confused in the zero visibility, abandoned their cars, and sought shelter wherever possible. Visibility officially remained at zero from 11:30 A.M., Friday, to 12:50 A.M., Saturday, January 29.

Thousands of people remained downtown; some did not reach home until several days later. Food and drink were shared, but shortages of milk for small children became a problem. Strict measures were enforced concerning unnecessary driving into the city. This official decision was criticized by some as being too severe a measure; yet, it appeared a necessary step. The situation became so serious that President J. Carter issued a declaration of emergency on Saturday, January 29. This permitted governmental agencies to move in and to assist with rescue efforts under the authority of the Fed-

Plate 46. During the Buffalo Blizzard of January 28, 1977, dunelike drifts of snow matched the height of many suburban homes as this one shown in Amherst, New York. (Photo by C.H.V. Ebert).

eral Disaster Assistance Administration (U.S. Army Corps of Engineers; 1977, p. i).

One major problem involved the enormous masses of snow. Dunelike drifts piled up against houses, in the streets, on parking lots, everywhere [Plate 46]. Cars disappeared under the snow; some people became entrapped in them. Nine persons died in automobiles either of hypothermia or carbon monoxide poisoning. Western New York suffered 29 snow-related deaths, a relatively small number of victims considering the bitter severity of this blizzard.

After the storm finally abated, the area slowly tried to struggle out of the paralysis that had gripped it for several days [Plate 47]. Soldiers from as far as Fort Bragg, North Carolina, had been flown in to help clear the streets with heavy equipment. But where to put the snow? One of the many ways, and rather unique, was to fill open freight railway cars with snow and take the "snow trains" south, away from the western New York region, so that the snow

Plate 47. For many days after the Buffalo Blizzard of 1977 had abated, hundreds of cars—abandoned during the storm—had to be located and extricated from the snowdrifts. (Photo by C.H.V. Ebert).

could gradually melt! The large quantity of snow, accumulated since December of 1976, created a considerable flood potential; however, favorable weather conditions slowed the melting and subsequent flooding was relatively minor.

The economic losses, as is usual in storms that strike large urban areas, were very large. Loss of production, sales, and wages amounted to at least $297 million for Erie and Niagara Counties. Some of this loss was incurred because of the emergency curtailment in the use of natural gas. Then there was also the physical damage to homes and cars.

Weather systems do not respect boundaries. Many severe storms, after leaving North America, pick up moisture and energy from the North Atlantic, and strike Western Europe. One of the worst blizzards of this kind blew in from the Atlantic into Scotland at the end of January 1978 and killed from 10 to 15 percent of all livestock (Whittow; 1979, p. 316). Two weeks later southwestern England suffered blizzard conditions from another storm with drifts as high as 20 feet (6 m).

As in most disaster situations, a great deal of harm to people, animals, and facilities could be avoided if people are informed, understand the nature of these storms, and use common sense. Many victims succumb because they overexert themselves and underestimate the danger of hypothermia. Some, as illustrated in the Buffalo experience, do not recognize the silent death brought on by **carbon monoxide.** Oxygen starvation and carbon monoxide buildup form a deadly combination. When staying in a car during a blizzard, one should use both motor and heater sparingly, and leave the downwind window partially open. This action would greatly enhance the chance of survival.

AVIATION HAZARDS

9

C H A P T E R

Key Terms		
aeronautical meteorology	wake turbulence	pressure-altitude
IFR (Instrument Flight Rules)	rotor clouds	sea level
	wing vortices	isobaric surface
heat islands	microbursts	great-circle
altimeter	temperature inversions	ice fog
air pocket	Doppler radar	sea smoke
albedo	rain shaft	diamond dust
lift surplus	airspeed	false relief images
lift deficiencies	groundspeed	whiteout
mountain waves	standard atmosphere	rime
	aneroid barometer	

Humans have ventured into space, walked on the moon, operated satellite stations, and used space vehicles, such as the shuttle *Challenger,* that bridge the gap between rockets and aircraft. However, everyday flying takes place in the lower atmosphere and must meet the challenges and hazards of weather.

Specially designed military airplanes have reached peak altitudes of about 124,000 feet (37,800 m), and sustained level flights are possible at over 85,000 feet (25,900 m). These flights leave the troposphere, which averages about 7 miles (11 km) in the midlatitudes. High-altitude flights by large jet aircraft reach cruise altitudes of 39,000 feet (11,900 m), or a little over 7 miles (11 km). The famous British and French-built *Concorde* operates at considerably higher elevations. Its cruising altitude, at Mach 2.04, is 51,300 feet (15,630 m); its service ceiling is 60,000 feet (19,290 m). The plane started regular passenger service in 1976 (Mondey; 1983, p. 10).

Meteorology, the science of the atmosphere and its associated phenomena, has a number of subfields. The one that deals primarily with weather, as it affects flying, is referred to as ***aeronautical meteorology.*** Just about every element of meteorology is directly, and indirectly, of some influence in aviation. The majority of aviation hazards have been thoroughly studied. In most cases they are well understood and are relatively easy to recognize. Some of them were treated in the preceding chapters covering hurricanes, tornadoes, weather fronts, dust storms, thunderstorms and lightning. Most pilots, and certainly professional fliers, are well acquainted with these storm types and associated weather conditions. Preflight weather briefings and in-flight weather updating enable the pilot either to avoid atmospheric problems or to deal with them effectively. This capacity is especially needed along well-traveled air routes in developed areas of the world.

In many ways sophisticated electronics respond faster than pilots; computers can fly and land planes and do this in an efficient, impersonal way. Pilots, on the other hand, may have to deal with disorientation, confusion, slow response, overreaction, and simple fatigue. Unfortunately, sophisticated electronic gear is very expensive. In some cases, because of military security restrictions, it may not even be available to the average pilot. Flight conditions may become quite hazardous because of either simple or inadequate equipment. Other hazards can be traced to geographical factors and to less well known atmospheric phenomena.

Geography, in both the physical as well as in the spatial context, constitutes a major factor in aviation hazards. Vast, isolated regions that are frequented by severe thunderstorms, such as the thousands of square miles of tropical rainforest, must be negotiated both in terms of distance and atmospheric conditions. Other problems, partially discussed in connection with dust storms (chapter 8) arise in desert flights. High mountains are known for their own local wind peculiarities. The Rocky Mountains, the Andes of South America, the Alps of Europe, and the towering Himalayas, pose at times very dangerous flight conditions. Flying in the high latitudes also entails very unique hazards; these will be analyzed later in this chapter.

But it is not only nature that creates problems for aviators. Many human-made factors have introduced new and sometimes unexpected complications. Industrial smoke and urban haze can close airports, slow down air traffic, and force pilots to follow **IFR (Instrument Flight Rules)** procedures in congested air lanes around urban centers. Moreover, the heat buildup of large cities, leading to the phenomenon called **heat islands,** can trigger thunderstorms and may produce severe turbulence [Plate 48]. This situation can interfere with the plane's glide path during the landing approach.

Nature itself has numerous surprises for the pilot. Several atmospheric conditions, often quite local in nature, can be exceedingly dangerous because they may arise spontaneously, may be invisible to the eye, and are too small to be detected by weather-monitoring instruments. Included in these problems are certain forms of thermal and mechanical turbulence, or rapid air pressure changes, which can result in **altimeter** errors, and several optical phenomena. The delicate interplay of air pressure, temperature, and moisture content may lead to very complicated weather situations. Their complexity and hazard potentials increase if we add the factors of topography, urbanization, instrument inadequacy or failure, and the human element called *"pilot error."*

The possibility of complex interactions of factors should not imply that all flying is exceedingly

Plate 48. As large urban centers warm up they may form heat islands that are capable of producing large convective cells and thunderstorms. This situation can severely interfere with aviation. (Photo by C.H.V. Ebert).

hazardous, nor should these factors discourage anyone from venturing into this activity. The overall record of aviation, especially in the United States, is very good and much better than other forms of transportation. But quite a number of people consider flying a risky business, and some develop a strong phobia when it comes to using an airplane. It is possible that the stark pictures of utter destruction associated with the air crashes contribute to this fear [Plate 49]. This fear exists despite the fact that, according to the Federal Aviation Administration, fatal accidents per 100,000 departures (U.S. flights only) between 1977 and 1990 averaged only 0.063 deaths a year.

Some years, such as 1982, were entirely free of fatalities, which averaged about 126 annually between 1976 and 1984. Of course, such statistics never tell the actual stories. One bad crash, such as the one of Turkish Airlines on March 3, 1974, at Orly airport, Paris, killed 346 persons (Nash; 1976, p. 572). This toll is more than twice the number of victims who died in civilian air crashes in the United States in 1978. Another terrible disaster, as the one in Dallas, Texas (August 2, 1985), killed 137 people. Only 27 passengers survived.

The majority of air accidents, about 94 percent of all U.S. Civil Aviation accidents, involve noncommercial general aviation. According to the National Safety Council, out of a total number of 2,930 accidents, 2,742 involved general aviation and claimed 937 deaths. This amounts to 75 percent of the total of 1,246 victims in 1985. Large U.S. airlines, in the same year, accounted for only 18 accidents, or 0.6 percent of the total, with 197 fatalities

or 16 percent of all fatalities. Statistically, flying appears to be rather a safe way to travel, especially when we look at the 45,600 deaths caused by car accidents during the same year. Yet, while there are about 26 more automobile accidents, per distance traveled (*passenger miles*), the probability of getting killed in airplane accidents, per distance traveled, is 13 times greater in general aviation than in car travel (Goldstein; 1984, p. 1.1).

In the preceding chapters we have studied the nature of hurricanes, tornadoes, thunderstorms, and lightning. In some instances aviation hazards were mentioned, especially those associated with lightning. Instead of reviewing this information, the following sections concentrate on the generally less well known hazards that pilots have to face.

Air Turbulence

Literally, *turbulence* means "full of restlessness." In modern atmospheric science turbulence refers to any irregular or disturbed air flow. Although major atmospheric disturbances can be detected quite readily, it is much more difficult to give precise information about turbulence. Because turbulence tends to be of local occurrence and frequently is transient in nature, prediction, detection, and reportage are each difficult. Thus, if compared to the total usable air of a given air lane, the volume involved in turbulence is very limited; but this phenomenon, especially when interfering with landing and takeoffs, can pose grave dangers. Turbulence may also develop at higher elevations and can present in-flight hazards. Some of the more general danger zones involving turbulence can be predicted with somewhat greater precision. With proper flight briefings and effective monitoring of weather reports such areas can be avoided. Strong and dangerous turbulence often develops around the jet stream. It frequently occurs under cloud-free conditions and is called *clear-air turbulence*. This can bring about severe buffeting of high-flying planes. Jet-stream, clear-air turbulence is quite variable, and its exact location may be difficult to predict (Battan; 1984, p. 139).

Turbulence can be produced whenever there is rapid air movement, when there is some interaction between differently moving air currents, or between the air and some obstruction. We recognize two basic categories of turbulence: (1) *thermal turbulence,* and (2) *mechanical turbulence.* In each category are specific types. It is difficult to sharply divide cause-and-effect in each incident because the

Plate 49. A person's fear of flying is reinforced by pictures such as the one shown above. This accident involved a jet aircraft that crashed near Dhahran (Saudi Arabia) because of severe heat turbulence and poor visibility caused by dust. (Photograph credit: Frank Gates).

different types may be interrelated or may combine.

The first type of thermally induced turbulence is frequently encountered when flying over convective clouds and rising air cells. Most air travelers have experienced "bumpy" flights when the aircraft suddenly lifts up and then falls into what is called, incorrectly, an *air pocket.* These sorts of disturbances are brought about by rising air currents so typical for late summer afternoons and in tropical regions. The sharp drop is the result of descending cold air that must compensate for the ascending warm air. In severe instances, and particularly when not using safety belts, passengers are in danger of being thrown against the ceiling of the aircraft during sudden downdrafts; these may reach velocities of up to 40 feet per second (12 m/sec).

Another type of convective turbulence develops near the ground in response to differential heating. Dark-colored surfaces, having a low *albedo,* reflect less solar radiation and therefore heat up more efficiently. The albedo, usually expressed in percentages, is the difference between the radiation received and reflected. A blacktop parking lot, for example, has an albedo of about 5 to 10 percent, a freshly plowed field about 14 to 17 percent, and new snow reflects about 75 to 90 percent of incoming solar radiation. Water, depending on the sun angle and wave conditions, averages about 8 percent (Miller; 1971, p. 42). In response to such differences in albedo, the earth's surfaces produce rising and descending air currents. Because for each updraft there must be a downdraft, turbulence develops.

Aircraft on their final low-level landing approach can be severely affected by this complex convective turbulence. Depending on the sum total of uplifts and downdrafts encountered along its glide path, an aircraft may find itself above or below the planned elevation at any given time. Crossing an urban complex, dark fields, or paved-over areas, would result in a *lift surplus* (fig. 9.1). In that situation the pilot must decrease the plane's speed. Downdrafts, resulting from flying over cool lakes and green fields, produce *lift deficiencies* and require an increase in the plane's speed. A mix of lift surplus zones and lift deficiency zones may result in abrupt changes in both altitude and flight speed, which could lead to overshooting a runway, or to stalls, respectively. These conditions can become especially serious when visibility is reduced by smoke or haze.

A good example is offered by the old Kansas City Municipal Airport, which frequently suffered

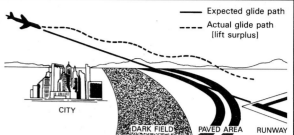

Figure 9.1. A plane's glide path is affected by convectional updrafts, which may be encountered over cities, above large fields of dark soil, and over extensive paved areas. Such updrafts can give an aircraft an unwanted lift, which placed the plane above the expected glide path. (Based on Federal Aviation Agency: AC-00-6A; 1975).

from sharply reduced visibility. This was caused by high humidity levels and severe air pollution from grain processing plants and burning waste dumps. The high humidity levels stemmed from the close vicinity of the Missouri River. The conditions markedly improved with the opening of the new airport located 18 miles (29 km) to the north of the old installation (Navarra; 1979, pp. 499–500).

Mechanical turbulence, caused by obstruction or friction, is the second major category. There are three situations involving mechanical turbulence:

1. *low-level obstructions*
2. *mountain waves*
3. *wake turbulence*

Any object in the path of wind produces some degree of interference and turbulence. Obstacles to low-level air currents could be a row of trees, a large building, a hill, or the sharp edge of an escarpment. In each case, eddies of variable intensity form on the downwind (*leeward*) side of the object. On a small scale, the well-known effect of a snow fence is a good example of this principle. Although there are great variations, brought on by different wind velocities, as well as by the shape and size of the obstruction, the average downwind-effect zone is about 10 to 20 times the height of the obstacle. The irregular whirls and the decrease in wind velocity are the main dangers to aircraft.

On a larger scale we find the *mountain waves.* When a wind reaches speeds of about 46 mph (74 km/h) turbulence develops on the leeward side of mountains (Federal Aviation Agency, AC-00-6A;

1975, p. 83). In many cases this turbulence is manifested by violent downdrafts. If these downdraft velocities are greater than the potential maximum rate of climb of a particular aircraft, the plane could crash into the mountain if it approaches toward the downwind side.

Although the most violent downdrafts are restricted to a relatively narrow zone near the mountains, *downwind waves* may extend for over 100 miles (161 km) away from the mountains. In the upper atmosphere these waves are not very dangerous and can easily be identified by bands of stationary clouds. Underneath the wave crests friction eddies may develop and produce **rotor clouds.** In the upper portion, the overturning part of the rotor cloud, turbulence is quite severe and can become dangerous to aircraft, especially to smaller planes (fig. 9.2).

The last type of mechanical turbulence to be discussed here is called *wake turbulence.* As the name implies, this is a set of normally invisible turbulent waves left behind an aircraft quite similar to the waves left in the wake of a ship.

This turbulence, also called **wing vortices,** develops from the wing tips of aircraft once the landing gear is retracted and the plane lifts off; at that moment the entire weight of the plane is carried by the wings. The strength of the vortices is proportional to the weight of the plane. For this reason, wake turbulence is more intense behind heavy transport aircraft as contrasted to light planes (Federal Aviation Agency, AC-00-6A; 1975, p. 89). The vortices are created as long as the plane is in flight; they trail temporarily behind the aircraft and may also drift with the wind. Danger to other airplanes arises when they fly into such vortices aloft, or cross them

while landing right after the takeoff of another plane.

The typical wing vortices generated by a Douglas DC-10, or the comparable Lockheed L-1011 TriStar, have a core diameter of 8 to 10 feet (2.5 to 3 m), with peak velocities of 150 to 225 feet per second (46 to 69 m/sec). This compares to a whirl blowing at 104 to 155 mph (167 to 217 km/h), or hurricane strength (Goldstein; 1984, p. 5.15). An incident that occurred during training flights involving two airliners illustrates this danger. Less than a minute after a DC-10 practiced touchdown, followed by an immediate takeoff, a DC-9 proceeded to land. It hit the vortex of the DC-10, and its left wing dropped sharply. The pilot corrected by righting the aircraft, but the plane suddenly tipped to the right, rolled over and crashed (Goldstein; 1984, p. 5.14).

Violent Downdrafts and Wind Shear

The development of cold-air downdrafts from thunderstorm cells was discussed in chapter 7, and also in chapter 6 in connection with the dynamics in the formation of tornadoes. These downdrafts, sometimes referred to as **microbursts,** can be very intense; they normally last less than about 15 minutes. They issue from the leading edge of the thundercloud and may reach the ground at speeds of 62 mph (100 km/h) and then burst radially outward along the ground (Moran, Morgan; 1986, p. 303). Violent turbulence may develop between the warm air rushing into the convectional updraft of the thunderstorm cell and the downrushing colder air. The pressure of the descending cold air may help to "feed" the thermal updraft into the storm, thereby intensifying the system (Eagleman; 1983, p. 81).

Two related but different aviation hazards arise from this situation: (1) *the downdraft,* or *microburst,* and (2) *wind shear.* The latter is a change in wind speed, or direction, or both, between two air currents. Wind shear can develop vertically or horizontally; it may be encountered under several very different weather situations.

The most frequently seen wind shears occur in connection with the following atmospheric conditions: (1) *along weather fronts* because winds always change across a frontal system, and the marked discontinuity in speed and direction may lead to wind shear; (2) *low-level **temperature inversions,*** which form during clear calm nights. Winds within the cold

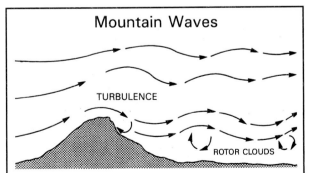

Figure 9.2. Mountain waves, a product of turbulence on the leeward (downwind) side of mountains, may result in violent downdrafts that can endanger low-flying aircraft (Based on Federal Aviation Agency: AC-00-6A; 1975).

surface air layer usually are calm or very light, but winds in the warmer air above it are stronger. Friction between the two unlike currents causes wind shear and turbulence. The resulting fluctuations in wind speed change a plane's airspeed resulting in possible stalls near the ground; (3) wind shear may arise in the vicinity of jet streams in connection with clear air turbulence (CAT), as discussed in the section on air turbulence; (4) wind shear occurs in association with cold-air downdrafts. Wind shear, especially when it forms near the ground and interferes with landings and takeoffs, has caused many airplane accidents. It was not until the development of the **Doppler radar** that these dynamic processes could be studied effectively.

The Doppler radar emits continuous radio signals that are reflected by raindrops. Depending on whether the raindrops are moving toward the radar transmitter, or away from it, the reflected radar waves show different frequencies. If they are moving away, the frequency lengthens; it shortens if the raindrops move toward the radar unit. Since the downdrafts are activated by falling rain, the cold-air downdrafts can be identified. A three-dimensional interpretation becomes possible by using two Doppler radars aimed at the storm from different sites (Eagleman; 1980, p. 214).

Downdrafts associated with rain falling out of a typical thundercloud, the cumulonimbus, can be spotted quite readily because they are pinpointed by the **rain shaft** descending from the cloud base. Some downdrafts, however, may occur without rain falling. This type is usually observed in regions that are dominated by dry air masses. In these kind of air masses, heavy cumulus clouds develop at higher elevations. As water droplets from these clouds begin to meet the uprising dry air, they begin to evaporate and cool the air around them. If sufficient evaporation and cooling take place, a powerful clear-air microburst develops and produces a dangerous and invisible wind shear. It is not until dust begins to swirl on the ground that the shear becomes detectable.

When a plane on its glide path during a landing approach encounters the deflected cold-air downdraft it will first experience a strong *headwind*. This will increase its *airspeed* but decreases its *groundspeed* (that is forward motion relative to the ground). The initial result is an unwanted lift surplus. This situation induces the pilot to put down the nose of the aircraft to overcome the lift effect. But, an instant later, the plane leaves the horizontal headwind of the deflected downdraft; this results in a loss of lift because the airspeed suddenly decreases. The situation is made even more dangerous because the plane is being pushed down by the impact of the descending cold air under the cloud. As the plane exits the downdraft, the wind direction changes abruptly and becomes a *tailwind*. This sharply reduces the plane's airspeed, lift becomes inadequate, the plane stalls and crashes.

A tragic event, involving such a violent microburst, occurred on June 24, 1975, at J. F. Kennedy International Airport in New York. During that warm summer afternoon several thunderstorms had developed over Long Island. Close to 4:00 P.M. three airliners approached the airport and were scheduled to land at about eight-minute intervals. The first (Allegheny 858) reported severe turbulence and some downdrafts, but it managed to land. The second plane, a Lockheed TriStar (Eastern 902) approached. Its speed suddenly reduced from 173 mph (278 km/h) to 136 mph (217 km/h), and the plane dropped from about 300 feet (91 m) to only 60 feet (18 m) above the runway surface! The pilot instinctively gave full takeoff power, miraculously managed to pull up, and subsequently landed safely at Newark airport. The third plane, a Boeing 727 (Eastern 66), on landing approach seven minutes after the aborted attempt by the Lockheed TriStar, was caught in a massive cold-air downdraft; it pushed the aircraft to the ground, just ahead of the runway, killing 113 persons.

This air disaster triggered intense research into downdrafts, or microbursts. One of the scientists primarily responsible for microburst research is Professor T. T. Fujita of the University of Chicago. He concluded that other plane crashes, including the Continental Flight 426 at Denver (August 7, 1975), and Allegheny Flight 121 at Philadelphia (June 23, 1976) were related to downbursts. It is estimated that more than 30 air disasters were connected with violent downdrafts since 1964 (Maybury; 1986, p. 251).

Altimeter Errors

In the introduction to atmospheric hazards it was pointed out that temperature and pressure fluctuations produce ever-changing atmospheric conditions. This variability poses problems for scientists who depend on standard conditions against which they measure deviations from the norm. To overcome this difficulty, meteorologists use the so-called **standard atmosphere,** which shows average values.

Atmospheric temperatures and pressure are intimately interlinked; both play a critical role in monitoring the altitude of a plane whenever an aneroid altimeter is used. The *aneroid altimeter* (aneroid = not using a fluid) and the **aneroid barometer** use the same working principle. A partially evacuated metal box is connected to an indicator scale via a coupling mechanism. As pressure increases or decreases on the outside, the metal box contracts, or expands, respectively. Atmospheric pressure changes with elevation, so that these changes can be correlated with the altitude of an airplane.

Standard pressure deviations, caused by changes in elevation, are measured from *sea-level pressure,* which is expressed in a number of ways: (1) 1013.2 millibars, (2) 29.92 inches of Hg, (3) 760 millimeters of Hg, or (4) 14.7 pounds per square inch. This system means that for about each 1,000-feet (305-m) increase in altitude the atmospheric pressure decreases by about 1 inch of Hg, or about 34 millibars. These values vary somewhat with latitude and above the lower few thousand feet of the troposphere; at higher elevations the rate of pressure decrease declines.

Without deviations from standard pressure, the pilot could rely on the altimeter for determining altitude. However, the aneroid altimeter simply shows **pressure-altitude,** which may be quite different from the plane's actual elevation above the ground. Whenever the pressure is different from the standard-atmosphere pressure, the aneroid altimeter readings are incorrect, and the instrument must be adjusted accordingly. Moreover, elevation above ground level does not mean elevation above **sea level.** If, for example, an airplane lands at La Paz airport (Bolivia), which is located at about 13,300 feet (4,050 m) elevation, the plane's altimeter would show that altitude while the plane actually is on the ground. These kinds of problems can be overcome by the use of more sophisticated instruments. One of them is the *radar altimeter,* which responds to electromagnetic waves bounced off the ground. This reflected energy is called the *radar echo.* Radar operates independent of sunlight because it provides its own screen images. This feature makes this instrument equally efficient during day or night flights (Avery, Berlin; 1985, pp. 168–169). Coupled with computer systems such devices can guide airplanes, at low-level flights, across variable terrain; however, thousands of planes still depend on pressure-altimetry.

Using standard pressure values, we could imagine the atmosphere as being composed of layers of decreasing air density (*lower pressure*) with increas-

ing height (fig. 9.3). Each layer is separated from the next by an imaginary surface called an **isobaric surface.** Pressure and elevation readings stay unchanged if (1) the isobaric surface remains parallel to the earth's surface, and (2) the aircraft would fly along this imaginary plane. Unfortunately, pressure conditions, comparable to terrain on land, result in the bending of these isobaric surfaces. In high-pressure zones the isobaric surfaces form domes, and in low-pressure zones they slope downward and may intersect the earth's surface.

Imagine a plane taking off at airport A with an altimeter setting of 1019 mb. The plane rises to 500 feet (152 m) where the pressure reads 1002 mb. The ground pressure at airport B, the flight destination, is 1002 mb, or 17 mb lower than that at airport A, because airport B is affected by a low-pressure weather system. If the pilot would rely entirely on the altimeter setting obtained at airport A, he or she would fly, indeed, at constant pressure along the 1002 mb isobaric surface. In reality, he or she would fly *downhill* along the sloping isobaric surface and would continuously lose altitude. This means that the plane would actually be at ground level by the time it reaches airport B. The altimeter, however, would still indicate 500 feet (152 m) (fig. 9.4). If such a situation is accompanied by fog or haze the results could be disastrous. Without ongoing *altimeter corrections*—obtained in flight by radio—the plane could fly at unauthorized altitudes and could endan-

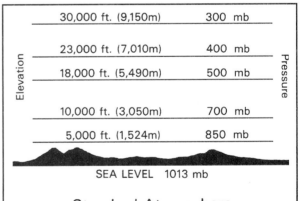

Standard Atmosphere

Figure 9.3. The standard atmosphere represents accepted values of average vertical pressure decreases with increasing altitude. Pressure values can be expressed in various ways. At sea level the pressure amounts to 1013 millibars, 14.7 lb/in^2, or 1.033 kg/cm^2. (Modified from: Air Force Manual 105-5, March 1951).

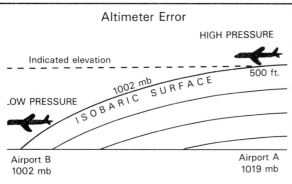

Figure 9.4. Isobaric surfaces (imaginary levels of equal pressure) slope toward areas of lower atmospheric pressure. A plane flying along a down-sloping isobaric surface will find itself at a lower altitude than that indicated by an aneroid altimeter unless altimeter corrections are made. (Modified from: Air Force Manual 105-5, March 1951).

ger other aircraft; it could strike hills, tall buildings, or other high structures.

The principles outlined above can be summarized as follows: When flying toward a low-pressure zone, or along down-sloping isobaric surfaces, the plane is lower than the aneroid altimeter indicates. The reverse is true when flying toward high-pressure zones. Since air temperature variations are directly involved in pressure changes (*warm air is less dense*) this factor must also be considered. The rule for temperature-pressure relationships is: When flying toward a warm air mass (upsloping isobaric surfaces), the plane is actually higher than the altimeter indicates. When flying into a reverse setting, the plane is lower than the altimeter reading. Many air crashes have resulted from underestimating the effect of strong temperature contrasts between air masses, especially in cold-weather flying by instrument (AF-Manual 105-5; 1962, pp. 6.10–6.11). Considerable pressure-altimeter errors can occur in the Arctic because of intense low-pressure systems and conditions involving air masses of much-below standard temperatures. Unless careful compensations are carried out, the sharply contrasting pressure and temperature conditions can lead to altitude miscalculations up to 2,000 feet (610 m).

High-Latitude Aviation

Aviation in the high latitudes, both in *arctic* and *antarctic* regions, began in this century; it posed,

from the very beginning, unique and often extremely hazardous flight conditions. These hazards are not usually encountered during high-elevation flights. They are prevalent near or on the ground and affect mostly takeoff and landing procedures.

From a geographical point of view it is easier to define the Antarctic because we are dealing with a distinct continent: Antarctica and its surrounding seas. The Arctic is more difficult to circumscribe unless one accepts that the Arctic encompasses the regions above the Arctic Circle (66 1/2 N). This definition, especially in terms of aviation hazards, is quite unsatisfactory; the continents, islands, and seas of the Northern Hemisphere interlace and thus present a complex climatic and topographic pattern. More realistic, from a physical-geographical point of view, is the demarcation of the Arctic based on land areas located north of the 50° F (10° C) *isotherm* (a *line connecting points of equal temperature values*) for July, or whichever month is the warmest month, and the coldest month with an average temperature not higher than 32° F (0° C). This demarcation also allows for short-range and long-range climatic fluctuations (Kimble, Good; 1955, p. 7).

The history of high-latitude flying has moved through a number of distinct phases. The early flights were designed exclusively for exploration. Some of these daring flights include the first successful crossing of the North Pole by Admiral Richard E. Byrd and Floyd Bennett on May 9, 1926. They flew a trimotor Fokker plane from Spitsbergen, north of Norway, over the Pole and back, a flight that lasted more than 15 hours. This was followed by Roald Amundsen's venture in the airship Norge on December 5 of the same year. Almost two years later the Italian, Umberto Nobile, crossed the North Pole in a small dirigible on May 24, 1928, but the airship crashed the next day northeast of Spitsbergen. Explorer Amundsen, attempting to come to Nobile's rescue, lost his life in this effort. These hazardous flights took place without the benefits of more recent technology: reliable and superior engine power, rescue helicopters, blind navigation equipment, radar, satellite communication and navigation systems, detailed maps and flying charts, and weather reports based on constant monitoring, almost instantaneous transmission, and analysis.

Although aviation in the arctic regions moved quickly into commercial, private, and military operations, especially during and after World War II, antarctic aviation continued to serve primarily exploration and scientific endeavors. The first flight into the realm of Antarctica was conducted by an Englishman, Hubert Wilkins, in 1928. This was followed by Admiral Byrd with a 1,600-mile (2575 km) flight on November 28–29, 1929. Modern aviation in the Antarctic mainly serves to supply the research stations operated by many nations. In recent years a few private explorers, and even some tourists, added to the slowly increasing flight operations to and within this last frontier continent.

Arctic Flying Hazards

High-altitude flights across the Arctic and low-elevation aviation within this region have increased dramatically. International airlines take advantage of *great-circle* routes; the shortest distance between two points on a globe, which interconnect cities of the Far East, North America, and Europe can be over the North Pole. These flights are hardly ever affected by arctic aviation hazards because they take place "above the weather." But the story of intraregional flying is quite different.

The pilots who have to face near-ground flying conditions, where arctic weather is fickle and at its worst, are the "bush" pilots, and those who serve scientific expeditions and supply remote settlements or weather or military stations. Many have to fly rescue missions under difficult conditions.

Contrary to the popular notion of ever-present howling snowstorms and hurricanelike gales, the overall flying conditions in the Arctic are probably less hazardous than for many low-latitude locations. But ground and near-ground weather conditions can change radically and often unpredictably. Two regions, southern Greenland and most of the Aleutians, are known for some of the most difficult and dangerous flying situations. These are primarily caused by the interaction of relatively warm and moist maritime air masses with very cold arctic air. Sudden winds, with velocities of over 100 mph (161 km/h), may spring up along the southern coast of Greenland.

Quite different in nature, and lacking the obvious perils of galelike winds, are the extremely dangerous low-level atmospheric phenomena that involve a sudden loss of *visibility* and *depth perception.* Such conditions can produce unnerving situations when pilots try to find small landing strips that lack sophisticated blind-landing instrumentation, or when pilots may have to put down on open snow fields under emergency conditions.

The sudden development of fog, both water droplet and *ice fog,* is quite common. *Steam fog,* which is also called *sea smoke,* is frequently encountered in the winter when very cold and dry air masses move from land areas to warmer sea surfaces. Fog layers may also form very quickly when somewhat warmer maritime air moves over colder water areas that contain ice-floes. This kind of fog frequently develops in sea inlets of Greenland, which tend to trap sea ice [Plate 50]. The upper limit of this fog is sharply separated by a warmer layer of air aloft. This air-layer structure is referred to as *temperature inversion;* it deviates from the normal situation where air temperatures decrease with altitude.

Pure ice fog, composed of tiny floating ice particles, forms within calm air, or when the wind is very light, and in air temperature of about −30°F (−34.4° C), or colder. In that case, the water vapor of the atmosphere passes directly into the ice crystal form. The development of this type of ice fog can also be triggered by the exhaust of any type of combustion engines and may linger for hours, or even days, if conditions are right (Federal Aviation Agency; 1975, p. 153). When sunlight strikes these suspended ice particles the pilot—especially when flying toward the sun—faces blinding conditions produced by the bright reflections and shimmering light. Severe *post-flash images*—quite similar to those following a photo flash—may further increase visual problems.

Arctic haze is another type of arctic fog. In this case the size of the ice particles is much smaller, and the crystals form in air layers that are in contact with the ground. Pilots face sharply reduced visibility in the horizontal mode and when looking at ground objects at an angle. Bright, shimmering lights

Plate 50. A dense fog layer—a typical arctic aviation hazard—develops as relatively warm and moist maritime air moves over an ice-covered sea inlet on Greenland's southeast coast near the village of Kusuluk. (Photo by C.H.V. Ebert).

and the intermixing of rainbow colors conspire to produce very confusing images. This ice haze is sometimes called **diamond dust** because of its sparkling characteristics (Neiburger, et al.; 1973, p. 150). The addition of color shades may generate **false relief images** on a level surface. A pilot may become very reluctant to set down his or her plane even though the snow is even and safe.

Another visual hazard, considered the most dangerous one by many experienced pilots, is the **whiteout.** Sometimes the term whiteout is used to describe the total loss of visibility during blizzards, but the arctic whiteout develops in calm air and in the absence of snowfall; it is a phenomenon based on the diffusion of light rays, which eliminates all shadows. The consequences can be a complete loss of depth perception and a dangerous disorientation for the pilot.

The basic requisites for the formation of this type of whiteout are: (1) a subdued terrain uniformly covered with snow or ice, (2) a relatively low, evenly thick cloud layer, and (3) the sun being at an angle of about 20 degrees above the horizon. It is believed that the following process is responsible for the whiteout development: The normally parallel rays from the sun become diffused as they pass through the cloud layer and thus strike the snow surface at various angles. The diffused light is reflected back to the underside of the cloud layer that, in turn, reflects it back to the snow surface. In this diffusion-and-reflection process the rays' angles change all the time. This phenomenon leads to the elimination of all shadows and to a loss of depth perception. If a few dark-colored objects are on the ground, they appear "detached" and seem to float in air. If fresh snow covers such objects, total *disorientation* may ensue (Federal Aviation Agency; 1965, p. 238). These whiteouts can also affect persons on the ground. Obviously, this effect could become exceedingly dangerous to those who are traversing an area that contains ice crevasses.

Many other hazards are associated with high-latitude aviation; some of these may also occur in other regions of the world. While flying through fog or clouds at below-freezing temperatures, planes can accumulate a type of ice called *glaze*. Droplets, striking the leading edges of the plane, freeze on impact. This clear ice can build up rapidly and is quite difficult to remove. Another type of ice deposit may form on an aircraft when it flies through a cloud composed of supercooled droplets that are too small to precipitate. These droplets, on striking the plane, build up rough, irregularly shaped ice, or **rime** (Riehl; 1965, p. 302). The accumulation of ice, glaze or rime, can radically change the *aerodynamic* characteristics of an aircraft in several ways: (1) lift is decreased, (2) friction, or drag, increases, and (3) the total weight grows. These effects may combine and decrease the plane's speed to a level where it stalls and becomes unmanageable.

Modern technology, higher-flying aircraft, better communication networks, in-flight deicing capabilities, and better weather forecasting have greatly reduced the hazards involved in high-latitude aviation. Moreover, techniques of cloud seeding have succeeded in temporarily opening cloud layers so that planes can land during whiteouts and under other poor-visibility conditions. But the remoteness of many small airstrips, the vast emptiness of the arctic and subarctic regions, the rapid local weather changes, and the often unpredictable ground conditions still make flying in these areas a potentially hazardous enterprise. This danger is a definite reality for aviation that employs small aircraft and engages in local flights. Although flying itself may become safer, the main problems associated with high-latitude aviation involve getting the aircraft off the ground and, above all, getting it back to a safe landing.

DISASTERS INVOLVING THE BIOSPHERE

The complexity of disasters—in terms of causes and effects—was repeatedly examined in the preceding sections dealing with the lithosphere, the hydrosphere, and the atmosphere. All disasters, regardless of their origin, have some impact on the *biosphere,* that part of the earth in which ecosystems function.

Ecosystems, interrelated systems composed of living things and of the nonbiological components of the physical world, can be as small as a local bog, or as immense as the tropical rainforest. Furthermore, during the eons of time they have been occupied by some life-forms, ecosystems on our planet have become more numerous and more complex. Each organism must find its *niche* within existing communities of plants and animals; each community must be able to cope with conditions set by the dynamic physical environment.

Our understanding of the earth's past, and of its interwoven natural history, is fragmentary and is primarily based on scientific deductions. Such knowledge derives from fossil evidence *(paleontology),* paleoclimatology, radioactive dating, structural geology, ocean sciences, and atmospheric sciences, as well as from the ever-increasing information base of the cosmos around us. As new evidence is gathered, new hypotheses are formed; more questions must be asked than ever before.

What we *do* know, and what *is* reflected in our present world, is that far-reaching changes have been going on in the past, are going on now, and will continue in the future. Some of these changes seemed to have occurred suddenly, at times in response to cataclysmic disasters. Others took place over long periods of time in response to changes of climate, or to modifications in ecosystems brought on by other living things, including man. Many species in the biosphere have disappeared, others approach the borderline of extinction. It appears that the key to surviving on Earth is not to resist change but to develop the ability to change with change (Wagner; 1971, p. 5).

Humanity's role in this changing world is controversial and in many instances very destructive. The role is controversial because humans, in a realistic sense, no longer consider themselves intertwined with the natural environment. Advances in technical knowledge, living increasingly in an artificial urban environment, and not having to interact with nature for daily subsistence, have given people the false impression that they are no longer creatures dependent on nature. However, the fact is that we as human beings are members of the biosphere. Unfortunately, some of our natural environments have already changed so drastically that we are barely capable of perceiving our role in them. This illusion of being free agents—apart from nature—is very dangerous because it induces humans to engage in reckless actions that may haunt future generations.

Disasters that affect the biosphere are so numerous that it is impossible to present even the majority of them. Some occur on a very small local scale. Others, such as the overwhelming threat of prolonged **drought,** have continental comprehensive dimensions.

In several cases it is impossible to identify clear cause-and-effect relationships. It is difficult to tell whether the disastrous droughts of the Sahel and other regions of Africa are "normal" cycles of climatic change, or whether they are short-cycle *anomalies* made worse by a growing population functioning in a highly sensitive environment.

Droughts have occurred throughout recorded history, and there is no reason to believe that prehistoric earth was free from this phenomenon. Ancient layers of wind-carried dust, deriving from the interior deserts of Asia, have been identified in undisturbed subsoils on the Hawaiian islands. Wind-deposited dust, millions of years old, was found in deep glacial ice layers on Greenland. Droughts have been, and still are, a scourge that threatens human existence. Ash layers from extensive prehistoric forest fires exist in many parts of the world. Fire, as will be discussed in chapter 11, is a natural and important element in the biosphere in general, and critical to the continuity of some ecosystems in particular. Obviously, when humans encroach on fire-prone areas, or when fires are set through carelessness, ignorance, or even with malice, then fire becomes a threat to the environmental equilibrium.

Other problems in the biosphere arose by transferring animals and plants into an environment to which they did not belong. Such actions have brought on *ecological stress* and sometimes disastrous consequences. The almost uncontrollable pest role of rabbits in Australia caused millions of dollars of damage each year. It was triggered by the importation of a dozen pairs of European rabbits—for sporting reasons—in the middle of the nineteenth century. Similarly, semiwild donkeys introduced to some of the Galapagos Islands have multiplied beyond the carrying capacity of the islands and threaten the existence of the giant Galapagos tortoises. The following chapters will explore some of the problems mentioned above.

DROUGHT AND DESERTIFICATION

Key Terms		
deserts	dendrochronology	marine food chain
semiarid regions	windbreaks	anticyclone
savannas	monolith	Trade Winds
desertification	evapotranspiration	Subtropical Highs
salinization	desertization	Satellite imagery
xerophytic vegetation	La Niña	overgrazing

The word *drought* invokes the feeling of dryness and the craving for water. In the archaic sense drought means "thirst." Thirst and drought very likely are some of the greatest scourges afflicting living things. In Africa alone—according to several reports issued by the United Nations Food and Agricultural Organization (FAO) in 1984—an estimated 150 million people were threatened by drought-induced starvation. But, drought is not a simple phenomenon subject to quick and effective counteraction. Drought enters the scene stealthily, like a disease. Once established, it takes a long time to assess the devastation brought on by drought and to stop further deterioration.

From the very beginning of this discussion it is essential to understand the differences between drought, arid lands, and deserts. Though scarcity of moisture, either temporary or permanent, is the common denominator of these terms, one has to carefully distinguish between them.

The *desert* is the most permanent feature. Definitions of the word "desert" do vary; however, one can state that deserts represent barren land that is not able to support anything except sparse vegetation—if any at all—because of inadequate average precipitation. The margins of deserts fluctuate, but the core of the presently existing great world deserts has existed for thousands of years [Plate 51].

True deserts, including some of their adjoining semiarid lands, probably cover in excess of one-third of the world's land; both are marked by low or very variable precipitation. Some deserts may go without any rainfall for a year, or considerably longer, and then receive a downpour within a few minutes. It appears unrealistic, therefore, to speak about *average annual precipitation* when it comes to a true desert. A true desert is in a state of permanent moisture deficiency.

The transitional areas, or the **semiarid regions** of the world, usually have sufficient amounts of precipitation to permit some forms of agriculture or grazing [Plate 52]. But these are the very places that are *high-risk* zones because they may act like traps. In those decades of more plentiful rainfall people were tempted to invade such lands. In some

Plate 51. The great desert on the Arabian Peninsula has existed for thousands of years. The scene above shows one of the very large migrating sand dunes in eastern Saudi Arabia. (Photo by C.H.V. Ebert).

Plate 52. Desert landscapes vary greatly. On their semiarid margins they may support scant, salt tolerant vegetation as illustrated by the coastal regions of eastern Saudi Arabia near the Persian Gulf. (Photo by C.H.V. Ebert).

cases these moisture-deficient zones have been involved in large-scale government projects. One of these developments was the so-called "Virgin Lands Project" of the former Soviet Union, which will be discussed later in this chapter. It is in these extensive *transitional* tracts of short-grass *steppes* of the mid-latitudes—as well as in the tropical *savannas* of Africa, marginal to the southern Sahara—where sporadic periods of precipitation failures occur. Such droughts can overwhelm millions of people and condemn them to starvation, diseases, and death.

In this context it becomes clear that the term *drought* should not be linked with true deserts. Marginal arid lands undergo periods of severe drought that can be defined as erratic moisture deficiencies; they have afflicted certain regions throughout history. Drought conditions, ranging in length from weeks to decades, may begin at any time and in just about any part of the world. As population pressure forces more and more people into marginal land, an increasing number have to face the vagaries of fluctuating precipitation. For example, in India alone, more than 80 percent of the country is chronically prone to drought (Wijkman, Timberlake; 1984, p. 43). But the pressure of population has forced people to try to live in drought-afflicted areas.

For many years climatologists, meteorologists, agronomists, hydrologists, and geographers have struggled to unveil the nature of arid-lands droughts. Opinions about their causes, and about the possibility of their cyclic occurrence, diverge greatly. Definitions of drought, and the development of *statistical*

boundaries of deserts and of arid lands, appear to be equally vexing. The borders—especially those of semiarid lands—fluctuate in response to some global influences. Research seems to indicate that correlations exist between the shifting of ocean currents and droughts. This relationship is not surprising when one considers the vast heat-storage capacity of the world's oceans and the ongoing exchange of energy between the oceans and the atmosphere (World Meteorological Organization; 1975, p. 9).

It is hoped that, once the planetary energy-exchange systems are understood more thoroughly, prediction of droughts will become a viable tool for dealing with these threats. In the meantime, despite various attempts at scientific definitions, the nature of drought could be summed up in the words of a minister of one of the African nations devastated by drought: "There simply is not enough water as people need."

Up to this point of the discussion it appears that drought is entirely a natural phenomenon—an "Act of God"—depriving the land of needed rainfall. Unfortunately, humans play an increasingly devastating role in turning much semiarid land into deserts, a process that is called ***desertification.*** This process, as is the case with drought, cannot be defined in simple terms because many factors can lead to desertification. Moreover, it is simply not yet understood to what degree the expansion of deserts is a natural process—going on for possibly thousands of years in response to global climate changes—or whether it is the outcome of abusing the land. Such misuse includes the overworking of sensitive soils, stripping land of vegetation, lowering the water table by excessive pumping of wells, overgrazing marginal pastures, and using low-quality water for irrigation, which leads to soil *salinization.* Excessive salt accumulation in soils, of course, kills the vegetation [Plate 53].

There is no question that humans are deeply involved in facilitating the spread of desert conditions. This problem is especially severe in Africa where, in 1984, about 26 percent of the 531-million population were fed entirely on grain. Since 1967 grain production has been declining, and in 1985 some 10 million people had abandoned their villages in search for food (Brown, Wolf; 1986, p. 177). The drought disaster in the *Sahel* region south of the Sahara will be discussed later in this chapter.

The Nature of Drought

The definitions of *drought* vary considerably, and it is just about impossible to find a universally acceptable one. If there is agreement, it centers on the fact that drought is accompanied by an abnormal decline in precipitation, and that severe environmental stress is brought on by *moisture deficiency.*

The onset of drought can be very subtle. Its initial impact is almost unnoticeable because there usually is a sufficient moisture reserve in the soils and in the vegetation. Subsoils, especially when fine-textured, have a considerable *moisture-storage capacity;* watersheds containing such soils may feed water into rivers and water-bearing rock formations long after normal rainfall has declined. *Xerophytic vegetation*—plants that have adapted to grow under predominantly dry conditions—also store moisture, transpire very little of it, and may endure prolonged droughts. But if droughts continue, either uninterruptedly or in intermittent ways, the progress of environmental stress and deterioration becomes inexorable. Drought may begin at any time, may last for highly variable time spans, and may occur in almost any climatic regime. The impact of drought may range from mere local inconvenience to an economic and political breakdown of a nation (World Meteorological Organization; 1975, p. 1).

The U.S. National Weather Service recognizes three distinct categories of drought:

1. An *absolute drought,* a period of at least 15 days without measurable rainfall.
2. A *partial drought,* or 29 consecutive days in which the mean daily rainfall does not exceed a trace (0.01 inch or 0.2 mm).

Plate 53. Excessive salt accumulation, caused by inadequate subsoil drainage, has killed off most of the irrigated cotton in this field near Piura in the Atacama desert of Peru. (Photo by C.H.V. Ebert).

3. A *dry spell,* a minimum of 15 consecutive days during which less than 0.04 inch (0.8 mm) of rainfall occurs.

These definitions may serve a useful purpose within the total cultural-economic and physical setting of the United States; however, one has to realize that there are many smaller countries. They are located in marginal lands and they do not have the great economic diversification that allows for alternatives. These vulnerable countries simply do not have the resilience to withstand the stresses brought on by drought. This is particularly so for those—as the ones located in the Sahel of Africa—that are situated in semiarid areas where boundaries with truly arid lands are not clear (White; 1960, p. 15).

In permanently arid zones humans and animals have adapted to the sparsity of water and engage only in those activities that are in basic harmony with the dry environment. In regions that normally have enough rainfall to support some types of agriculture and pastoralism a temporary drought is much more disruptive. Here, human factors—the ability to cope physically and economically with such disruptions—decide for the most part how severe the impact of the drought will become.

Some regions of the world have so-called *seasonal droughts.* These are climatic zones which typically have a distinct dry as well as a wet season. This is the case in the savanna lands of Africa that are located between the Sahara to the north and the rainy equatorial regions to the south. Other world areas, especially those parts of Asia that are affected by the seasonal monsoon winds, also experience a seasonal drought. These seasonal dry periods, although more predictable in nature, can result in severe stress if (1) the preceding wet season brought below-average precipitation, or (2) the dry season is abnormally long. Many droughts in India fall into these categories.

In contrast, a quite unpredictable drought is the one that results from marked *precipitation variability.* The concept of variability is a controversial issue because it is based on a deviation from the "*normally expected*" amount of precipitation. The problem is that in most cases the "*normal amount*" is very difficult to determine. In arid and semiarid climates—to reiterate this important fact—the total amount of precipitation, its spatial distribution, and the time of its occurrence vary more than in any other zone in the world. Whether a short-term deviation is merely a momentary fluctuation or the beginning of a new climatic trend is almost impossible to determine either at the time of change or over a relatively short time span. This problem will be treated later in this chapter in connection with questions related to desertification.

Based on its record in many parts of the world, drought must be viewed as one of the most serious threats to humankind. Droughts must have exacted terrible casualties in past ages, but estimates of the number of deaths are vague. Even in our times it is often quite difficult to obtain accurate counts of drought victims. In many instances of modern drought disasters—as that of the Sahel region—we will never know for sure how many people died, or how much livestock perished or had to be destroyed.

In some parts of the world, especially in China, two extreme conditions, flood and drought, alternated over centuries and claimed untold victims. The orderly bureaucracy of that nation—dating back further in history than for any other country still in continuous existence—tells us that China suffered 1,029 floods and 1,056 droughts since records were kept on these topics since 200 B.C. (Whittow; 1979, p. 280). For the past century alone some estimates of deaths attributable to famine or malnutrition, brought on by either drought or flood, go as high as 100 million (Cressey, 1963, pp. 66–67).

Droughts and floods repeatedly ravaged the semiarid provinces of Kansu, Hunan, and Shansi-Shensi, where wind-carried *loessial* materials choke the Yellow River. Both disaster types caused staggering casualties because they were frequently accompanied by diseases and prolonged famines. The severe droughts between 1876 and 1879 probably were the worst that struck China in modern times. Estimates of deaths for this three-year period exceed 10 million people.

Drought can easily turn into disaster when drought conditions are abetted by humans' reckless abuse of the land. In some cases it is the sheer number of people and their livestock that exceeds the land's capacity to support them. However, we do not have to look at India or China, where the precarious physical conditions, in addition to the tremendous population pressure, led to repeated calamities. Unwise farming practices, overgrazing, and years of lower-than-normal rainfall conspired to create the American "*Dust Bowl*" (Bennett; 1939, p. 16), an expression that has found application to this phenomenon ever since its inception in 1933 (Dasman; 1984, p. 199).

Droughts must have struck North America in prehistoric times. This is reflected in tree-ring studies, or **dendrochronology,** carried out by scientists of the Laboratory of Tree-Ring Research at the University of Arizona and going back as far as A.D. 500 (Bark; 1978, p. 12). Extended periods of drought are typical in the southern Great Plains of North America; but, as long as these lands were protected by natural grass cover, such dry spells probably did only limited harm to the environment.

Some *soil blowing* apparently took place in the Great Plains even before plowing the land. The sandhills of Nebraska attest to such dust storms. Furthermore, archaeological excavations revealed that severe wind erosion has affected the prehistoric Great Plains in prehistoric times. This situation is only possible if a great amount of vegetation disappeared as a result of drought of great intensity and considerable length (Carter; 1968, p. 410).

How often did such dry periods occur? Tree-ring studies carried out by H. E. Weakly in 1965—going back to the year 1220 and extending to 1952—showed average intervals of 24 years (Bark; 1978, p. 13). The shortest drought lasted 3 years, and the longest resulted in 38 years of below-average rainfall. Whether the drought intervals are true *"cycles"* or convenient statistical groupings can be debated.

The American "Dust Bowl"

The region that was called the *American Dust Bowl* includes parts of Oklahoma, Texas, Kansas, New Mexico, and southern Colorado. Annual precipitation in this area does not exceed 15 inches (38 cm), and in many sections it is even less. When early pioneers began to farm these lands, precipitation was relatively abundant. The favorable weather conditions, and the fact that the soils had not been touched before, combined to paint an attractive early picture as far as agriculture was concerned.

But only two years later a severe drought, lasting well into the decade, struck the region in 1890. Periods of good rainfall and short drought spells followed. In their virgin state, soils of the sub-humid regions contain considerable amounts of organic material, or humus. Once the soil is disturbed, however, the organic content declines rapidly. This decline is accompanied by decreasing soil consistency as was pointed out in the discussion of dust storms in chapter 8. In 1910, another drought arrived, and by that time the soil began to blow (Dasman; 1984, p. 199).

With the outbreak of World War I, the United States became the major supplier of wheat, corn, and cotton. Greed and ignorance helped to promote the plowing up of at least 6 million acres (2.4 million hectares) that should have never been touched by the plow. The stage was being set for the worst drought disaster experienced in the United States.

The ravaging of the land continued more or less until 1931, when a severe drought hit the area. The dry spells continued in 1933, 1934, and again in 1936. The 1931 drought set the background for untold human suffering triggered by another blow, the Great Depression. The ominous *black dusters* that darkened the skies matched the black mood of thousands of farmers and workers. Watching their soils blow away, facing rapidly shrinking markets for their products, and unable to find work, many desperate people engaged in a mass migration to California or to the great urban centers of the east. The mood of those years is captured in stark reality in John Steinbeck's book, *The Grapes of Wrath.*

In response to the disastrous drought and the destroyed lands of the Dust Bowl the government established the U.S. Soil Conservation Service in Washington in 1935. Later, in February 1936, soil conservation was broadened under the Soil Conservation and Domestic Allotment Act. This legislation allocated $5 million to stop the wasting of soil, to conserve fertile land, and to develop flood control projects aimed at preventing soil erosion (Barck et al.; 1950, p. 879).

Millions of acres had been damaged beyond repair, but returning rains—and the pressure for food and uniforms in World War II—introduced another cycle of irresponsible land abuse. It has been said that with the first rains, following a drought, comes a sense of security. Conservation plans are discarded, and plans for coping with the next drought are set aside until a new cycle begins (Easterling; 1987, p. 1).

The post-World War II droughts of the 1950s, and the recovering economies in other parts of the world, brought the plunder of the precious land resources to a partial halt. Ironically, violent rainstorms in 1957 caused severe flooding because the water-holding soils had eroded away. Under the guidance of the Soil Conservation Service, and with federal aid funds, further destruction of the land was halted. Portions of the damaged land was restored, other sections were placed under permanent grass cover; but the soil that blew away, and with it the innate fertility of the land, is gone forever.

⍟ "Virgin Lands" Project, Khazakhstan (Former Soviet Union)

After the death of Joseph Stalin on March 5, 1953, Nikita Khrushchev became the First Secretary of the Soviets' Central Committee. He had risen from humble origins and soon became a vigorous and controversial leader. When he visited the American National Exhibition in Moscow, Khrushchev said to Richard Nixon—later president of the United States—that "when we catch up with you, in passing you by, we will wave to you" (Nixon; 1982, p. 185). At that time Khrushchev had started a gigantic agricultural project in the semiarid lands of the Khazakh Republic and surrounding areas. He was convinced, in his pragmatic way, that the Soviet Union would eventually catch up with the agricultural production of the United States and even surpass it.

Khrushchev approached the economic problems of his country as a "reckless gambler does a roulette wheel" (Nixon; 1982, p. 188). Breaking up millions of acres of virgin steppe soil—up to that point used for marginal grazing only—to produce spring wheat was, indeed, a gamble. A total of about 104 million acres (42 million hectares) was opened up during the seven-year period from 1954 to 1960.

These subhumid lands normally receive about 8 to 16 inches (20 to 40 cm) of precipitation, but the amount fluctuates greatly; moreover, droughts are common in that part of the former Soviet Union. Some Soviet climatologists were worried that these droughts, in combination with the often fierce winds that blow during the growing season, could lead to an environmental disaster paralleling the experience of the American Dust Bowl. Agroclimatologists, however, claimed that climatically the new wheat areas of the virgin lands would complement the old wheat-growing regions of the Ukraine. Based on an evaluation of data, covering a time span of 60 years, it appeared that droughts had not occurred simultaneously in both areas (Lydolph; 1977, p. 351).

Initially, wheat production increased remarkably. The soils still had a good humus content, and the reserve of subsoil moisture, combined with fairly good precipitation, proved adequate. However, as the soils declined in quality, and when the first drought began to plague the area, matters went from adequate to bad, and from bad to worse. Dusty topsoil began to blow away and traveled far to the west (Durrell; 1986, p. 170).

Soviet scientists became aware of the fact that this project had exacted a high price, and further plowing of new land was halted by 1965. Windbreaks were planted, some lands were allowed to rest for extensive fallowing intervals, and crop stubble was left in the field. Several million acres were taken out of production and were allowed to revert to grass. But at least 7.4 million acres (3 million hectares) are still subject to severe wind erosion. During dry spells, sandstorms and dust have become problems where there were none before (Wijkman, Timberlake; 1984, p. 45). Mr. Khrushchev's dream did not come true: in subsequent years the Soviet Union had to import about one fifth of its annual grain supplies (Worldwatch Institute Report; 1986, p. 19). Even though future generations may condemn this environmental gamble, wheat production—although fluctuating greatly in yields—goes on. On average, about half of the Soviet grain-crop production came from the still productive lands of this region (Symons; 1983, p. 128). Whether this drought-prone region will eventually collapse, similar to the American Dust Bowl, remains to be seen.

Desertification and Desertization

The preceding sections showed that drought is a highly variable phenomenon that can occur in just about any part of the world. Moreover, the examples of the American Dust Bowl and of the Soviet Virgin Lands project illustrate that unwise handling of drought-prone land can lead to ecological disaster. Such disasters do not just refer to the disruption of the economy, but also point toward an ominous word first coined at a worldwide conference on arid lands held in Nairobi, Kenya. The word is *desertification*.

Desertification implies that deserts are spreading, and that they are spreading as a result of *human* activities, including overgrazing, clearing of forests, and overworking sensitive soils. A 1984 study—carried out by an Australian geographer, Jack Mabbutt—pointed out that more than a third of the world's land, with about 850 million people living in that area, is threatened by desertification (Worldwatch Institute Report; 1986, p. 72). Stated in different terms, each year the expanding deserts claim more than 14 million acres (5.6 million hectares) of grassland and other productive land (Moran et al.; 1980, p. 118).

The analysis of desertification is difficult and also controversial because of at least three complications: (1) there is *no real agreement* in defining the word "desert," (2) there is *no finite answer* to the question of whether we are facing long-term or short-term climatic changes, and (3) there is *no clear understanding* to what degree, and how fast, the land deteriorates in direct response to humans' activities [Plate 54].

The word *desert* evokes the image of barren wasteland in which precipitation is so slight and erratic that such an environment is incapable of supporting vegetation and any significant number of people. But, looking at the various types of deserts in the world, we find an astounding variety of plants, animals, human cultures, and landforms. The greatly contrasting desert landscapes include the fog-shrouded sections of the Namib, on the southwestern coast of Africa, the reddish sand mountains of the Rub' al Khali (the Empty Quarter) of the Arabian Peninsula, the drifting white sands of the Libyan Desert, and the lone ***monolith*** of Ayers Rock in the desert heart of Australia.

Because of the multifaceted aspects of this environment, the word desert is used in a very inconsistent, and at times almost glib way. For example, the pioneers referred to the dry-land western tracts of America as deserts. It would be interesting to learn how that area would be viewed by the nomadic desert dwellers of North Africa! But, whenever we look at any type of desert, it is the lack of available water, the sparsity and unpredictability of precipitation, and the sensitivity to human abuse that offer the common denominator of the arid-lands environment. Moreover, there simply are no clear boundaries of arid lands (White; 1960, p. 15)!

Whatever factors one wishes to use to delineate deserts, rainfall alone cannot be used as a criterion. Seasonality and duration of precipitation, moisture-holding capacity of soils, groundwater location, presence and distribution of vegetation types, and temperature regimes must be considered in assessing a desert environment. No agreement exists here because opinions diverge about what elements should be included—and what weight each would have—in any aridity equation (McGinnies et al.; 1968, p. 3).

Plate 54. Salt encroachment from nearby irrigated land and overgrazing conspire in the destruction of the once thriving floodplain vegetation near the Murray River at Renmark, Australia. (Photo by C.H.V. Ebert).

Any atlas, on the other hand, that shows the physical features of the world does show the locations of the world's deserts. In most instances the sharp boundaries are drawn for statistical convenience. True climatic deserts—to differentiate this concept from others—generally have been defined by sta.... on the average, annual *evapotranspiration*... precipitation, thereby leaving a moisture... ost cases the 4-inch (10 cm) annual rainfa... *isohyet,* serves this purpose.

Th... d major question, whether actual climatic c.... s are taking place, is equally complicated. All climatologists agree that long-term climate changes have occurred and will occur in the future. The problem is that it is almost impossible to distinguish between *secular* climatic changes—those that take place over centuries—and short-term climatic oscillations (Raikes; 1971, p. 147). The short-term variations, sometimes called "drought cycles" or even "weather cycles," also are *statistical* entities. Their recurrence is unpredictable and the term "cycle" appears inappropriate (Glantz; 1977, p. 24).

The... d aspect, to what degree arid lands deteriorate... result of human-induced factors, is equally problematic; both natural changes and human-made changes overlap in cause-and-effect relationshi... [Plate 55]. Obviously, there is no way to dra... dividing line between what could be calle... desert" and a *"false desert."*

T... ion of typical desert landscapes, throug... al processes only, to areas where they did not exist in the recent past is referred to as *desertization.* This term is advocated by H. N. LeHouerou, coordinator for UNESCO's Man and the Biosphere

Project No. 3 (Glantz; 1977, p. 17). In other words, long-range and persistent change in climate is believed to be the central element in the explanation of why deserts expand. Climates, of course, are not static; but why they change, and how fast they change, is a major subject of contention. It is suggested that climate changes take place in response to (1) internal oscillations within the climate system, and (2) external influences, which may include continental drift, variable solar output, changes in the makeup of the earth's atmosphere (dust; greenhouse effect), and alterations in the reflectivity of the earth's surface (Bryson, Murray; 1977, p. 138). It is particularly the shifting of ocean currents and of planetary wind belts, considerably beyond their normal seasonal positions, that sheds some light on the occurrence of droughts, on the spreading of arid conditions, and on temporary periods of abnormally high amounts of precipitation in otherwise dry regions. In this context the phenomenon of *El Niño* deserves further attention.

⊚The El Niño-La Niña Phenomenon

Short-term climatic oscillations, which result in either abnormally great amounts of precipitation or may also be responsible for very low rainfall, seem to be in part linked to the phenomena called "El Niño" and "**La Niña**", respectively.

The name "El Niño" (meaning child) refers to the disappearance—usually around Christmas time—of cold offshore waters which are known for their high concentrations of plankton and higher marine life along the low-latitude coasts of Peru and Ecuador. During the occurrence of El Niño the cold waters are overrun by warm equatorial water masses. Air which passes over the warm water surface picks up large amounts of moisture. The higher vapor content of the air in this way increases both its instability and precipitation potentials in these normally dry coastal regions of South America, Mexico, and in the southern regions of California.

The main environmental impacts of these changes are: (1) a breakdown of the **marine food chain** because the warm water contains less dissolved oxygen and fewer nutrients, and (2) the heavy rains bring about severe flooding, soil erosion and mudflows in areas that tend to remain dry for as long as a decade or longer (Ebert; 1978, pp. 349–350). On the average, El Niño events take

Plate 55. The excessive pumping of groundwater and the removal of vegetation encourage the relentless advance of the Atacama desert in northwestern Pery. (Photo by C.H.V. Ebert).

Plate 56. During years of low rainfall—or "La Niña," Rio Rimac in Lima, Peru, is a sluggish and mud-choked river. This situation changes drastically when "El Niño" prevails as shown on Plate 57. (Photo by C.H.V. Ebert).

place at intervals ranging from 3 to 7 years [Plates 56 and 57].

In March 1995, one of the worst flood disasters struck the central coastal sections of California.

Abnormally high ocean temperature, the El Niño, was one of the major contributing factors leading to the severe flooding. The rains broke a 6-year drought cycle but the destructive power of the floods completely overshadowed that benefit.

The weather maps—covering the period from Thursday to Saturday (March 9 through 10)—showed a major cyclonic depression over the eastern Pacific. The rain-producing fronts were expected by Wednesday, but the system stalled and then moved sluggishly into the coastal regions. The storm's progress was prevented by a large high-pressure cell, an **anticyclone,** which dominated the interior of North America.

The almost stationary storm picked up large amounts of moisture from the warmer-than-normal ocean surface and subsequently produced incredible amounts of rainfall in the Santa Barbara and San Luis Obispo regions. Precipitation amounts for a 24-hour period ranged from 8 to 10 inches (20 to 25 cm).

In addition to the El Niño and the unusual meterological conditions, several factors exacerbated the situation: (1) the steep topography of the

Plate 57. When "El Niño" brings heavy rainfall to coastal Peru, the usually peaceful Rio Rimac changes into a raging torrent causing widespread flooding and destruction. Compare to Plate 56. (Photo by C.H.V. Ebert).

Plate 58. The heavy rains in March, 1995, caused numerous flash floods in California. These powerful and dangerous floods are able to carry debris and large objects—as this car—over long distances. (Photograph credit: Steve Malone; Santa Barbara News Press).

coastal mountains produced massive runoffs and flashflooding [Plate 58], (2) the amount of moisture exceeded quickly the soils' absorptive capacity, (3) unstable soil conditions resulted in massive mudslides, and (4) large amounts of eroded soil and debris choked the streams and canyons [Plate 59].

Many theories have been advanced dealing with the development of El Niño. It was not until the early 1990s that two international research efforts (Tropical Ocean Atmosphere Program and the Coupled Atmosphere—Ocean Research Experiment) produced enough data that allowed for some prediction

Plate 59. Throughout the coastal mountains of California narrow valleys and canyons were hit hard during the severe floods of March, 1995. (Photograph credit: Steve Malone; Santa Barbara News Press).

of an El Niño event. Also, an extensive series of Goesat measurements gave a synoptic view of sea level fluctuations for the entire Pacific basin. For example, in the 1987 El Niño occurrence, the Trade Winds slackened and sea level throughout the eastern tropical Pacific rose. These winds normally blow seaward along the equatorward side of large high-pressure cells referred to as the Subtropical Highs. This situation prevents the cold nutrient-rich subsurface water from welling up along the western shores, and the warm water temperatures result in the El Niño problems discussed previously.

In 1988, the Trade Winds resumed their normal strength and drove the warm surface waters westward into the Pacific and thus permitted cold-water upwellings. The sea level became depressed in the eastern tropical Pacific and rose correspondingly in the western parts (Koblinsky; 1993, p. 230). These upwellings generate the favorable conditions that lead to a high biomass production in the coastal oceans. The cold surface waters also stabilize the air by giving off less water vapor and thus account for dry atmospheric conditions. These periods of drought, in contrast to the rainfall associated with EL Niño, led to the designation of "La Niña" a term which originated in Peru and in Ecuador.

Many questions remain concerning the complex interactions between the shifting ocean waters and the prevailing global winds, but it is now well accepted that the effects of El Niño are encountered worldwide (Gross; 1995, p. 82). The capability of predicting the occurrence of both El Niño and La Niña has become more realistic by the use of remote sensing; furthermore, the employment of anchored buoys in the tropical Pacific, which report fluctuations in ocean water temperatures, assist in assessing oceanic conditions conducive to the formation of these events.

Many unknown factors play a role in the cyclic occurrence of EL Niño and La Niña a pattern which seems directly linked to changes in ocean temperatures. The study of these phenomena deals statistically with their past, but the past picture may not give an answer to the time intervals in the future. If, for example, global warming will supply a greater amount of heat energy, the occurrence of El Niño could become more frequent. Once it will be possible to create a more reliable predictive model it will become feasible to deal more effectively with such devastating flood events of those in Mississippi in 1993 and the ones that created havoc in California in 1995.

◎ The Sahel Disaster

The Sahel of Africa is more a concept than a clear-cut geographical region that lends itself to easy definition and mapping. The Arabic meaning of the word sahel implies that it is a "shore" or a "border" where something ends. It would be far too simplistic to say that it is here where the great Sahara desert ends, and a different "climate zone" begins. The Sahel is a semiarid, drought-prone belt of lands that roughly straddles a line drawn from Cape Verde, on Africa's west coast, to Cape Guardafui, the point of the "Horn of Africa" on the east coast (fig. 10.1).

Rainfall fluctuations, both in amount and in yearly occurrence, are as difficult to define as the boundaries of the Sahel. At best one can say that it is a transition zone about 500 to 600 miles (800 to 1,000 km) wide that receives some fickle rainfall. Normally, and we want to use this term with considerable caution, the Sahel derives its precipitation from the seasonal northward shift of moist equatorial air masses; on the other hand, the long, rainless season is dominated by the dry northeasterly trade winds blowing southward out of the Sahara.

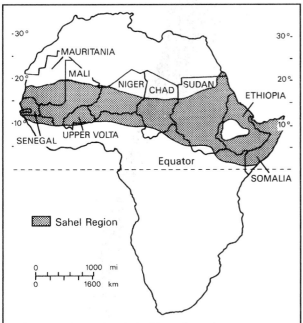

Figure 10.1 The Sahel is a drought-prone region to the south of the Sahara. This region, although fluctuating in size in response to climatic cycles, straddles the continent from the west coast to the "Horn of Africa" on the east coast.

In response to the unreliable and marginal moisture conditions a basic cultural-economic pattern established itself in the Sahel: the northern sections, grading into the Sahara, support nomadic herding and a few oases; the southern belt allows some cultivation—mostly drought-resistant grains—and contains permanent settlements. Supplementary trade between the two different regions benefited both systems.

After World War II, a host of European colonies in this zone, mostly French, obtained their independence. That time of political transition was tenuous because the bulk of the population had not acquired adequate governmental, political, and managerial skills during colonial times. But, the wave of joyous emotions, combined with years of good rainfall in the mid-1960s, supplied sufficient momentum to about 21 million people to face the future optimistically.

Drought has never been a stranger to these lands, especially not in the drier and landlocked nations such as Mali and Chad. Two severe droughts—preceding the more recent disastrous period from 1972 to 1974—had ravaged this zone this century. So it was no surprise for the people when the rainfall became sparse in 1968. Crop yields declined in the south, and the northern nomads had to trade more than the usual number of livestock as pastures became sparse. The following year brought worse conditions. Dust filled the air, wells went dry, saltwater from the ocean began to creep into rivers on the west coast. The region braced itself against the impact of another drought, but this time the resilience of the Sahel had been undermined: Thousands of wells had been drilled in the drier sections, water storage tanks were built so that the herds of livestock had increased far beyond the capacity of the land to support. The efforts to commercialize livestock raising by restricting the movements of the nomads—begun under the colonial regimes—resulted in severe overgrazing (Ehrlich et al., 1977, pp. 627–628).

The destructive **overgrazing,** the excessive and often radical removal of trees for the use of firewood, or just as an operation to clear the land, and an annual population growth rate of 2.5 to 3.0 percent, put pressure on the Sahel that this ecologically sensitive environment could not tolerate. Even normally productive oases turned dry because the removal of the vegetation had lowered the water table (Collinson; 1977, p. 153); thus the stage was set for a disaster.

The rains continued to fail the next year, and the year after brought no change. The humid air from the equator did not make it far enough north, and the dry desert air masses prevailed. As it was already pointed out in connection with the discussion of desertization, there is no agreement as to why these rain-failures occur. Are they just unpredictable "cycles" or are they the harbinger of a slow, insidious climate change?

By 1972 the world began to realize that disaster had struck the Sahel. Fear of bad publicity—potentially revealing incompetence—inefficient communications, and probably a certain stoicism about accepting stress and suffering delayed an earlier detection and response.

Within a year widespread starvation became rampant. The nomads had abandoned their former grazing lands, now denuded and buried under dust and sand, and a trek of desperate people, accompanied by their emaciated livestock, moved southward toward the more humid regions and permanent villages. The local economy could not handle the influx of helpless millions, and thousands died of exhaustion and dehydration. Massive crop failures had also befallen the southern Sahel, and the refugees were received with fear and often with open hostility.

The United Nations and numerous charitable organizations launched major food and medical shipments; but the death toll climbed rapidly. How many of the people died of starvation, of dehydration, and of diseases will never be known. Numbers ranged from a low of 50,000 to highs exceeding 200,000 victims.

A brief respite followed in 1974 and 1975 when rainfall improved. Some of the nomads, despising the life in crowded refugee camps, began to move north again with a few remnants of their livestock. They planned to resume their former life, but their land had turned into a desert. The topsoil, denuded of all vegetation cover, had either blown away or was deeply buried under dust or drifting sands. The wells, where they could still find them, were without water, or they had filled in with sand (Bryson, Murray; 1977, p. 96).

The ancient empire of Ethiopia, on the far eastern side of the disaster area, had crumbled under the stress of the drought. The rains of 1974 and 1975 had come too late to save the emperor, Haile Selassie, who was overthrown in September 1973. This revolution was in reaction to both political and economic problems that arose in connection with the drought. It is believed that the prolonged drought had cost Ethiopia at least 100,000 lives. A similar rebellion had already toppled the president of Niger, Diori, in April of the same year.

Since these events many areas in Africa have been subjected to more drought and desertification. As both people and earth are weakened, each subsequent drought impact becomes more severe. The affected areas expanded; the number of suffering people is on the increase. It becomes quite clear that shipments of emergency food, and drilling more and deeper wells, is not the answer to this staggering problem. Reaction to disaster—although necessary under the conditions described—must be replaced by predisaster planning. This planning requires a thorough understanding of the ecological parameters set by the environment. Many useful techniques (reforestation, building terraces, stabilization of migrating sand dunes by the use of vegetation cover) are available. These are, of course, long-range methods that cannot change the situation within a few months. But, if used systematically, and on a large scale, they will assist the severely damaged region to regain some degree of resilience needed to cope with the future droughts.

In addition to the physical repair indicated above, the weather and climates of drought-prone regions must be carefully monitored to develop an answer to the haunting question: variability or long-term trend? The consequences of the answers to this question could be immensely important in preventing mistakes of the past and to plan, intelligently, for the future.

FOREST FIRES

Key Terms		
Devonian	ground fire	fire vortices
Mississippian	surface fires	valley breeze
slash-and-burn	chaparral	mountain breeze
fuel load	crown fire	summit aridity
monocultures	running crown fire	firebreaks
kindling temperature	firestorm	fuelbreaks
pyrolysis	area fires	triangulation
ignition temperature	fire wind	backfiring
canopy fires	conflagration	

Eons before the dawn of history thunderstorms raged in the atmosphere of the primeval planet Earth. But vegetation fires could not exist until a sufficient plant cover developed on the lands. We do not know when lightning started the first forest fire, but forest-like plant associations began to spread in the late **Devonian** and early **Mississippian** periods, or about 300 million years ago. Some of these treelike plants, the *cordaites,* probably reached heights of close to 60 feet (18 m).

Fire has always been a powerful force in the shaping of ecosystems. Areas that are repeatedly visited by fire tend to develop plant communities which could be called a *fire climax,* as a balance between the vegetation and endemic fires (Laubenfels; 1970, p. 77). Fire must be viewed as a natural process, but one should also realize that the role of fire changed drastically once humans learned how to use it. Fire in the hands of humans has evolved from the sporadic use by primitive peoples to the opposite extreme of a devastating tool of war capable of incinerating cities and nature.

Early peoples received fire as a gift because humans did not invent fire, but in short time they learned to become users of fire. The art of *making* fire came long after volcanic eruptions and lightning set trees afire and demonstrated the properties of fire to early humankind. Collecting burning branches and glowing embers probably represent the earliest attempts to obtain fire from nature.

Fire gave humans the ability to hunt more efficiently by frightening and containing animals with firebrands. Later, fire was used to clear the land and to supply the soil with mineral nutrients deriving from the ashes. Shifting agriculture, still using the ancient **slash-and-burn** technique, is widely encountered nowadays in tropical forest areas [Plate 60]. Wherever agriculture evolved it seems that fire, in one way or another, took part in this activity. Humans became the only living beings to use fire, and fire played a critical role in the evolution of myths, superstitions, crafts, and in science.

Fire, of course, is an *oxidation* process; so is the forming of rust. It is the speed and violence of this type of oxidation—*combustion*—sets it apart. The roaring of a fully developed forest fire, the bluish-green hissing flames of a magnesium incendiary bomb,

Plate 60. The slash-and-burn method demonstrates the use of fire in agriculture. The forest vegetation is burned so that the nutrient-rich ashes are returned to the soil, which is cultivated for several years. The process of burning plant residues is shown in this photograph taken in the Polochic valley of eastern Guatemala. (Photo by C.H.V. Ebert).

the searing heat radiation from an open furnace door, and the angry red of a burning town—these are the elements of fire that spread terror.

The abuse of fire (and the lack of understanding of fire as an inherent component of natural systems) placed the seeds of distrust and fear in the minds of people. Many think that fire is basically bad, destructive, and unwanted. These thoughts are quite prominent when it comes to *forest fires.*

There is no question that uncontrollable forest fires are terribly destructive. Pictures of burned wildlife within the charred remains of a forest speak a clear language. People, while viewing such a devastated landscape, will find it hard to believe that the majority of the animals were not hurt by the fire and that, in a few years, new vegetation will start a vigorous new life cycle. Moreover, most people find it unbelievable that fire can be an important and desirable event in the evolution of a healthy forest. They have to learn that fire can be used as an effective tool in *forest management,* and that the excessive protection of forests from fire can be ecologically detrimental. Above all, total exclusion of fire lays the foundation for the worst kinds of forest fires.

Ecological Assessment of Fire

For many years it was accepted that plant associations, including forests, evolve over time into a *steady-state* climax vegetation. This kind of thinking ignores the role of *natural disturbances* that occur in ecosystems. Such disturbances include windstorms, droughts, floods, diseases, insect invasions, and fire. Humans, the most powerful modifiers of the natural environment, certainly must be added to the list of disturbances. But of the natural factors it is fire that is the greatest single instrument in plant community change (Laubenfels; 1970, p. 76).

The steady-state concepts are being replaced by the knowledge that disturbances play a critical role in the development of ecosystems, and that fire is an integral part in the growth process of a forest (Borman; 1981, p. 153). Studies have revealed that forest fires occurred at regular intervals in the northern hardwood forests of North America long before the regions were settled by Europeans. What, then, explains the traditional fear of fire and the fervent desire to ban it from our forests?

Throughout recorded history humankind faced the varied role of fire. Whether fire is feared or viewed as a beneficial factor seems to depend primarily on its size and intensity. The warmth of a well-tended fire was welcome to the cave dweller; bonfires spread joy and festivity; the glow of candlelight is peaceful. But the flames of war, the fierceness of a running forest fire, and fire in the hands of an arsonist are the aspects of fire that people dread. Fire can easily turn "wild" and uncontrollable. The feeling of not being in control could be the basic reason to fear fire (Wright, Bailey; 1982, p. 2).

In addition to these attitudes about fire one must also consider cultural and environmental factors. The understanding of the role of fire in forest ecology, and especially as a managerial tool in forestry, came quite late to the United States. *Excluding* fire from our forests is a deeply engrained perception in America. The main reason for this traditional attitude can be found in the fact that the European concepts of strict *fire control* in forests were introduced to Americans during the nineteenth century. At that time the Europeans, particularly the French and Germans, were the pioneers of modern forestry education.

Many of the western European forests were carefully tended. In many cases their trees represented replanted species grown on clear-cut land, and these forests included species that were in greatest demand (Dasman; 1984, p. 316). The forests were carefully thinned out, litter was removed, and undesirable undergrowth was eliminated. Compared with the huge, unbroken wilderness forests of North America some of the European forests looked more like tree plantations. From the European point of view it was desirable, and relatively easy, to eliminate fire from these well-tended forests. Under the existing circumstances this policy seemed reasonable, but it proved unworkable in the ecologically diversified North American wilderness forests [Plate 61].

The most critical difference between the European and the American forests was that the latter accumulated large amounts of *litter:* branches, dead trees felled by wind, and dead undergrowth. An occasional fire would burn out this debris. A fire of this sort would move quickly through the forest without touching off a major fire. On the other hand, if fire was kept out of the forest, or was quickly extinguished, the amount of forest floor debris would become excessive. Prolonged accumulations of such materials generate a high *fuel load.* Once set afire the massive fuel load, now capable of sustaining an intense fire, will ignite the trees and start a full-fledged, serious forest fire.

As public forest reserves and national parks were established in this country the desire to protect them became the official policy. But as early as 1902 protests surfaced against the increasingly rigid guidelines of the Forest Service concerned with the protection of the land from fire (Pyne; 1982, p. 102). The fight against fire received a popular boost in 1945 when Smokey the Bear admonished the public: *"Remember, Only You Can Prevent Forest Fires."* The continued elimination of fire from national

parks allowed shade-tolerant species to form a thick undergrowth, which made it difficult for the trees to reproduce. In addition, the fire hazard intensified (Wagner; 1971, p. 64).

It was not until 1963 that a breakthrough was achieved. At the Twenty-Eighth North American Wildlife Conference A. S. Leopold demonstrated the inherent dangers of the *fire-exclusion* policy. The dangers included, in addition to the fire hazard of fuel buildup, the stagnation of young trees, the encroachment of unwanted species, and the development of tree *monocultures,* which invited an increase in diseases and insect damage (Wright, Bailey; 1982, p. 2).

Controlled burning (referred to as *prescribed fire*) has established itself in modern American forestry. In contrast to uncontrolled fire, controlled burning has a number of beneficial results:

1. Reduction of potentially dangerous accumulations of both litter and unwanted undergrowth
2. The return of organically immobilized nutrients in the form of ash. This is of significance in arid regions where vegetation litter becomes so dry that bacteria and fungi cannot decompose it (Odum; 1963, p. 74).
3. The sorting out of mixed forests if the more fire-resistant trees are the desirable ones
4. The assistance to species that actually benefit from occasional fire. The recycling of nutrients in connection with fast-moving low-intensity fires benefits some of the African savanna grasses. Unburned grasslands are also subject to the invasion by unwanted woody plants (Phillips; 1959, p. 60). The North American black spruce also needs fire to open its cones; unless scorched by flames the cones would remain closed. Thus nature has designed a species that needs fire as a part of its life cycle.

Fire has a specific effect in different *biomes,* or plant communities. Many of these effects, as discussed above, are needed and beneficial; it is uncontrolled fire that poses the danger to humans and nature.

Causes of Forest Fires

Many campers have experienced the frustration of being unable to build a camp fire when the available wood is damp. Moist wood is difficult to ignite because a great amount of heat is used just to vapor-

Plate 61. Excluding fire from a wilderness forest, as illustrated by this forest tract in North Carolina's coastal region, allows, excessive undergrowth to develop, which increases the fuel load. Occasional controlled burning would reduce the potentials for a very intense forest fire. (Photo by C.H.V. Ebert).

ize the water before the wood burns efficiently. For the same reason a forest becomes more susceptible to fire once it has dried out in response to dry spells, heat, and wind. These conditions—especially when prolonged—could be called *fire weather.* When fire weather prevails, the forest rangers worry about campers and hikers in the forest.

Various agents are involved in starting forest fires, and it is difficult to say how many fires result from each factor. *Lightning* starts about 6,000 forest fires in the western United States alone (Critchfield; 1960, p. 330), but it is doubtful that more than 5 to 10 percent of all forest fires in this country are started in that way. Some of the worst fires are set by arsonists and by individuals who may gain from them; deliberately set fires may account for almost a third of all wildland fires! Contrary to popular belief, lumbering accounts for less than 10 percent of the fires. The largest share must be ascribed to carelessness on the part of inexperienced campers and hikers. A single cigarette butt flipped out of a car window can spell disaster.

Any fuel must reach its specific **kindling temperature** before it can burn. When subjected to high temperature, fuel will undergo a process called **pyrolysis,** or the decomposition of organic compounds. As the temperature increases to the critical level many substances in wood become volatile, come in contact with air, and will start to burn. Combustion takes place at the fuel surface; any increase in the surface area speeds up the burning process.

The **ignition temperature** of wooden materials depends on their density, moisture level, total surface exposed to heat and air, and on their chemical composition. Wood types that contain large amounts of resinous and oily substances generate large amounts of highly inflammable gases. The kindling temperature of wood is about 600°F (315° C), although this may vary. Rotten and dry wood may have a kindling temperature as low as 482°F (250° C). It is important not to confuse the kindling temperature, the critical temperature of initial combustion, with the much higher temperature levels that are generated during forest fires. Full-fledged **canopy fires** are reported to have reached temperature levels of 957 to 1,497° F (500 to 800° C), while heath fires generated temperatures of $1,750° F (950° C) in England (Moore; 1982, p. 6).

Types of Forest Fires

Woodland fires, according to their location within the biomass and based on their behavior, can be grouped into four categories. In some cases these are stages in the development of forest fires, but they may also occur as individual events.

The first type, the **ground fire,** is primarily restricted to thick layers of organic materials, such as old rootwork and peaty deposits. Such a fire may also spread in heavy accumulations of compacted litter, or duff, which has undergone considerable decomposition. Ground fires can develop through *spontaneous combustion* or can be started by surface fires and lightning (Albini; 1984, p. 591). Spontaneous combustion is caused by the continuous accumulation of the heat of slow oxidation within the compacted mass of organic materials. The heat generated in this fashion cannot dissipate efficiently and thus attains the kindling level.

Ground fires spread by the glow of combustion and may smolder underground for months. Although they are difficult to extinguish, they spread quite slowly. How long such a fire prevails depends on the amount of fuel, the location of the water table, and on rainfall. Despite the lack of visible flames a ground fire can be very damaging to tree roots and to the living matter in soils. In less-compacted layers, especially when they contain branches, ground fires burn more intensely and can trigger **surface fires.**

As the name implies, surface fires feed on fuel near and on the ground. Their intensity depends on the amount of vegetative debris and on wind conditions. Surface fires tend to move with considerable speed, which may attain 6 mph (10 km/h) in fire-prone brush. Such speed is reached in the **chaparral,** the evergreen shrubs of southern California (Albini; 1984, p. 592). Grass fires are capable of moving much faster [Plate 62].

Unless litter accumulation is extensive, surface fires move rapidly and therefore will not damage mature trees. In the absence of wind a high-fuel surface fire becomes very intense. Then the flames extend upward and can reach the lower branches of trees. The newly added fuel further intensifies the fire, and a new type of fire, the **crown fire,** is in the making.

The initial crown fire depends on the heat energy from the surface fire. This situation is especially true when the trees are relatively far apart and are in good growth condition. Under these circumstances an isolated crown fire may be short-lived; it cannot produce the heat energy to sustain itself.

The susceptibility to develop an *independent crown fire* increases with greater tree-stand density and also with drier climates. An area of exceedingly

Plate 62. Surface fires, as this quickly moving grass fire near Georgetown, South Carolina, are of low intensity and assist in removing unwanted vegetation. Such fires are readily controllable because of their low fuel load. (Photo by C.H.V. Ebert).

high fire risk of this kind is the southeastern portion of Australia, including the island of Tasmania, where even living trees are partially dried out during the long dry summers (Butler; 1976, p. 86).

A well-developed crown fire is difficult to control [Plate 63]. The fire does not move forward at a steady rate; it stabs forward and surges. The fact that large fires produce their own wind systems is partially responsible for this behavior. Fire-generated hot wind gusts may form powerful *fire whirls* that can raise burning debris high up into the air. In many instances such debris is dropped far ahead of the burning area and will start new fires. Furthermore, intense radiation generated by the wind-fanned flames can ignite trees that are not yet reached by the fire itself. *Heat radiation* from a raging crown fire can produce skin burns at distances exceeding 300 feet (100 m). Studies by C. Ahlgreen (1970), cited by P. D. Moore, indicate that a temperature of 1,497° F (800° C), and lasting for one minute, were recorded in crown fires of jack pines in Minnesota (Moore; 1982, p. 6).

The fourth type of forest fire is not really a different category but could be called a culminating stage in a large-scale crown fire. A variety of names have been given to this totally uncontrollable fire. Its vast dimensions justify the term *mass fire* [Plate 64]. Its high speed of advance led to the designation of **running crown fire.** Its fierce wind system, including tornadolike vortices, produced the name **firestorm.** After World War II the concept of firestorms was mainly associated with the enormous

Plate 63. This crown fire envelops all levels of the forest from the surface litter to the high crowns. Intense crown fires, such as the one shown above, are very difficult to control and may get out of hand under high wind conditions. (Photograph credit: U.S.D.A.; Forest Service).

area fires caused by the firebombings of German and Japanese cities (Mullaney; 1946, p. 99). These events are described in the last chapter of this book. However, this ominous word was in usage long before the fiery death of cities in World War II. It was used, in a descriptive way, in connection with the hurricanelike *fire wind* that incinerated Peshtigo, Wisconsin, in 1871, and earlier than that, when the City of Hamburg, Germany, was destroyed in 1842.

The transition from a large-scale crown fire to a running crown fire usually is quite brief. Whether a mass fire or a firestorm develops depends on a number of factors. Wind fans the flames of any fire, but wind also tends to drive a fire quickly across an area. Research seems to agree that mass fires respond to

Plate 64. This scene depicts an uncontrollable mass fire, which is also referred to as a "running crown fire." The high intensity of such fires, which produce their own wind system, can result in devastating firestorms, (Photograph credit: U.S.D.A.; Forest Service).

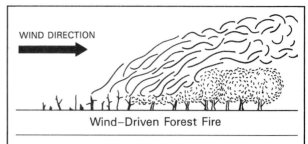

Figure 11.2. A wind-driven forest fire, in contrast to the more stationary firestorm shown in figure 11.1, results in a rapidly progressing mass fire.

their own energy system (Pyne; 1982, p. 24). The heat from a mass fire produces a powerful convectional updraft. This massive uplift of hot air and gases results in violent winds that blow into the burning area (fig. 11.1). There seems to be a distinct pattern to area fires that results in true firestorms. In addition to a very high fuel load there is also an absence of *ground wind.* This makes firestorms more or less stationary (Ebert; 1963, NFPA, p. 9). A mass fire that is driven forward by preexisting ground wind should be called a *general* **conflagration** (fig. 11.2). But this distinction, under the conditions of a working forest fire, may not be as clear as the terms

indicate.

In addition to the powerful convectional wind system, mass fires also produce local inconsistencies in wind direction. Such wind variations are the result of *ground turbulence* (Ebert; 1963, NFPA, p. 8). These high-velocity *fire whirls* act much like tornadoes. **Fire vortices** also form during forest fires in mountainous terrain because topography is also an influential factor in fire behavior [Plate 65].

Forest Fire Behavior

How forest fires establish themselves, how they develop, and whether they get out of hand, depends on a number of factors. *Fuel load, weather conditions,* and *topography* are the most influential ones. It is difficult to rank such factors, or to state the control each holds over the nature of a given fire. In most cases it is the combined interaction of all that creates

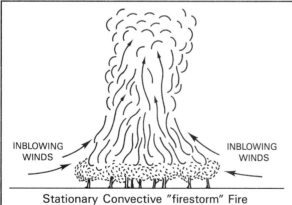

Figure 11.1. In the absence of ground wind, a major forest fire develops a generally stationary firestorm pattern, marked by inblowing winds and a powerful convective column of hot gases and smoke.

Plate 65. Some intense forest fires produce powerful convective wind systems. This convection can develop tornadolike fire whirls called fire vortices. (Photograph credit: U.S.D.A.; Forest Service).

the specific dynamic characteristics of each fire.

In wildland fires it is the fuel load, represented by the specific type of vegetation involved, that determines the basic intensity of the fire. There are marked differences in fire behavior because of vegetation characteristics. The major variations in fire response are the result of plant size and its chemical makeup.

The size of plants (grasses, shrubs, trees) influences the height to which flames will reach and how much fuel mass is available. The chemical composition, especially the waxes, oils, and terpenes, determines the flammability of the gases that emanate from the plant when the kindling temperature is reached. This chemical make-up is also a factor in the *fire resistance* of certain vegetation. But the chemical makeup by itself is not always an indicator of fire resistance. The Australian Eucalyptus initially is quite fire resistant, but it burns fiercely once fully ignited. Dry grass has an extremely low resistance to fire. But, because of both its low fuel mass and its chemical composition, it is incapable of sustaining a high-intensity fire.

Forest fires are absent, or very rare, in both tropical as well as in temperate rainforests of the world (Dasman; 1984, p. 310). This fact illustrates how significant the water content is in the biomass and the presence of moisture on its surface. How influential the *outer* water factor is, as represented by the action of rainfall and by *relative humidity,* is presented in the role of weather as a factor in fire behavior.

Before the advent of computers and information systems, the U.S. Forest Service depended on the analysis of local and regional weather data. This data was obtained from weather stations and other observation points. The collection and dissemination techniques of such data have changed dramatically, but the weather components, leading up to fire weather, have remained the same. Critical information includes the present and projected atmospheric conditions, fuel moisture content, days since the last rainfall, and the amount of rainfall recorded. This data is correlated with all other fire factors to establish the potential danger level for the development of forest fires.

Lack of rainfall, combined with high temperatures, has a desiccating effect on both forest litter and living vegetation. As air warms up—at a given moisture level—its relative humidity is reduced. This process means that the air can pick up additional moisture. Relative humidity thus can serve as a valuable indicator of fire hazard potentials. During hot days with low relative humidity the forest floor dries out very quickly. Studies have shown that coniferous litter readily ignites when its moisture content, by weight, drops below 10 percent (Whitaker, Ackerman; 1951, p. 246).

The influence of wind—not to be confused with a fire's own wind system—was mentioned before in the differentiation between firestorms and conflagrations. Wind feeds fresh oxygen into a fire and raises its temperature. This process occurs in any blast furnace. As the flames are fanned by the wind the fire's radiation temperatures increase, and this accelerates fire spread. Wind also lengthens the flames so that they may reach out far ahead of the *fire front;* this also facilitates the forward migration of the fire.

Wind also plays a major role in the direction of fire propagation and in shaping of the fire area. Steady winds, blowing consistently from one direction, allow for a more precise prediction of the fire's path than under fickle or changeable conditions that produce erratic and very dangerous fire behavior [Plate 66].

Another weather factor, although mostly of consequence in mass fires, is the *atmospheric temperature stratification* over the burning area. A steep *temperature gradient* develops when ground temperatures, intensified by the heat energy of the fire, are high and much cooler air is encountered aloft. This arrangement of air layers, as already discussed in connection with tornadoes (chapter 6), triggers strong convective currents. This convection, in turn, attracts large quantities of fresh air into the

Plate 66. Wind is a major factor in the behavior of forest fires. Steady winds, blowing consistently from one direction, make it easier to predict the direction and speed of a fire's path. (Photograph credit: U.S.D.A.; Forest Service)

mass fire and may set the stage for a firestorm. A similar atmospheric stratification was a prime factor in the formation of the firestorm in Hamburg, Germany (July 27, 1943), which totally burned out an area of about 5 square miles (13 km²) and generated street-level temperatures of about 1,400° F (760° C) as will be presented in the last chapter (Ebert; 1963, AMS, p. 71).

The third major factor in fire behavior is *topography,* or *surface configuration*. The role of topography in fire is about as complex as topography itself can be. On flat terrain it is primarily the nature of the vegetation, its moisture content, and the existing wind factor that exert a major control over fire spread and intensity. But in mountainous topography the influence of *slope* and *elevation* becomes significant. Slope characteristics are expressed in *slope length* and *slope angle;* they combine to form the *topographic gradient.*

Because fire has a natural tendency to rise, a long uniform slope allows a fire to move upslope without hindrance; in fact, such a slope aids in fire spread because warm air rises upslope. This up-slope wind is well known as the **valley breeze.** The valley breeze is most active during the hottest part of the day, when the air is warmed in the valley, expands, and begins to rise. Whenever such solar heating is sharply reinforced by the energy of a forest fire, the gentle valley breeze becomes a fierce wind, which draws the fire uphill (fig. 11.3). Upslope fire spread may be slowed down during night hours when colder air flows downslope from the higher elevations. This downslope wind is called the **mountain breeze.**

Less clear is the effect of topography in severely dissected areas, because the resulting wind patterns are much more variable. Each valley, and each mountain gorge, produces a different set of wind currents; they may either reinforce or counteract each other. This makes the prediction of fire behavior much more difficult.

Elevation and slope orientation, the direction in which a slope faces, also affect fire. North-facing slopes—as a general rule in the Northern Hemisphere—are cooler and hold more moisture because they are less exposed to sunshine. Northern slopes, therefore, tend to be less fire-prone. Moreover, there are distinct differences between *wind-facing* and *downwind slopes.* The windward slopes receive more orographic precipitation and generally are more resistant to fire. But such advantage can be offset by the more frequent occurrence of thunderstorms and lightning on the windward side.

The effect of higher elevation is mainly one of lower temperatures. This means lower evaporation from soils and plants so that moisture levels are kept high. Also, at higher elevations clouds and mist prevail and tend to protect high-altitude forests from fire. At lower elevations the very tops of hills and ranges are usually more prone to fire development. Here the soils are thinner, more eroded, and tend to be drier as they are exposed to wind. These factors result in a condition called **summit aridity,** a state that favors vegetation fires. Moreover, peaks may induce triggered lightning and attract cloud-to-ground lightning.

Fighting Forest Fires

Understanding the *spatial* and *temporal* behavior of fire is the key to successful counteraction; but theories developed under controlled laboratory conditions can fail under natural circumstances. Too many variables are involved in each fire, and each control factor can change rapidly during the progress of a fire. Nevertheless, a number of fire-fighting techniques have proven themselves. Such techniques can be very successful when combined with alertness and common sense.

Time is the most critical factor in dealing with forest fires. Firefighters must get to them while the fires are still small and controllable. Unfortunately, many fires begin in remote and often in accessible places. This can be partially countered by the construction of access roads. In some regions already existing lumber trails can be improved to facilitate the deployment of firefighters and equipment near

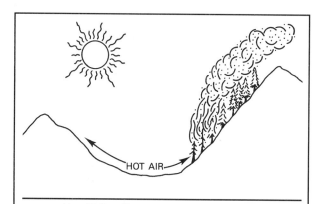

HOT AIR

Hot Valley Breeze and Upslope Fire

Figure 11.3. Daytime valley breezes, caused by the heating of the valley floor and subsequent convection, reinforce forest fires as they move upslope.

the fire's location. Such roads must have numerous *turnarounds,* places wide enough for trucks to turn around to avoid being trapped by fire.

The construction of *firebreaks* also helps to check the spread of fires. These breaks must be cleared of vegetation, and they must be carefully maintained to prevent new undergrowth. In some regions of relatively constant wind direction the layout of such firebreaks should make a right angle with the prevailing winds. This design anticipates the general direction of fire spread.

The maintenance of firebreaks requires considerable efforts and expenditures. In 1957 the U.S. Forest Service introduced a new idea, the *fuelbreak technique.* The fuelbreak method is based on planting fire-resistant vegetation in the firebreak lanes. *Fuelbreaks* are beneficial in several ways: (1) they reduce soil erosion, (2) very little maintenance is required, and (3) they may offer a good habitat for wildlife.

One of the most decisive factors is to spot fires quickly and accurately. Look-out towers, placed strategically throughout the region, are of great help. These towers are equipped with sighting devices and with communication equipment. The location of a young fire is achieved by establishing its *geographical coordinates.* The position of the look-out towers is shown on the map. The precise point of the fire, as seen from at least two towers, can be found by simple *triangulation.* During periods of threatening fire weather—especially in vast forested regions—small spotter planes are also used to pinpoint fires.

Whenever the location of the fire is accessible, firefighters and equipment can be rushed to the area. If the fire starts in a remote place it may be necessary to deploy *fire jumpers,* highly trained firefighters. They parachute into the immediate vicinity of the fire to control it in its early stage. Another technique uses large airplanes capable of releasing either water or fire-retardant chemicals during low-level flights [Plate 67]. These high-risk flights are necessary to achieve two goals: (1) to concentrate the drop into a relatively small area, and (2) to spot the heart of the fire that otherwise may not be visible from either

Plate 67. Low-flying airplanes, dropping fire-retardant chemicals, have become a valuable tool in spotting and fighting forest fires in inaccessible areas. (Photograph credit: U.S.D.A.; Forest Service).

high-level flights or from look-out towers.

Under favorable circumstances the method of **backfiring** is used to stop a fire. This potentially very dangerous practice consists of burning the vegetation ahead of the advancing main fire. This method can be used only when wind velocities are fairly low and wind direction is consistent.

The use of a backfire serves to deprive the approaching fire of fuel, and the main fire actually assists in this effort. A fully developed fire—as discussed with fire behavior—produces a rising column of hot air over the burning area. In the absence of strong ground wind this convection draws ground air toward the fire, and with it the backfire (fig. 11.4). By the time the two fires merge, the fuel load in the path of the main fire is consumed. But this situation can become very dangerous when the wind changes abruptly and the backfire moves in the wrong direction.

Catastrophic Forest Fires

Few natural sights are more sobering than the charred remains of a forest devastated by fire. The absence of color or signs of life is depressing to all who love nature. At that moment it is difficult to appreciate the powerful regenerative ability of the biosphere. The forest we once knew with its tall mature trees and its cooling canopy may not return in our lifetime; but a new life cycle will emerge quickly to cover some of the scars. The loss of human lives,

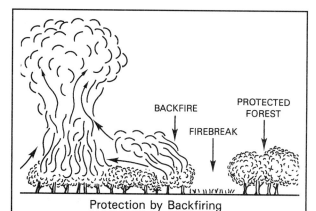

Protection by Backfiring

Figure 11.4. The strong convective wind system of a major forest fire draws ground air, and with it the backfire, toward the main conflagration. This action creates a burnt-out zone that protects the remaining forest against the advancing main fire.

however, is a more permanent wound.

It is not easy to "rank" catastrophic forest fires. But, as far as human casualties are concerned, the 1871 Peshtigo fire in Wisconsin must be viewed as one of the worst on record. Although the concurrent Great Chicago Fire received then, and today, much greater publicity, the Peshtigo disaster, which caused in excess of one thousand deaths, caused about five times the number of casualties as the Great Chicago Fire.

The hot and dry summer of 1871 left the huge forests of Wisconsin and Michigan in a dangerous dry state. Lumbering operations, aided by the construction of railroads, had penetrated the great forests leaving behind large amounts of debris, or *slash*. Scattered fires infested the region by the end of the summer, and a smoky haze hung over the entire area. It was reported that, by the beginning of October, the smoke had become so dense that steamers on Green Bay used their foghorns and had to navigate by compass (Wells; 1968, p. 41).

Fires in the interior began to merge, and an ominous glow reddened the night sky. By October 7, the people of Peshtigo could hear the eerie rumbling sounds of a growing mass fire. Many farm families and lumbermen from the surrounding countryside fled to Peshtigo. The great fire arrived there with a roar on October 8. Burning branches and glowing embers rained from the sky, and a fierce wind pushed the flames into the doomed town. One survivor, John Cameron, saw, while sitting on the wooden steps of a boarding house, a large piece of burning wood falling out of the sky. It landed on the street covered with dry sawdust. More pieces followed, and in an instant the pine sidewalks burst into flames; the great fire was under way (Holbrook; 1944, p. 66).

Most people were mowed down quickly by the fire; others fled in panic to the banks of Peshtigo River, but even there the flames caught up with them. Some stood up to their necks in the water, but they were burned by the wind-driven flames that extended over the water (Wells; 1968, p. 119).

Detailed accounts by survivors appeared in the contemporary papers and were retold in various publications. Many stories became distorted and some aspects probably were exaggerated; this is understandable considering the impact of the deadly conflagration and the panic. But the howling *fire wind* was real, the victims were real, and destruction was total. The total death toll, including 600 who were buried in Peshtigo, is not known. Holbrook, using minimum figures, estimates the regional total at

1,152 (Holbrook; 1944, p. 72).

Brief mention should be made of two other highly destructive forest fires that occurred in the United States in 1933 and 1967, the Tillamook and the Sun-dance fire, respectively.

The Tillamook fire, in northwestern Oregon's coastal ranges, started in connection with lumber operations. Just when the initial fire seemed to be under control a strong, dry wind blew into the fire area. It had been an unexpected shift of the wind. Within a short time the burning area increased from 40,000 to 240,000 acres (1,620 to 9,700 hectares), and a full-fledged firestorm consumed the mountain forest. The wind velocities toppled mature stands of Douglas firs, and flames were reported to have reached heights of 1, 600 feet (480 m) as they were fanned by winds of hurricane force (Pyne; 1982, p. 330). The fire burned more than 311,000 acres (125,860 hectares) of which 87 percent, or 270,000 acres (109,920 hectares) were destroyed within 20 hours on August 24–25 (Holbrook; 1944, p. 141).

The second mammoth conflagration, the Sun-dance Mountain fire, occurred within a 150,000-square-mile area (388,500 km²) of *wilderness forest* shared by Oregon, Washington, Montana, and Idaho. Persistent hot and rainless weather had dried out the forest. The first sign of a fire—probably started by lightning—was spotted from a look-out tower on August 23, 1967. The strong northeasterly winds quickly spread the fire. Debris was carried far ahead of the fire front so that it could not be contained. Soon the growing fire made its own wind system with velocities as high as 80 mph (128 km/h) and got totally out of control. At the peak of the fire its smoke plume reached 25,000 feet (7,600 m) into the sky (Page; 1983, p. 18).

Twenty years later several western states, mostly in northern California and southwestern Oregon, suffered 1,900 fires, which destroyed 837,000 acres [338,730 hectares] of forest. Starting on August 30, 1987, and for the following four days, 4,520 known lightning strikes ignited over 1,500 fires. Strong winds accompanied by low relative humidity resulted in the most severe fire problem on National Forest lands since 1929 (Gale; 1987, p. 1).

In 1968 this author visited Tasmania, Australia, the site of another devastating fire. This fire raged around Hobart, the capital, burned 12 townships, killed thousands of sheep, and destroyed extensive forest tracts on the slopes of Mt. Wellington. About 60 people lost their lives in this fire [Plate 68].

Plate 68. In early 1967, hot and dry weather produced a strong sea breeze near Hobart, Tasmania. This wind drove an intense forest fire up the slopes of Mt. Wellington. About sixty people lost their lives in this devastating fire. (Photo by C.H.V. Ebert).

In that part of the world rainfall is quite variable from year to year. The abundant precipitation of 1966 had fostered a heavy growth of grass and brush. As is customary, the sheep ranchers of that area burn the pastures in the dry season to get rid of excess vegetation. February 1967 was a hot, windy, and dry month. The high temperatures—some days in excess of 100° F (38° C)—induced low atmospheric pressure to develop over the land area. This produced an intense *sea breeze* with wind blowing from the sea into the interior. Local residents recalled strong winds of about 45 to 60 mph (72 to 96 km/h), with gusts of much higher velocities, throughout the area.

The burning of the dry grass, the bone-dry bush vegetation, and the rising wind conspired to trigger a chain reaction of erupting fires; soon, a glowing wall of flames raced across the region. The fire spread within a radius of about 40 miles (64 km) around Hobart, but the city itself was spared. Eventually the conflagration moved up the dry slopes of Mt. Wellington where the dessicated forest exploded with fire. The conflagration finally subsided when the tree cover thinned out and gave way to bare rock.

As we have seen, forest fires can be exceedingly damaging and especially so in overprotected forests. But their periodic occurrence is a part of nature to which we have to adjust. Living with forest fires—and at times actually using them as a tool in forest management—becomes easier with a better understanding of their nature. Such knowledge makes it more feasible to prevent disastrous fires or, at least, to control them more effectively.

SELECTED THREATS TO THE BIOSPHERE

CHAPTER 12

Key Terms	
molting	plague
Black Death	non-native species

Dynamic changes mark our planet. We have already viewed some of the processes that shape the lithosphere, are active in the hydrosphere and atmosphere, and those that interact with the biosphere. It was also stated that the biosphere—*the world of living things*—is an ever-evolving system of animal and plant life. How such changes have taken place in the past and how long it took to achieve them is to a large extent a matter of speculation. But many changes can be observed today, and from them one can infer to some degree how changes occurred in the past.

Some alterations in the biosphere take place in response to slow changes in climate as well as a result of *mutations,* or a change from the parent type. Others occur rather quickly under the impact of diseases or fire. Or changes may be caused by the invasion of other species and by the ever-increasing role of humans. The total story of changes in the biosphere, and the potential threats to it, is far too large and complex to permit a complete review. The following accounts, though highly significant by themselves, reflect only small segments of the immensely interwoven system in which we live.

The Locust Plague

"For they covered the face of the whole earth, so that the land was darkened; and they did eat every herb of the land, and all the fruit of the trees which hail had left; and there remained not any green thing, either tree or herb of the field, through the land of Egypt." This famous passage of *Exodus* 10:15 describes one of the oldest plagues known to humankind—the ravages of the locust.

Locusts, also called grasshoppers, exist in most parts of the world, but their devastating impact is particularly damaging in the semiarid regions. While different species are known—different in terms of appearance and life-styles—the common characteristics center on two facts: (1) locust populations explode in cycles—achieving incredible numbers, and (2) locusts become a double threat because the plague seems to occur most effectively when drought also occurs.

Drought, as discussed in chapter 10, has its most serious implications in the marginal semiarid zones that border on the deserts. It is here, in the drier grasslands, where the locusts do their greatest damage. The coincidence of drought and a population explosion of this insect has resulted in some of the worst famines the world has known.

Locusts exist in small numbers at all times. Their numbers are held in check—as it is normally true for all organisms—by *predators* and by *diseases*. Certain parasites, such as the *Opsophyto opifera,* can kill off large numbers of grasshoppers. This event was reported in New South Wales, Australia, where between 60 and 70 percent of the insects were killed by the parasite (Essig; 1965, pp. 316–317). The killing of locusts by predators is especially effective when rainfall is sufficient for predators to remain in the area and to multiply. Moreover, if high moisture conditions remain too long, the eggs of the locust, laid in the soil, may rot. To ensure successful hatching the soil should be warm, sandy, and endowed with some moisture. The sandy nature of a soil not only facilitates the placing of the eggs but also enables the locust *nymphs,* a young locust that has not yet undergone complete *metamorphosis,* to reach the surface easily after hatching.

Subsequent drought conditions seem to trigger a locust explosion. When excessive soil moisture is reduced, the soil may crack open allowing for easy egress of the nymphs, and the predators leave the region because of the dry conditions. Several studies have shown a close correlation in the occurrence of droughts and locust outbreaks (Rudd; 1964, p. 234).

The newly hatched locusts are wingless and move along the surface, walking and hopping, in search for food. Their wings develop after five stages of *molting,* which takes about five weeks. When excessive numbers of young locusts survive they quickly destroy the local food; then they must move toward regions that offer more vegetation. This situation is typical for the Sahel, where mature locusts migrate across the Sahara and as far northward as the coast-lands of the Mediterranean (White; 1960, p. 56). It is here, mostly along the Nile, where irrigated crop areas are exposed to these ravaging insects. Locusts also breed in favorable areas of the Arabian peninsula, in the Sudan, and in Somalia. From there they migrate throughout southwestern Asia and as far as Iraq. In the past, crop losses in Iraq have gone as high as 70 percent (Cressey; 1960, p. 396). In recent years Saudi Arabia has invested heavily in developing new agricultural areas devoted to producing irrigated wheat. In 1986 production reached two million tons and made that country not only self-sufficient but allowed some export. To safeguard this crop Saudi Arabia has developed a large-scale program to monitor and

to combat locusts in their suspected breeding grounds (Hobson, Lawton; 1987, p. 10).

Major locust outbreaks have also occurred in Australia (1972 and 1974) and in the Great Plains of the United States. Grasshopper swarms plagued the western regions of the United States in the 1860s correlating with droughts, and massive swarms were observed migrating eastward in search for food. In 1865 General A. Sully, bivouacking with his troops near Sioux City, Iowa, reported that one of his soldiers, while sleeping in the tall prairie grass during the midday hours, was severely bitten by grasshoppers (Barker; 1960, p. 3). One mammoth swarm was reported in Nebraska; it took six hours to pass a given point. The swarm was said to have advanced on a front 100 miles (160 km) wide and containing an estimated 124 billion locusts (Page; 1985, p. 69).

Various studies indicate that *overgrazing,* as well as dry-land cultivation, produces conditions favorable to the explosive increase of grasshoppers. Such research was conducted in the Sahel, in Morocco, and in North America. R. E. Smith pointed out that few grasshoppers existed in ungrazed grasslands in Kansas compared to the numbers in nearby intensively grazed areas (Rudd; 1964, p. 234). This harmful effect of overgrazing, again, is a good illustration of how one disturbance in the biosphere can trigger disastrous responses.

A grasshopper consumes its own weight in food every day. The collective destruction caused by a large swarm can be total. Should such a swarm contain about 60 billion locusts it would weigh nearly 100,000 tons! With densities as high as 1,000 individual insects per square yard (0.83 m²), this mass of insects could cover an area of 600 square miles (1,550 km²) and leave bare ground behind.

Many steps have been undertaken to combat the locust. These efforts require international cooperation because of the large-scale migratory habits of this pest. After World War II the Food and Agriculture Organization [FAO] of the United Nations launched a massive international locust control program. Pesticides were applied by airplanes, by truck-mounted sprayers, and by thousands of local workers. Though these efforts could be called successful—as far as the fate of the locust was concerned—the poisoning of the environment soon raised voices of protest. Moreover, political upheavals and war in the regions of Ethiopia and Somalia, both breeding grounds of locusts, interrupted the operations. Giant migrating swarms were reported again in 1978.

Other methods less damaging to the environment have been recommended for many years, and new techniques are being researched. It was shown that early-seeded crops tolerate damage much better than late-seeded crops because early crops make considerable growth before the grasshoppers hatch. Grain crops, which have already developed seed heads, can withstand extensive defoliation without serious reduction in yields (Fleming; 1957, p. 328). *Deep-tillage* can also discourage the grasshopper; the eggs may be buried so deeply that the newly hatched insects cannot get to the surface. Plowing also moves the eggs to the surface where they are picked up by birds or are destroyed by exposure to sun and wind. Controlled flooding of fields can also be used to kill off locust eggs as well as newly hatched nymphs.

Since the early 1980s a new technique, based on *remote sensing* from satellites, has been added to the international arsenal to combat the locust. Satellites can spot areas in semiarid lands where moisture conditions are favorable to the breeding of grasshoppers. The FAO subsequently sends out locust alerts to the endangered regions to provide for timely countermeasures (Page; 1984, p. 148). As long as the locusts are in the wingless nymph stage they can be destroyed on the ground. But once the swarms darken the sky, the disaster on wings cannot be held back.

The Destruction of Tropical Rainforests

In 1541 Francisco de Orellana descended the Napo River, which joins the mighty Amazon about 60 miles (96 km) below Iquitos, Peru. It was not until 1542 that he reached the Atlantic Ocean more than 2,000 miles (3,220 km) downstream. Orellana, and later Alexander von Humboldt—who explored similar regions bordering the upper Orinoco and Rio Negro—probably did not comprehend the size of the Amazon Basin and the immensity of the then unbroken primeval rainforest. They must have been deeply impressed by the stillness of this forest and by the apparent absence of people [Plate 69].

Looking at a world population map today, vast regions of tropical rainforests still show a population density of less than one person to the square mile (2.6 km^2). The Amazon Basin, confidently colored green in atlases and mostly classified as *Tropical Rainforest,* appears unchanged. But drastic changes *have* occurred in this vast domain, which measures about 66 percent of the size of the United States

Plate 69. This primeval tropical rainforest in eastern Guatemala reveals a great variety of trees of different shapes and heights. The density of the canopy is such that relatively little sunlight is able to reach the forest floor. Excessive deforestation is now endangering this majestic forest. (Photo by C.H.V. Ebert).

excluding Alaska. Similar changes are going on as well in other tropical rainforest regions in Africa and Asia. Of the 7 million square miles (18 million km2) only about half of the tropical forests remain. According to the Washington-based Environmental Policy Institute, about 100 acres (40 hectares) of tropical forests are being destroyed, or seriously degraded, *each minute* [Plate 70]. This is about the time it may have taken you to read from the beginning of this section to this point!

At one time about 85 percent of the island of Madagascar's area of about 230,000 square miles

Plate 70. Large tracts of tropical rainforest are being permanently destroyed by this mining operation in eastern Guatemala. Exposed to heavy rainfall the soils of this area will be quickly eroded. (Photo by C.H.V. Ebert).

(596,000 km²) was densely forested; today less than 8 percent of the forest remains. Thailand was 75 percent forested as recently as at the end of World War II; only a small percentage is left, and of the Philippines, only 38 percent has remained in forest cover (Koopowitz, Kaye; 1983, p. 82). Even in small Costa Rica, forests are being cleared at a rate of nearly 160,000 acres (65,000 hectares) per year (Durrell; 1986, p. 46), and in Brazil the total forest area has declined by nearly 30 percent.

What forces are responsible for this reckless onslaught on one of the most sensitive and ecologically one of the most critical natural resources on our planet? It is a combination of greed, ignorance, and wanton unconcern, combined with what some developers call *"economic necessity"* [Plate 71].

Over the last decades the world's appetite for pulp and lumber has dramatically increased. During the past 30 years world paper use grew from about 40 million tons annually to 180 million tons, it may reach 400 million tons by the year 2000 (Rainforest Action Network; March, 1987). The tropical rainforests presently produce about 7 percent of the world's paper and paperboard. This percentage is rising quickly because new processing techniques permit the use of *hardwood* species for pulp production for which predominantly *softwood* had been used. The lavish annual consumption of 600 pounds of paper per person in the United States is almost 40 times the amount used by developing nations. The search for new lumber resources has also intensified. Japan, trying to conserve her own forests, buys vast

Plate 71. To clear the land for agriculture, the tropical rainforest was totally removed from these hills in eastern Guatemala. Compare this picture to Plate 69 which shows the same region before deforestation took place! (Photo by C.H.V. Ebert). ————————

quantities of wood from Southeastern Asia, from Tasmania, and from Indonesia.

A major experiment, financed and planned by an American billionaire, Daniel K. Ludwig, aimed at removing the ecologically diversified rainforest in Brazil and replacing it with a fast-growing tree species (*gmelina*) from Asia. But unsuitable soil conditions (sandy low-nutrient soils as well as compaction by heavy machinery) reduced expected yields by 50 percent. Other tree species were selected, including sand-tolerant pines and eucalyptus, but the financial losses had been too severe. Of the total 4 million acres (1.6 million hectares) only about 250,000 acres (101,000 hectares) had been cleared. Recently the land was bought back by Brazilian interests.

Another reason behind the disastrous destruction of the tropical rainforest is the growing beef cattle industry. Virgin rainforest land is cheap, and labor cost in developing countries is low. Grass-fed cattle can be raised in Latin America at 25 percent of the cost at which they can be produced in Colorado (Ehrlich, Ehrlich; 1981, p. 163). A great portion of this beef is used—the world over—in the fast-food industry where tough beef is made into hamburgers!

Pioneer farmers turn to cattle ranching once cultivation has exhausted the soils. This change takes only a few years because the nutrients in tropical rainforests are stored in the biomass, not in the soils. Once the trees are removed, the remaining soil nutrients are readily *leached* out of the soil; they are no longer within reach of the crop roots.

In some tropical areas the tropical rainforest is being destroyed because the land is given to poor people or developed by governments to relieve the population pressure of overcrowded regions. One of these projects involves the western Brazilian state of Rondonia. Here, an area of the size of West Germany (95,970 square miles or 248,570 km²) was being cleared of its forests and was denuded by 1990. According to the Environmental Policy Institute (Washington, D.C.) about 200,000 new migrants arrived there in 1985. But only 50,000 were able to obtain government land. The massive removal of forest and the construction of access roads spell ecological disaster: unstoppable soil erosion and a decline in soil fertility within about three years. A similar project involves the moving of millions of rural people from Java and Bali to other Indonesian territories on Borneo, New Guinea, Sulawesi, and Sumatra. Though this scheme will not even make a dent in the population problem, it will destroy precious forest reserves and also place local populations under severe economic and social stress. Java's birthrate is 10 times the number of people involved in this *"transmigration"* project!

In many countries the collection of wood for fuel and for other purposes exceeds by far the *sustainable yield* of remaining accessible forests. A 1980 World Bank study of West Africa showed that the demand for fuelwood exceeds the estimated sustainable yields in 11 out of 13 countries (Brown, Wolf; 1986, pp. 23–24). In the same report it is pointed out that the world's tropical forests, on an annual average, are disappearing at a rate of about 2 percent, but much more rapidly in West Africa and in southeastern Asia. In a very different geographical setting, as in the Himalayan foothills of Nepal, radical deforestation has left behind an utterly destroyed landscape. Deep erosion gullies crisscross the abused uplands that once were covered by a protective highland forest [Plate 72].

Will the massive removal of our planet's tropical rainforest have some far-reaching effects on climate and water resources? Only the future can answer this question for certain. The enormous amount of water falling as year-round precipitation, and the matching *evapotranspiration* from soils and plants, account for the fact that an estimated one-half of the world's water vapor is held in the atmosphere of the wet tropics. If this supply is cut off, many areas may experience acute moisture deficiencies and possibly an accelerated desertization.

It has been said that human progress has been built on the ashes of forests (Durrell; 1986, p. 46).

Plate 72. At one time this highland area of central Nepal was covered by forest. Radical deforestation exposed the land to the heavy monsoon rains. Deep gullies now traverse the ravaged landscape. (Photo by C.H.V. Ebert).

Will we learn our lesson and save the tropical rainforest before it is too late? It was estimated that by 1980 one or two plant species became extinct each day. By 1990 this rate was expected to accelerate to one species every hour, so that between 15 and 25 percent of all higher plant species may be gone forever by the turn of the century (Koopowitz, Kaye; 1983, p. 6).

Introduction of Non-Native Species

In its youthful stage the volcanic island of Surtsey, risen from the sea just south of Iceland in 1963, was a barren wasteland of lava and ash. About two decades later a few simple plants and marine birds had made their home there.

The violent explosion of Krakatoa in 1883 practically destroyed all life on the island. Although it might be possible that some insect larvae and earthworms survived the eruption, the *fauna* and *flora* had to reestablish themselves. Researchers visited Krakatoa as early as nine months after the catastrophe, and the presence of a small spider was reported by E. Cotteau. Subsequent visits in 1908 and in the 1920s discovered an increasing variety of animals and plants (Simkin, Fiske; 1983, pp. 431–432). How did the species get to these isolated islands? The wind and birds carried both plant seeds and insects from other islands. Some animals arrived by swimming; some floated ashore on driftwood and flotsam. The point is that both animals and plants, actively or passively, can move over considerable distances and establish new varieties of ecosystems.

Such migrations have been going on for millions of years. The *diffusion* of species can be triggered by many factors: population pressure, climatic change, forming of land bridges between continents, glaciation, and mountain building. Both *stress* and *opportunity* supply the momentum for diffusion. However, eventually it was *humans* who proved to be the most powerful agent in promoting the movement of both plants and animals across the globe—at times with unexpected and dire consequences.

As humans migrated they carried with them items of trade and of necessity, such as grain and other food items. Later on they herded domesticated animals over great distances. Both animals and people carried insects on their bodies, and these spread at the point of arrival. A good example of this is the disastrous impact of the *Black Death,* or the *plague,* which reached Europe in the 1340s and killed an estimated one-third of the population between India and Iceland.

After A.D. 1200 the climates of Greenland and northern Europe turned markedly colder. Hardship struck the continent, and many Viking settlements lost their lifeline with Europe as the growing sea ice enveloped the shores of Greenland and Iceland (Boorstin; 1983, p. 215). The summers in Europe turned cooler, the winters were unusually cold, and famine spread. The people were weakened and were not capable of dealing with the much greater disaster brought into their midst by an almost invisible invader—the *flea* (Burke; 1978, p. 98). The actual killer, however, was the *plague bacillus, Pasteurella pestis* (now classified as *Yersinia pestis*), which was not identified until about 500 years later. The deadly bacillus, living either in the flea or in the bloodstream of a rat, was transmitted to people, or to animals, by the bite of the flea or *rat* (Tuchman; 1978, p. 101).

The flea had been transported along Asian caravan trade routes, being fed by both human and beast, and reached the port of Caffa (now Feodosiya) on the Black Sea. From here ships carried both rats and fleas to the Mediterranean ports of Messina and Genoa. Infested rats scurried off the ships and spread the disease onshore.

Shipping and trade, as well as human migration, diffused the Black Death to just about all corners of Europe. Death overwhelmed smaller towns in less than six months; larger cities struggled longer. The victims showed large black swellings in their armpits and groins. Internal bleeding spread black blotches on their skin, hence the name *"Black Death."* Death rates ranged from 20 to 100 percent. How many people were killed no one knows, but estimates indicate that about 20 million people, or *one-third* of Europe's population, fell victim to the plague (Tuchman; 1978, p. 94).

Another small organism, a tiny insect (*phylloxera*), which lives on the rootstock of grapevines, did not cause death and disaster as did *Pasteurella pestis,* but it had the power to ruin the famous French vineyards in the mid-19th century. This change, in turn, gave the fledgling California wine industry the incentive for massive growth.

Infested rootstocks from American grapes had been transferred to France. The American vines were practically immune to phylloxera, but the French vines, as well as the European vines in general, could not tolerate the insect and died, and the French vineyards were threatened with extinction. The news of the dying French vineyards reached the West Coast and offered California the opportunity to become a leading wine grower. By 1855 the boom was on, and

California was caught by the wine fever (Balzer; 1964, p. 87). Land prices soared, and by planting a few thousand vines new growers could become prosperous. However, the road to success was not that simple. Resistant American rootstocks were shipped to France, Germany, and Italy. The high-quality European vines were grafted to these rootstocks, and the continental vineyards regained their dominant position.

Accidental introductions of insects—often with catastrophic results—have affected many parts of the world. The havoc raised by the *Mexican fruit fly* is well known in the orchards in the southern United States. Equally notorious is the recent introduction of the *Asian cockroach.* Increasing world trade, and especially food shipments, multiplied the import of unwanted insects.

Earlier incidents, such as the escape of the *gypsy moth* from a laboratory in Medford, Massachusetts, in 1869, caused severe damage. Silk-producing moths were to be cross-bred with the gypsy moth to make them more resistant to disease. But within 20 years the escaped gypsy moths evolved into a major environmental threat to the forests in the eastern United States.

Another accident occurred in the 1870s in connection with shipments of potatoes that had been infected by the *Colorado potato beetle* from Colorado to the East Coast and subsequently across the Atlantic. By 1930 this pest had spread throughout Europe. The beetle did not harm the native wild potato found in the Rocky Mountains but was harmful to the European potatoes that were introduced to Colorado in 1859 (Wagner; 1971, p. 286).

Insects have also been introduced intentionally. This was the case when Australian ranchers wanted to get rid of the prickly pear—a rapidly spreading cactus—which was crowding out pasture land. After a comprehensive study by entomologists, about three billion eggs of an Argentine moth, an insect that destroys the prickly pear, were introduced in 1930. Seven years later the prickly pear was eradicated and grazing on reclaimed pastures was resumed. This was achieved at a cost of less than a penny per acre (Carson; 1962, p. 83).

However, the voluntary introduction of another animal to Australia, the *European rabbit,* proved to be a disaster. In 1859 a rancher, Thomas Austin, had a dozen rabbits imported from England. In the absence of native predators the rabbits quickly spread throughout the large estate in Victoria Province. Six years later about 20,000 rabbits had been killed; an estimated 10,000 remained and migrated to many sections of the continent. Soon vast areas became overrun by the pesty animals, which destroyed valuable pasture grasses. In 1887 about 20 million rabbits were killed just in one area, New South Wales (Laycock; 1966, pp. 176–177).

After the turn of the century the systematic battle against the invader began: Rabbit-proof fences were designed; poison was spread, affecting sheep as well as rabbits; and a virus was introduced after World War II. Although the virus nearly eradicated the rabbit (only five out of a thousand survived) the effectiveness of the virus lessened with time as the remaining rabbits built up their immunity. There may be a billion rabbits left in Australia. Their consumption of grass could support about 100 million sheep.

The European rabbit struck another devastating blow in a different setting, namely on Laysan in the northern islands of Hawaii. After they had been released in 1903 they nearly turned the island into a wasteland. More than two dozens of known vegetation species became extinct, and several bird species disappeared (Laycock; 1966, p. 171).

Accidentally or intentionally introducing ***nonnative species*** into an ecosystem is a risky business, which may result in unpredictable chain reactions with frequently catastrophic results: Trading vessels accidentally brought rats to Jamaica in the 18th century. To control the rats a vicious ant (*Formica omnivora*) was introduced from Cuba. Some baby rats were actually killed by this ant, but subsequently the ants became a widespread pest over the entire island. Then the *mongoose,* a ferretlike carnivore, was imported from India. Unfortunately, the mongoose usually hunts during the day while the rat tends to be nocturnal. Moreover, the mongoose also killed many birds, reptiles, and eventually poultry, young pigs, and small lambs. The rat, of course, survived.

Another highly destructive animal is the *goat.* The once heavily timbered island of St. Helena in the South Atlantic was virtually denuded after the goat was introduced and spread uncontrollably. Goats also ravaged Santa Catalina Island offshore California and caused a sharp decline of humans, reptiles, birds, and small mammals (Ehrlich, Ehrlich; 1981, p. 165). A similar threat to Santa Fe Island, a member of the Galapagos offshore Ecuador, was eventually halted by exterminating all the goats on that island.

The threats to the biosphere are frequently unpredictable. But many modern technologies, such as long-range cargo trucks, all-terrain and four-wheel-drive vehicles, and ship container packaging make it likely that new hazards from the introduction of unwanted species will endanger our future.

MAN-INDUCED DISASTERS

The human mind finds it quite difficult to comprehend scales—proportional sizes or measurement relationships—that extend beyond everyday dimensions. This problem is especially true when we look at scales of the universe about us, or when viewing, under a microscope, an entity of a few microns in size. Similarly, *time scale* is another very elusive dimension.

Could we show the total geological history of Earth within a 30-minute time span, the role of humans would probably occupy the final 3.5 seconds of this spectacle. Expressed in a linear scale this tiny fraction equals about 10 feet out of a mile (5280 feet), or slightly less than 2 meters out of 1 kilometer (1000 m). Humans very likely existed more than 2 million years ago. In prehistoric times they were simply an integrated part of the then prevailing ecosystems but did not play a dominant role in them. Humans *shared* the environment with other living creatures by preying on other organisms for sustenance (Ehrlich, Ehrlich, Holdren; 1977, p. 8). It was much later, about 10,000 years ago, when humankind made its first major impact on the environment. This event was the beginning of the *Agricultural Revolution,* which triggered a rapid increase in world population.

The Agricultural Revolution was not a sharp break with the past but rather a *slow* development over many centuries (Stoddard, Blouet, Wishart; 1986, p. 12). But this phase in human history probably marked the sharpest transition in humans' ability to affect and to disturb the existing ecosystem.

The second phase, and a much more dramatic one, took place in the middle to late 18th century and is generally known as the *Industrial Revolution.* This step, again, was accompanied by an enormous acceleration of the population growth rate; it was also the beginning of an increasingly damaging impact on the environment. It is generally believed that humankind placed more stress on nature during the past 80 years, or so, than during its entire prior existence on this planet.

As world population increases further—and we added about 87 million people in 1986 and 91 million in 1991—the pressure on resources, water, air, and livable space increases accordingly. With this pressure grows the probability for *human conflict.* There is a good chance that the human race and its ever-widening activities may exceed what may be called the natural environment's carrying capacity. However, the exact knowledge does not yet exist to precisely define that limit. Moreover, we do not know what consequences will arise once this *global carrying capacity* is exceeded (Holdren, Ehrlich; 1981, p. 15).

The subsequent discussions will focus on some of the more ominous concerns we must face. We have to do this not only to become fully cognizant of them, but we must also think of ways to meet these challenges and, if possible, to avoid catastrophic consequences.

SOME MAJOR GLOBAL CONCERNS

13

CHAPTER

Key Terms		
system dynamics	topsoil	nonrenewable
population growth rate	hydroponics	aquifers
Third World	fisheries	fossil fuels
genocide	mariculture	coastal ocean
demographic transition	renewable resource	biodegradation

Humans' relationship with the ecosphere proceeds along a J-curve; that is to say that at first humankind was an integral part of the living world, existing in harmony with the total system, and did not exert any excessive influence. But, over time, humans assumed a different role and attained the position of changing the environment consciously, not instinctively (Dorfman, Dorfman; p. xiii). While early people's niche of survival was narrowed by the forces of nature, modern people widened their sphere of action and became a dominant factor on our planet. Science and technology raised Homo sapiens to a potential master role that removed humankind, step by step, from the more balanced relationship to nature as it existed in the past. In their influence today, humans are becoming comparable to, and possibly exceeding in many aspects, many natural processes (Holdren, Ehrlich; 1981, p. 11).

As a result of the new role of modern humans several *global concerns* have emerged. Four of these are especially critical in nature: (1) *world population growth,* (2) *food supply,* (3) *depletion of natural resources,* and (4) *avalanches of waste materials.* These factors by themselves, plus many associated ramifications, conspire to bring about an increasing stress on the environment; they are accompanied by many specific dangers and a general deterioration of nature and quality of life. This aspect will be discussed in detail in chapter 14.

The above-mentioned global concerns tend to grow exponentially; they grow by a more or less constant percentage of the whole entity (e.g., population; food base; resources, etc.) over a constant period of time. Each of these problems assumes a different role in different parts of the world so that no hierarchy of *universal importance* can be readily assigned. But the problems are all interrelated and form a complex system. Each component grows at a given rate, and all components make up the total structure of the system. A technique called **system dynamics,** developed at Massachusetts Institute of Technology, is based on the concept that the total structure of any system many times is as important in explaining the system's behavior as is that of the individual component, or variable (Meadows et al.; 1972, p. 31). However, it is clear that the world's population growth represents the common denominator underlying just about all human-induced environmental problems.

World Population Growth

War, floods, famines, volcanic eruptions, and earthquakes readily attract media and world attention; however, an ongoing major threat to the quality of life—if not to our very survival—is a pressing issue: *world population growth* (fig. 13.1).

Most critical problems that we are facing today, and will be facing increasingly so, are directly or indirectly related to the steadily growing number of people who all need food, clean water, housing, as well as economic and political stability. The same people, of course, generate rising amounts of waste and garbage, create pollution, and in many instances degrade and even destroy the natural environment. In most quarters these issues are recognized, but many people are not fully aware of the exponentially increasing danger of overpopulation and its potentially dire consequences.

The United Nations Population Fund labeled October 12, 1999, the "Day of Six Billion" as the

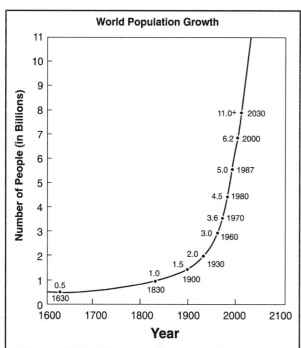

Figure 13.1. The world's population growth progresses along a J-curve and is expected to reach about 11 billion by the year 2030! (Population Reference Bureau; Washington, D.C., 1994).

world population passed this number. It took only four decades since 1960 to reach this level when the United Nations estimated the world total at 3 billion. Where are we going from here?

It is very difficult to reliably predict population growth for the world as a whole. Many variables and yet unknown factors blur long-range projections. What will be the effect of HIV infections, of famines, of epidemics, and of war? Locally available data may not be dependable or up-to-date; however, some basic premises exist: (1) world population will continue to grow, (2) developing countries will for a long time account for the largest share of this growth, and (3) the overall make-up of the world's population will shift toward older age (Lutz;1994, pp.32–34).

The aging of the world's population, paralleled by lower fertility, is a major factor in the worldwide declining growth rate. In fact, the United Nations' projections (Internet Communication; 10/18/99) for the year 2050 have been scaled down from 9.4 to 8.9 billion. Thirty-two countries, which account for about 12 percent of the total, report either steady or declining birthrates. Even Brazil, Mexico, and South Korea, countries with traditionally high rates, now have declining growth rates. This trend reflects the influence of a number of conditions: (1) favorable economic development, (2) better achievements in education, and (3) more effective efforts in family planning. On the other hand, there are considerable differences from country to country and from continent to continent. Of great concern is Africa with a total population of about 840 million and a birthrate estimated at around 2.7 percent. At this rate the population could double within less than 30 years; however, war, genocide, and diseases could change these projections.

It is not surprising that the greatest concern about rapid population increases centers on what is often referred to as the *"Third World"* or still developing countries. This term may not be quite realistic; technologically most world regions are now closely intertwined. World economic events, both positive and negative in their impacts, cause well-known *ripple effects*. Thus it is obvious that there is no easy solution to the urgent problem of world population growth. Education, religion, politics, as well as economic competition prevent patent solutions despite the many efforts made in many countries and by concerned world organizations.

China, with her present policy of "one child per family" experiences a sharply declining birthrate which at this time is just over 1 percent annually.

This trend means that her population, now at about 1.28 billion, would double in 60 to 70 years. About 9 percent of China population is over 60 years old; and this percentage is projected to rise to 20 percent by the year 2030 (Lutz; 1994, p.32).

China's long-time first rank in population is challenged by an ominous development in India where the population grew from nearly 350 million in 1950 to the present mark over 1 billion. With an annual birthrate estimated by the United Nations at 1.76 percent, realizing an increase of 18 million a year, India's population could pass that of China by the year 2037.

World population growth gained momentum in the 1800s, when industrialization and scientific achievements in medicine and general sanitation significantly lowered death rates. This change was especially true for the technologically advanced countries of Europe and North America. Birthrates initially remained high so that population growth accelerated dramatically. In modern times, the lowering of birthrates and death rates, at least in Europe and North America, completed what is known as the **demographic transition.** This cycle begins with balanced birth and death rates, changes into high birthrates accompanied by lowering death rates, and ends with falling birthrates in overall balance with death rates.

Thomas R. Malthus (1766–1834) expressed his deep concern about world population growth in his "Essay on the Principle of Population as It Affects the Future Improvement of Society" (1798). He was concerned that population, growing at a *geometrical* rate, would eventually outstrip the world's food supply, which grows *arithmetically.* To some modern demographers Malthus's ideas may appear "unsophisticated," but the basic premises in his essay are just as important now—though debatable—as they were in his days (Woods; 1986, p. 141).

One of the controversial aspects is that there is no agreement as to what " *overpopulation*" really means. Are we talking about the total population on planet earth, or do we mean that there are too many people in certain places? Are we referring to crowding, or do we worry about the earth's capacity to sustain life? It becomes obvious that talking about population pressure takes on different meanings, and implications, under different circumstances. These differences in meaning, however, must not detract from the basic gravity of the problem.

Whatever the possible answers, and whatever science and technology may have to offer to deal with world population growth, the sobering facts remain: if we allow the 6-billion mark or the 8-billion mark to pass by as casually as we have the 5-billion mark in the mid 1980s, our children will have to pay the price (Meyer; 1986, p. 1).

We must foster an awareness of the fact that poverty, famines, health care delivery, general sanitation, and the integrity of our environment are all connected with uncontrolled population growth: the world community must show an increasing commitment and determination to face this problem which eventually will affect all of us.

Food for How Many?

Hardly a day passes when we are not reminded of children dying of hunger and disease in Africa, in Asia, and in some parts of South America. Many appeals—usually accompanied by harrowing photographs—reach us by mail or by television and ask us to contribute to the fight against hunger. Is the world running out of food? Can Planet Earth feed the ever-increasing masses of people?

These questions cannot be easily answered, and it seems that the literature on this subject is more confusing than helpful. Some scientists, such as land use specialist Colin Clark, believe that the earth could support almost 50 billion people, or 10 times the number reached in 1987 (Stoddard, Blouet, Wishart; 1986, p. 54). But we hear also that Africa's food sufficiency ratio declined from 98 percent in 1961 to 78 percent by 1978, and that food imports rose at an annual rate of 8.4 percent between 1970 and 1980 (Bradley, 1986, p. 89). We also learn, according to FAO of the United Nations, that the world output of wheat, corn (maize), and rice for the same decade increased by 40, 50, and 30 percent, respectively.

The above statements appear to be contradictory, but they are not "wrong" by themselves. The situation is much more complicated than that. Statistically, there is enough "food" in the world, but it may not be *the right kind,* in *the right amount,* in *the right place,* at *the right price,* and at *the right time.* World agriculture is a maze of confusion! In some world regions per-capita food production is rising and falling in others. But in many places there is far too much food for existing market conditions. Obviously, problems of production and distribution are interrelated.

What is meant by *food* production? The types of food produced and eaten vary greatly in different cultures and in areas with varying standards of living. Thus it is not possible, except for a given group or for a specific location, to express what "food" means unless the term is stated in *caloric* values or types of nutrients. According to L. R. Brown and E. P. Eckholm (*By Bread Alone*), as quoted in Ehrlich et al., cereals represent about 56 percent of human food energy, while livestock and fish products account for 11 percent (Ehrlich, Ehrlich, Holdren; 1977, p. 286). For this reason grain output and consumption frequently serve as an index for the food situation of the world. But whether or not people starve is not directly tied to this index. *Famines* are rooted in an incredibly interwoven array of factors that include the physical environment (soils, water, climate), politics, economic systems (policy, cost), culture (priorities, values, dietary laws and preference, religious concerns), technology (energy, machinery, fertilizer, land management, transportation systems), and, of course, a host of *unpredictables,* which include natural disasters, social upheavals, and war.

The scope and purpose of this book does not allow a discussion of all these elements, but since our existence depends basically on the land on which we live, the *physical world* is of prime importance. Of the physical factors soil erosion, and with that land exhaustion, is one of the most serious ones. Much of the world's agricultural land is already severely overstrained, and it is questionable whether enough food can be grown, under the present trends of population growth and consumption patterns, without destroying our soils (Passerini; 1986, p. 15).

Soil erosion depends on many variables, such as the nature of the soil, the amount of slope, wind conditions and precipitation intensities, and how the soil is worked. Thus no uniform picture exists as far as soil erosion is concerned. However, as a general rule an annual average soil loss of more than 4 tons per acre (10 tons per hectare) is considered serious (Hausenbuiller; 1985, p. 226). The problem is made even more severe because soil erosion removes the most fertile and irreplaceable portion of the soil: *topsoil.*

When using soil erosion statistics one has to realize that such data is very difficult to obtain, especially in developing nations. The reason for this is that the data collection needed to form a reliable information base is quite demanding. There are simply not enough domestic resources available in developing countries to carry out such a program without considerable outside financial and technological aid. Also, there are many uncertainties how the various symptoms of soil erosion evolve through

time (Blaikie; 1985, p. 15). The projections are, therefore, probably not exact.

It is expected that world per-capita cropland will decline by 19 percent between 1984 and 2000. At the same time at the present rates of soil erosion the amount of topsoil per person could decrease by 32 percent. This loss will offset food production trends (Brown, Wolf; 1984, p. 8).

Although it appears plausible that soil loss, caused by erosion, and declining crop and livestock yields are interlinked, the statistical appraisal of this problem carries considerable complications. R. Dumsday and J. Flint (1977) and M. J. Amos (1982), as presented by Piers Blaikie, point out that there are different research models for erosion processes and for crop growth processes; the functional link between the two still is underexplored (Blaikie; 1985, p. 16).

The amount of soil loss, even in technologically advanced nations such as the United States, is ominous. Tolerable rates of erosion are exceeded on 44 percent of U.S. cropland. India, just as one example for a less-developed country, is losing about 6 billion tons of soil annually. Some estimates indicate that total world soil loss carried to the oceans nowadays amounts to 24 billion tons a year. Unfortunately soils form so slowly that their destruction can be viewed, for all practical purposes, as *permanent* [Plate 73].

Are there alternate ways to solve the food problem? Of course there are some schemes that appear workable; however, in many instances the techniques are experimental, or they prove to be too complex, or too costly, and they may result in foodstuffs that are unacceptable for a variety of reasons. One of

Plate 73. Lack of erosion control and neglect have allowed this field in Georgia to erode down to its hard, stony, and totally unproductive subsoil. (Photograph credit: Marlene Johnson)

the more exotic plans was the *"Hyperion Process,"* which involves the production of high quality vegetable protein by way of a blue-green alga (*Cyanophyte*) of the genus Spirulina (Smith, Newsom; 1976, pp. 1–2). Partially successful, except for *cost* and *scale* of production, is the process of **hydroponics,** the growing of food plants in nutrient solutions. Growing more tree crops is also advocated because yields (fruits, nuts) are high per unit, marginal land can be used, and the soil is protected (Passerini; 1986, pp. 17–18). In a realistic sense, however, it seems that the greatest hopes, as far as agriculture is concerned, are rooted in developing more productive crops, using mineral and organic fertilizers prudently, and managing the land responsibly. Can we look to the oceans for additional help?

The information about the ocean's role in solving the world's food problems is just as contradictory as that available on *overpopulation and food supply,* and the potential of the oceans to provide food has been and still is the subject of considerable debate (Alverson; 1977, p. 194). As important as ocean-supplied food is to many millions of people, the total amount of protein provided by fish in recent years is only around 3 percent and about 10 percent of all animal protein. But, how important ocean protein can be in individual cases is illustrated by Bangladesh where of all animal protein consumed about 80 percent comes from fish (Gross; 1987, p. 325).

How much food could *fisheries* eventually produce? By 1984 the total world catch amounted to about 74 million tons. The FAO estimates that the maximum annual sustainable yield—based on *traditional* fisheries—would be about 120 million tons (Thurman; 1985, p. 321). Other estimates go as high as 250 million tons (Alverson; 1977, p. 195). Obviously, if presently underused or unused species (octopus, squid, cuttlefish, krill, and many others) would be intensively harvested, the total food contribution by the ocean could be markedly increased. Moreover, if systematic **mariculture,** the systematic cultivation of *aquatic* organisms, could be greatly expanded, the watery environment of oceans, ponds, and lakes could become a significant contributor of food. But, none of these ways represents the magic road to the ultimate problem of hunger, nor are they the answer to the all-important question: *Food for how many?*

Depletion of Natural Resources

Anything that is needed to sustain human existence can be called a *resource.* This, by definition, must also include people themselves, the human resource, who are needed to perform work, who represent a market for goods and services, who build civilizations and achieve certain standards of living. Standards of living represent a composite of (1) individual wants and needs, and (2) those that are representative of the social group to which the individual belongs. Once such standards are established, there develops a strong resistance against any force that threatens to lower them (Zimmermann; 1964, p. 31). Said in different terms: There is a considerable resistance in some quarters to part with wealth and to share it with the less fortunate!

Some resources supply the *necessities of life,* such as food, clothing, and housing. Beyond these, humankind struggles to obtain material *wealth* and *comforts* that *exceed* basic needs. Moreover, people strive for better education, cultural achievements, health services, and, whenever possible, for some degree of luxury. Within reason this is good because these goals supply energy and ambitions, the ingredients to cope with life.

Unfortunately, resources are not evenly distributed over Planet Earth; however, most nations have some major resource that can be developed, used, or traded. But, modern-world money constraints, differences in economic and political philosophies, cultural priorities, glutted markets, unbearable costs of development, remoteness in location, distribution problems, price fluctuations, all these and other factors play a role in whether or not a resource can be used.

In a more restricted sense the word *resource* usually is applied to commodities that are either used in manufacturing, including food production, or that supply the energy base for such production. Some of these resources such as *wood, water, air,* and *crops* are considered **renewable.** But, whether or not a resource is truly renewable can also be determined by the time interval, or *lag time,* between its first use and the renewal process. This time span can be so long that, for all practical purposes, a **renewable resource** may take on the characteristics of a **nonrenewable** one (Brookins; 1981, p. 1). This is especially true for some water supplies.

The water that is used in the oases and industries on the Persian Gulf in eastern Saudi Arabia could be called nonrenewable *"fossil"* water. This water moves slowly through **aquifers** from the Tuwayk Mountains, about 300 miles (480 km) in the interior of the Arabian Peninsula, to the wells of the coastal oases. Precipitation in the source area averages about

2 inches (5 cm) a year. It takes an estimated 18,000 years for the water to make this migration, and it is used up faster than it is being replaced. For this reason it must be viewed as nonrenewable (Ebert; 1965, p. 496).

Wildlife, especially the big game of Africa, represents another valuable resource both from an aesthetic and economic point of view. Depending on conservation efforts such wildlife could remain renewable, or if hunted, poached, or displaced to the edge of extinction could easily turn into a nonrenewable resource [Plate 74]. How much wildlife can become a critical economic resource is reflected by the approximate $50 million left behind by tourists in East Africa in 1967 alone (Leakey; 1969, p. 189).

Other resources are definitely *finite.* These are the nonrenewable **fossil fuels** (oil, natural gas, coal) and minerals such as *metallic ores* and hundreds of natural substances that are used in modern technology; thus at some time in the future these finite resources will be gone. At that point and probably long before that moment humankind must be ready to get along without them or to find substitutes—unless we want to face universal disaster.

The use of resources increases with a growing world population. This growth has led to an energy use that by 1979 was about 20 times larger than in 1860, the first year for which overall energy consumption could be reliably assessed (Odell; 1986, p. 68). Whether this trend continues, and with it the total impact of world population growth on the earth's resource base, is difficult to analyze; both *population growth* and *rising affluence* are factors in this process (Brown; 1979, p. 6).

Energy is the critical requisite for sustaining life whether it is solar energy, metabolic energy, fuel combustion, or nuclear energy. A quick look at the more conventional energy sources, coal and oil, and their ultimate depletion will illustrate the massive problems future generations will have to solve.

Coal, of all grades, is the most abundant fossil fuel. This holds true for the United States as well as for the world as a whole. This energy source was used thousands of years ago (Bronze Age) and was also well known to Native Americans. Coal was first mined on a large scale in Newcastle, England, in 1233 (Grun; 1979, 1233G). Five hundred years later coal made its entry into the Industrial Revolution. By 1750 England was the first major user of coal; by 1850, this fuel supplied about 7 percent of all energy in the United States (Coates; 1981, p. 140). In 1980 coal supplied about 20 percent of all U.S. energy,

Plate 74. Excessive hunting in the past and the encroachment on their natural habitat have reduced the number of cheetahs to a critical few. The cheetah has been placed on the list of endangered wildlife. (Photo by C.H.V. Ebert).

whereas oil accounted for 74 percent. More than 60 percent of the coal will continue to be used in electric utilities (Shell Oil Co.; 1980, p. 16).

How long will coal last on a worldwide basis? Many unpredictable factors cloud this question. But, assuming the present annual world consumption of about 4 billion tons, the *retrievable* coal (estimated to be near 6 to 8 trillion tons) could last over 200 years; however, this very abundant energy source is not renewable.

Oil or *petroleum* is one of the most critical energy sources on our planet. The world economies virtually rise and fall on waves of oil. Modern empires have been built with it, wars were fought for its possession, and oil played a key role in the two world wars of this century. And how much tension focuses on the Middle East—and on the Persian Gulf in particular—because of oil! Oil plays a highly *strategic role* in the economic and political arsenal of nations. Over two-thirds of the (former) Soviet Union's supply of Western "hard currency" derives from rising oil exports after the oil crisis of 1973 (Flavin; 1986, p. 83).

How much longer can humankind count on worldwide oil supplies? Decade after decade the available references predicted that oil would be depleted by the 1980s, by the 1990s, by 2000. New oil fields are being discovered, and old fields go out of production. This situation creates, indeed, a confusing picture, but the bottom line is that oil resources are not renewable and are therefore finite. At the annual production rate of 21 billion barrels in 1985, the ultimate depletion appears to be between 50 to 88 years away (Flavin; 1986, p. 89). This means that we are talking about the years 2035 and 2073, respectively. The reference books to be printed in those years should be interesting! Obviously, but apparently not so obvious to many people and governments, the world must plan to meet and to function beyond the post-oil era or it will face a general breakdown.

Another critical resource is *natural gas;* it is frequently associated with regions that also contain oil. Natural gas is an efficient and clean-burning fuel that supplies almost a third of the energy consumption of the United States. It can be transported by pipelines over great distances from its place of production to where it is used. It can also be shipped in *liquefied* form, which presents very serious potential risks.

Liquefied natural gas (*LNG*), as well as petroleum gas (*LPG*), is transported in specially designed tankers. LNG, with a boiling point of −259° F (−162° C), must be shipped at extremely low temperatures and in concentrated form. In that state it occupies only 1/600th of its normal volume (Couper; 1972, p. 128). However, the potential danger involving such tanker ships is enormous. A tanker that carries 4.4 million cubic feet (125,000 m³) of LNG has an explosive power of about 20,000 tons of TNT (Coates; 1981, p. 174). This is equivalent to a 20-kiloton atomic bomb of the type dropped on Hiroshima (U.S. Dept. of Defense; 1962, p. 6). Loss of temperature control, structural failure of the metal gas containers, or a collision—all of these factors could result in a terrible disaster. It is not surprising that many ports ban such ships!

Planning for the potentially disastrous effect of the depletion of energy supplies includes the systematic exploration and development of *alternative* sources. But there is no easy solution both in terms of *technology* or in *cost.* Theoretically, total world *tidal power* could supply nearly half of the world's energy requirements, but—in reality—only a small fraction of the potential lends itself to development. *Wind power,* in theory, amounts to about 12 percent of the world's energy needs, but only a small part of this power could be used. Similar barriers, far too complex to allow for a quick analysis, confront *geothermal energy* (recovery of earth heat) and largescale *solar energy use. Nuclear power* is, of course, a very real power, real both in usability—and in dangers. With the memories of Hiroshima and Nagasaki, and the nuclear accidents at Kyshtym (1950s) and Chernobyl (1986) in the Soviet Union, as well as the threat of disaster at Three Mile Island in 1979 in this country, the overall trust of people in atomic energy is low. But in many countries, especially in France, this type of energy supplies already a considerable amount of the total power generated.

The development of additional *hydropower,* beyond the present level, is limited because of restrictive factors, such as (1) the *remoteness* of the site of power generation from the places of utilization, (2) the *high cost* of construction and of long-distance power transmission, (3) the *limits* imposed by severe climates or irregular river regimes, and (4) the desire to maintain the *beauty* of some of the world's greatest waterfalls. Niagara Falls is an example of this consideration.

Many critical minerals and other materials show signs of decline both in terms of *quality* and *quantity,* and an increasing number of industrial nations depend on rising amounts of import. No nation is truly self-sufficient in raw materials. The United

States must import around 40 essential ones. Some of these are highly strategic in nature, such as manganese, cobalt, titanium, chromium, nickel, and tungsten, in addition to about one-third of our oil needs.

Although there may not be an ultimate solution to the depletion problem of the world's resources, there are critical steps that can and must be undertaken to avoid a disaster. Such steps include at least the following:

1. Realistic conservation programs
2. Systematic recycling
3. Finding suitable substitutes for scarce materials
4. Lowering the excessive demands for luxury goods
5. Slowing down population growth with the aim of stabilizing it at ecologically and economically acceptable levels.

These goals can only be achieved if there is a *worldwide awareness* of the urgency of the situation, and if the efforts are based on effective international coop-eration. But common sense is not a very common resource!

The Waste Avalanche

To discard something means to cast something aside, that is, to get rid of it. This action implies the uselessness of what is thrown away. Unfortunately, there is no universal rule as to what is regarded as discardable, or what could still be used after repair, modification, or recycling [Plate 75]. The ease with which things are thrown out seems to be directly correlated with the wealth of a person, group, or nation engaged in this action (Plate. 76). In the poorer regions of the world many items are being repaired over and over again; they may serve for tens of years. In the affluent and technologically "*sophisticated*" countries, especially in the United States, a repairable carburetor tends to be discarded instead of being repaired. This is faster and considering contemporary labor cost much cheaper. But this practice is terribly wasteful.

Plate 75. The thoughtless discarding of refuse reflects a disregard for the environment and for the right of people to enjoy the beauty of nature. (Photo by C.H.V. Ebert).

Plate 76. Enjoying an outdoor festival is great, but leaving behind thousands of plastic cups and litter contributes to our seemingly unstoppable waste avalanche and degrades our environment. (Photo by C.H.V. Ebert).

Plate 77. The waste disposal problem shows no geographical restriction. This picture depicts a vast, desolate refuse dump site just outside of Cairo, Egypt. (Photo by C.H.V. Ebert).

The dimensions of refuse disposal are staggering, and they are growing almost uncontrollably as both populations and industrialization increase all over the world. It is frequently difficult to visualize statistical data, and to translate this information into comprehensible concepts. Some of the following accounts may illustrate the problems associated with waste removal.

Each day about 240,000 tons of materials—adding up to 87 million tons a year—are being discarded by the people of New York City. This refuse, according to a study made by the Environmental Defense Fund in 1985, contains many valuable metals and reusable items. Many unspoiled edible foodstuffs are also contained in this refuse. Moreover, this mass of waste also involves *hazardous* materials, such as mercury from old batteries and cadmium from fluorescent lights (Pollock; 1987, p. 101). Other large world cities, Beijing, Shanghai, Calcutta, Mexico City, Cairo, Tokyo, Milan, and many others, face the same problem [Plate. 77]. In the United States municipal refuse, not including sewage, makes up about 15 to 18 percent of the total waste.

The *waste avalanche* can be partially explained when we look at the often totally unnecessary amount of fancy packaging that is thrown at us. Every item from ballpen refills, batteries, household goods, toys, to thousands of other items are wrapped in cardboard and plastics. Every American discards about 660 pounds (300 kg) of packaging materials each year (Pollock; 1987, p. 103). Obviously, some packaging serves a useful purpose and is even essential, espe-

cially in the food industry when longer display or storage is required. But there is no question that our "throw away" mentality, reinforced by the very expensive policy of *planned obsolescence,* greatly contributes to the mountains of waste [Plate. 78].

The problems of getting rid of waste are complex and expensive. The proper discarding of waste in an urban area can run up costs of $50 to $70 per ton in this country. There are ways in which the volume of refuse can be reduced; one of them is *incineration.* Burning waste reduces the volume anywhere from 70 to 90 percent (Pollock; 1987, p. 105). But many valuable items and reusable materials are also being destroyed in this process.

Currently, 65 garbage incinerators operate around the country with more than 30 under con-

Plate 78. The "throw away" mentality and policy of planned obsolescence—unfortunately still a part of the American way of life—account for the unending stream of discarded items. (Photo by C.H.V. Ebert).

struction; another 250 are proposed. Many of these are mass burn incinerators, the worst possible design: they burn waste without any separation or recovery of materials. Simple incineration will not solve the solid waste crisis. Instead of leaking landfills polluting groundwater, incinerators pollute the air with *dioxin, furans* (a substance used in making certain plastics and as a solvent in the refining of petroleum-based lubricants), *heavy metals* and *acid gases.*

The dioxins are an important argument against incineration because they produce their *toxic* effects at extremely low levels, and they may form after incineration is complete. Work undertaken by Dr. Barry Commoner's research group on Long Island, New York, discovered that dioxins and furans form on *fly ash* particles in the cooler parts of the incinerator as the particulates leave the furnace and pass out through the stack (Lester; 1987, p. 8). There is also the problem of what to do with the ash residue. Fly ash from industrial power plants may contain valuable amounts of boron, potassium, and other nutrients that could be used as fertilizer (Thompson; 1978, p. 326), but ash residue from mixed municipal refuse may contain unpredictable amounts of toxic substances that preclude its further use.

The most common method for discarding waste is to place it in *landfills.* In the past many natural gullies or artificial pits were used to bury refuse without much concern about the waste materials' composition or the safety of the containment. Since the creation of the Environmental Protection Agency (EPA) in 1970 many new rules and guidelines were drawn up to protect public health and, we hope, our natural environment. But many old dump sites now come to haunt us as they are beginning to pollute groundwaters, streams, and lakes. In some countries, especially in the less-developed regions of the world, many cities discard mountains of waste outside the urban boundaries; they are creating *garbage landscapes* that attract insects and rodents. The most degrading aspect of these waste dumps is that many poor people are flocking to them; they hope to extract a few usable items and—a pitiful sight indeed—to find scraps of food.

Modern, planned *sanitary landfills* are placed in environments that are sufficiently remote and relatively safe. Factors in locating such dump sites involve many *physical factors,* such as the type of soil, groundwater levels, and the depth to bedrock. The depths of landfill pits vary but range from 7 to 43 feet (2 to 13 m). A soil layer is placed on each compacted waste layer—at a ratio of about 1:4—and the last refuse is covered with at least 24 inches (60 cm) of soil to avoid excessive infiltration of precipitation or snowmelt (Keller; 1979, p. 292). Various sources estimate that about 90 percent of the refuse generated in the United States still is discarded in landfills. New methods for handling waste materials must be found soon.

World population, urbanization, and industrialization continue to expand relentlessly, and the amount of trash increases accordingly. This situation is illustrated by Japan where the amount of refuse increased annually by about 3 to 4 percent since 1985. By 1990, the amount of waste material reached about 51 million tons.

Tokyo, with a population of over 12 million produces approximately 5 to 6 million tons a year, and the capacity of landfills and incinerators can no longer keep up with the ever increasing amount of disposals.

In view of this development, Japan is looking at the success story of one city, Numazu, which started a trash-handling system as early as 1975. Trash is divided into three basic categories: (1) *combustibles,* (2) *non-combustibles,* and (3) *recyclables.* Subsequently, the recyclables are separated into cans, bottles, sheet glass, metals, old cloth, and dry batteries (Arai; 1994, pp. 20–22).

For untold years the oceans have served as receivers of refuse and sewage. At one time this appeared safe, but the massive increase in population and industries makes this no longer possible. The sheer amount of waste materials, and the emergence of new and often *nondegradable* products, represents a serious threat especially to the **coastal ocean** (Wenk, Jr.; 1977, p. 258).

For several reasons the *coastal waters* are particularly sensitive to the dangers of pollution. This part of the ocean—representing only about 10 percent of the total ocean area—is relatively shallow because it generally coincides with the waters overlying the continental shelves. Here, the ocean depth hardly ever exceeds 600 feet (180 m). Furthermore, the coastal ocean is most directly exposed to pollution because rivers empty into it, estuaries and harbors mix their contaminated waters with that of the coastal ocean, shipping is concentrated here, and many waste dumping sites are located there. What makes this situation worse is the fact that an estimated 90 percent of all marine life is concentrated in these waters and on the bottom of the continental shelves (Heyerdahl; 1973, p. 447).

Many countries engage in dumping waste, including treated and untreated *sewage,* into the coastal waters [Plate 79]. This action is done with the erroneous idea that such materials will be quickly diluted and broken down by bacteria. It was even hoped that the marine environment would benefit from the additional "nutrients." But the rate of **biodegradation** in the coastal ocean is slower than when waste is buried on land or when exposed to the atmosphere (Davis, Jr.; 1987, pp. 402–403). Unfortunately, the United States is one of the offenders in this activity, although new legislation is beginning to restrict the worst aspects.

In the United States about 80 percent of the waste carried out to sea—mostly on special barges—is composed of harbor and river spoils derived from *dredging;* the remainder is mainly industrial waste and *sewage sludge* (Moran et al.; 1980, p. 392). In 1986 alone about 28 million tons of dredge sludge was dumped into the coastal ocean. Some of the sewage and industrial sludge contain considerable amounts of heavy metals and also have a high bacterial count (Bascom; 1982, p. 237).

In the past there were a number of equally irresponsible practices, such as the dumping of obsolete military supplies and munitions. In the 1960s a number of Liberty ships, loaded with about 53,000 tons of military scrap, were sunk offshore. Moreover, dangerous chemicals, including *rocket fuel* and *nerve gas containers,* were disposed of in the ocean; some of these ended up in the area of the Florida Current (Whipple; 1983, p. 160).

Plate 79. Several beaches and coastal waters of Mediterranean Sea are severely polluted by untreated urban sewage. In many areas valuable recreational beaches and fisheries are being degraded. (Photo by C.H.V. Ebert).

In 1972 the Marine Protection Research and Sanctuaries Act represented the first step to exercise some control over dumping waste into the ocean, at least as far as the United States is concerned. Although this is a step in the right direction, this problem is encountered in all parts of the world; it can only be solved through strict international legislation and enforcement. The mobility of the ocean waters makes this essential: *"There is no such thing as territorial waters for more than days at a time"* (Heyerdahl; 1973, p. 446).

As populations increase, and as technology and demands for "the good life" spread across the globe, the waste avalanche seems unstoppable. It appears essential that in addition to much stricter regulations on a worldwide basis a great portion of humankind's discarded waste materials must be recycled, and the wasteful use of resources must be curbed. Resistance to such policies is considerable, especially in the United States and in other technologically advanced countries. Apathy, planned obsolescence, yearning for conveniences, the lack of efficient facilities, and the apparent absence of economic rewards: These elements seem to be some of the major obstacles that stand in the way of a nationwide, and worldwide, intensive *recycling* program.

Some progress has been made abroad as well as in the United States. Examples are the *bottle return laws* first initiated in Oregon and later in New York. The result is cleaner public parks and a landscape mostly freed from empty bottles and aluminum cans. How essential this law was, just in New York State alone, is illustrated by the fact that about 400 million cases of beverage containers are sold there each year (Pollock; 1987, p. 114). That progress can be made is also shown by the Netherlands where about 40 percent of used aluminum, 46 percent of used paper, and more than half of used glass is successfully recycled. The lack of progress, on the other hand, is dramatically demonstrated by the 1987 accidental fire that involved more than 2 million old tires in a dump site near Denver, Colorado. Each ton of scrap rubber could yield more than 100 gallons of oil if tires were systematically recycled. This fact was recognized by the Oxford Energy Company near Modesto, California, which located a power plant next to a dump site where millions of tires had been discarded over the past decades. These tires will serve as fuel for the plant and will be converted to useful energy.

This chapter gave an overview of humans' increasingly detrimental role in the natural environment. This role is no longer in harmony with the ecosphere because humankind has removed itself, step by step, from a much more balanced relationship to nature as it existed in previous times.

Humans' unwise actions and attitudes have led to a number of global concerns, such as the ominous growth of population, the question of adequate food supply, the potential shortages of resources, and the ever-increasing impact of waste disposal. All these concerns grow *exponentially* and could drastically lower the *quality of life* on our planet and bring about a general environmental deterioration. The broader and some specific problems—at times interrelated—are presented in the subsequent chapter.

ENVIRONMENTAL DETERIORATION

Key Terms		
environmentalism	sea level	"ozone hole"
thresholds	eustatic process	Montreal Protocol
soil salinization	colloidal dispersion	phytoplankton
ultraviolet radiation	photochemicals	nerve gas
sea breezes	buffering capacities	deep-well injection
prevailing westerlies	ions	half-life
hygroscopic particles	cations	vitrification
pH	chlorofluorocarbons,	leachates
acid precipitation	CFCs	swales
greenhouse effect		

The preceding chapter pointed out that humankind has placed more stress on nature during the past 80 years, or so, than during its entire prior existence on earth. This statement should not be taken to say that humankind's interaction with the natural environment is a very recent phase in earth-human history. What it does state is that technologically advanced systems have become a major force in the shaping of the total ecosystem of our planet.

By itself this fact cannot be labeled "good" or "bad"; it exists as a *new integral component* of Earth's evolution. The evaluation of this relationship between humankind and the environment depends on the specific issues in question, and on how reasonably *Homo sapiens* will act in view of the persistently growing global problems presented in the preceding chapter. How will we deal with world population growth, world food supply, potential exhaustion of natural resources, with the waste avalanche, and with the deterioration of the environment?

When the first departments of geography became established in this country, shortly after the beginning of this century, their self-proclaimed charge was to function as a bridge between the natural and the social sciences. Many geographers tried to find their professional arena within this theme but, as is true for most trends, the initial enthusiasm faded and "*innovators*" took their place. A general retreat from **environmentalism** began in the 1920s, and by the end of the 1930s almost the entire research program initiated before World War I was abandoned (Mikesell; 1974, pp. 2–3).

For one reason or another the 1960s sparked in the United States an awareness—and one of unprecedented intensity—of what could be called the "*environmental crisis.*" Many factors contributed to the general mood of "*malaise,*" a feeling of frustration and discontent. It vented its tension through vehement protests against the "*establishment*" as well as against the perceived—although quite real at times—disregard for human welfare and disrespect for nature. The literature, as well as the contemporary speeches, centered on themes such as *global doomsday* and *national guilt.* Events such as *Earth Day* helped to direct the public's attention to these issues. Though some of these outcries degenerated into mere rhetoric offered by self-appointed "experts," most of the concerns and feelings were real and honest; they often acted as the driving force and catalyst for more appropriate, focused, and more effective action. Responsible scientists from all over the world presented serious, professionally sound, and sobering data and analyses. The scientists recognized the potential threats to the natural environment and warned humankind about the real dangers that exist around us; such dangers were easy to see for all who wished to see them. One major problem with threats to the environment is that many processes and resulting dangers are not always visible.

The partial reason for this is that *natural systems*—such as the atmosphere, the oceans, soils and vegetation, as well as the human body itself—can tolerate stresses for a long time without revealing overt symptoms of injury or recognizable changes. The oceans, which cover about 71 percent of the earth's surface, seem so immense that it is hard to imagine that dumping sewage or chemicals into this limitless expanse of water could do any harm. Each individual polluter thinks he or she has done an acceptable act. The blue skies above us look so pristine and unspoiled that the release of "a little" sulfur dioxide or submicroscopic dust particles does not seem to be detrimental. However, as we will see, many final effects are the result of inconspicuous *cumulative* processes, and minute amounts of toxic substances are capable of doing irreparable damage. The critical point is reached when conditions suddenly deteriorate. Scientists do expect such sudden changes, called **thresholds** or *jump events,* but they can rarely pinpoint when they will occur (Postel; 1986, p. 7).

Another general and dangerous perception is that humankind, through technology and scientific advancement, has achieved a *"master-over-nature"* position, which seems to validate an increasing degree of independence from nature. This viewpoint is conceived in arrogance, born of the Neanderthal age of biology and philosophy, when it was supposed that nature exists only for the convenience of man (Carson; 1962, p. 297). The intimate dynamic interrelationship between humans and the natural environment is inseparable and therefore must remain mutually beneficial.

This interrelationship was clearly stated: "Technology and natural environmental processes together—not technology alone—have permitted the human population and its material consumption to reach their present levels. Today human beings continue to depend on the nontechnical environment for a variety of services that technology not only cannot replace, but without which many technological processes themselves would be nonfunctional or unaffordable" (Ehrlich, Ehrlich, Holdren; 1977, p. 9). In other words, *we need nature*—air, water, plants, soil—to help us in many ways, and at a scale that we could never achieve alone. Plants and oceans absorb excessive carbon dioxide, soil particles adsorb harmful substances; bacteria, practically invisible helpers, decompose waste and restore available nitrogen that subsequently is cycled through the food chain. But how long can nature handle a growing input of waste materials—many being highly toxic—without reaching the threshold of destruction or sterility that will lead to a disastrous breakdown of the systems that we need for survival?

Previous parts of this book demonstrated the more overt "wear-and-tear" on our planet: excessive deforestation and depletion of water resources in the Sahel; the destruction of the tropical rainforests; the erosion of agricultural land; the careless introduction of alien animal and plant species; the massive extraction of mineral resources; and the overwhelming avalanche of waste materials. Some of the processes that will be discussed in the following sections deal with somewhat less tangible aspects of environmental deterioration. But in some cases these processes have already proven to be deadly phenomena; some events caused disasters that destroyed many lives. Environmental deterioration runs the entire gamut from mere discomfort and annoyance to lethal threats. The first important step is to understand the causes of this deterioration, but the greatest need—and possibly the last effort that can be made before it is too late to reverse the trend—is to develop the determination and the means to stop it or at least to contain it within survivable thresholds.

Humankind: A Powerful Force of Change

On a typical acre of pasturing land in the midlatitudes live an estimated 300 to 400 million insects that are part of the existing ecosystem. Most of their activities such as feeding, breeding, and dying are invisible to us, but their existence is part of a larger natural process on which total Earth life depends.

For the world as a whole, there may be as many as one million insects for each human being, and the total accumulated weight of insects probably amounts to nearly 12 times as much as all of humankind. Just as we cannot see most of the insects that populate our planet, so the occupants of a spacecraft, circling the earth, could not observe any of the 5 billion humans that are engaged in their life activities. But humans—now more powerful in influence than some of the natural processes—are marking the surface of our planet to a degree that seems to be totally out of proportion to our size. The human brain spawned this ability; human ingenuity has led to the technology that may determine the ultimate fate of our earth and its inhabitants. In less than a minute pesticides, possibly sprayed from a low-flying airplane, can wipe out most of the insects on the acre of grassland mentioned above!

Humans' constructive and destructive works can be seen from space. Astronauts have marveled at the glittering sea of lights emanating from the great cities of the world. Some of the space travelers were able to identify the Great Wall of China, while others looked at the veil of dust extending westward from the Sahara over the Atlantic Ocean; a great amount of this dust originates from the drought-stricken Sahel. Astronauts also observed immense scars of open-pit mining, burning forests, and also the browning and dead expanse of millions of trees in Southeastern Asia; their life-giving foliage had been destroyed by tons of *herbicides,* toxic defoliants, that had been sprayed on their canopies. Humankind, indeed, has become a powerful agent on Earth.

The preceding chapters have already illustrated some of the dramatic events in the biosphere that were partially, or fully, the result of humans' activities. But there are many more ways in which humankind has altered the natural environment in the past and is doing so at the present. In modern times humans have engaged in construction works that not only brought on major ecological changes, but in some cases totally obliterated the existing landscapes.

Immense irrigation projects are diverting water from rivers at such a scale that some bodies of water, previously sustained by these rivers, are beginning to shrink dramatically and could be doomed in the foreseeable future. The massive diversion of water from the Sea of Galilee (Lake Kinneret) in Israel, and from the Yarmuk River in Jordan, has drastically reduced the flow of the Jordan River; this has led to an accelerated lowering of the Dead Sea [Plate 80]. In a similar way, the large-scale use of water taken from the Amu Darya and Syr Darya is endangering the Aral Sea in the Soviet Union. As huge amounts of water are used for irrigation, the Aral Sea is slowly changing into an unusable body of salt water.

According to I. P. Gerasimov, a Russian scientist, the level of the Aral Sea, a *landlocked* body of water with an area of about 24,700 square miles (64,000 km^2), dropped about 10 feet (3 m) between 1960 and 1975 (Symons; 1983, p. 80). If the present climatic conditions and the projected use of irrigation water continue, the water level in the Aral Sea could drop an additional 28 feet (8.5 m) by the year 2000. The shallowness of this water body, averaging from 33 to 66 feet (10 to 20 m), accelerates salinization. This increased salt content had already a very detrimental influence on the local fisheries, which declined from an annual total catch of 40,000 tons in 1962 to about 6,000 tons in 1970 (Lydolph; 1977, p. 355), a decline of 85 percent in eight years!

The impacts of other projects have already proved to be equally severe or forecast bad results for the future. Well known among these is the deepening of the Suez Canal after 1956. The canal was deepened to allow larger tankers to pass through this waterway. As more water began to flow through the enlarged canal, the concentration of salt in the Bitter Lake—formerly ranging from 8 to 10 percent as

Plate 80. The Dead Sea is shrinking steadily because of excessive evaporation and too much water being diverted from the rivers that flow into this landlocked body of water. Heavy salt encrustations on the shore reflect the very high salt content, which ranges as high as 27 percent. (Photo by C.H.V. Ebert).

compared to the average of about 3.5 percent for the world oceans—was markedly reduced. This allowed unwanted fish species and other marine organisms to migrate from the Red Sea into the Mediterranean. Another problem arose from the deepening of the Welland Canal, Canada, which connects Lakes Ontario and Erie. After the canal was deepened in 1932, the *lamprey eel* engaged in a massive migration into Lake Erie and endangered the local fisheries. Chemicals were later used to kill the lamprey larvae, a questionable step from an environmental point of view.

The construction of the Aswan High Dam on the upper Nile in Egypt is another example revealing the unfavorable impact on an ecological system. The author visited this region in 1985 and discussed some of the problems with local officials. The huge dam was once labeled "*protection against hunger*"; it was designed to hold adequate amounts of water for a period of seven years. This storage is necessary because of the considerable fluctuation of precipitation in the highlands of Ethiopia, which is the source region for the Blue Nile. The envisioned holding capacity of the reservoir, Lake Nasser, was about 247 billion cubic feet (7 billion m³), which was able to boost greatly Egypt's output of electricity and increase vastly the irrigated areas along the Nile river (Chalaby; 1981, pp. 111–112). However, several problems had not been anticipated: The reservoir could not be filled to the designed capacity because water seeped into *pervious sandstone* formations; fertile silt that was traditionally deposited on the agricultural lands bordering on the lower Nile became trapped behind the dam. Accordingly, sardine fisheries offshore the Nile Delta were deprived of mineral and organic nutrients; high evaporation rates from Lake Nasser, with an area of about 1,500 square miles (3,880 km²), increased the mineral concentration and lowered the quality of the irrigation water used downstream.

The author observed another side effect of bringing almost 1.5 million additional acres (607,000 hectares) under irrigation: The water table, including that of the immediate Cairo area, rose markedly. This effect has already caused flooding of basements in some districts and also makes proper soil drainage more difficult. There is another severe danger: **soil salinization** that will be followed by declining crop yields.

The examples presented above could be augmented almost indefinitely and clearly illustrate the far-reaching and sometimes unpredictable effects that resulted from humans' tampering with nature. However, one of the most drastic and continuing modifications of the natural environment is the conversion of rural areas into a new and totally different form: *the city.*

Urbanization: Creating a New Habitat

Early cities, the places where "*citizens*" live, evolved about 7,000 years ago in the region of Mesopotamia, the "*land between the rivers,*" which is located between the Euphrates and the Tigris. Cities served many purposes: the worship of gods; the concentration of power; the protection of people, wealth, and food stores. The early cities were often a focal point of trade. The city, probably from its beginning days to the present, acted as a magnet. The flow of rural inhabitants toward the urban environment has always been fostered by either perceived or by real opportunities for economic or social advancement (Stoddard, Blouet, Wishart; 1986, p. 276). Nature may offer shelter for a few, but nature has no cities: They are entirely human-made, and all aspects of the urban center, for better or for worse, are of human creation. A city is the greatest result of construction (Menard; 1974, p. 354).

Nowadays in the United States the city seems to offer a good life, unless one is poor. Clean water flows from the tap, air conditioning and heating keep people comfortable, and a great variety of food fills the stores. But urban safety and comfort hang on a thin thread, although most urban dwellers do not perceive this. How thin this thread really is can easily be seen during power failures, when a city is paralyzed by a blizzard, or when a strike by garbage workers allows growing waste piles to rot in the streets. A rather sobering question is: What would happen to huge cities—such as New York, Los Angeles, Chicago, London, Tokyo, Moscow, Mexico City, or Cairo—if for just two weeks the supply of water, electricity, fuel, and food were cut off?

It is obvious that cities represent an *artificial environment* that contains an incredibly concentrated mass of people. This concentration constitutes a totally *new habitat* with unique demands and conditions. All environmental problems seem exacerbated by cities (Wagner; 1971, p. 360). Moreover, no city can sustain itself. The requirements for food, water, and fuel are of a massive scale not found anywhere in nature. Nor can nature disperse the waste produced in cities (Brown, Jacobson; 1987, p. 50).

The concentration of people in cities represents a market for both goods and services; arts and crafts focused early on the urban environment. It was quite natural that the Industrial Revolution, with its rapidly growing demand for raw materials and labor, triggered urban growth. As less-developed nations begin to industrialize, *urbanization* gains momentum, and with urbanization come all the problems associated with cities. How will Mexico City, unable to handle its present problems, brought on by almost uncontrollable growth, deal with an estimated influx of 1,000 people a day and with a projected population of near 40 million by the year 2034? How will other rapidly growing cities fare, such as Cairo, Djakarta, or Shanghai? They may reach populations of 17 to 39 million by the same year. The role of these urban giants, as far as political dominance and the setting of national policy are concerned, will be increasingly disproportionate compared to the rest of the country. Will the "vision of urban development" turn into a true nightmare?

According to AID (Agency for International Development), internal migration (*rural to urban*) is more critical in the growth of urban population compared to international migration (Echols; 1981, p. 20). By 1800, less than 3 percent of the world's population lived in cities. By 1950, within 150 years, the urban population had increased to about 38 percent; it is to reach at least 51 percent by the year 2000.

By the 1970s, it was North America that led the world in urban population with almost three-fourths of its people living in urban areas. This, according to AID, compares to 62 percent for Europe, 57 percent for Latin America, and a low of 20 percent for Africa. Because of a growth rate of 2.5 percent per year (half again as fast as for world population growth) the number of people living in cities will double between 1986 and 2014 (Brown, Jacobson; 1987, p. 38).

In addition to the physical problems encountered in large cities, such as proper housing, food supply, heating, and sewage removal to mention just a few, there are also growing social problems brought on by the urban life-style. David Clark expresses this well by saying: "In contrast to the close-knit, intimate, and stable relationships which were found in rural society, lifestyles in the rapidly growing twentieth century metropolis were impersonal, segmented, and transitory" (Clark; 1985, p. 19). It is not surprising that stress, mental health problems, social discontent, and crime have become universal characteristics associated with the basically unnatural and often overwhelming population densities of large cities!

Urban Weather and Climate

The protective screen set up by air conditioning, heating, air filtration, humidification and dehumidification, tinted glass panes, and by airtight windows shelters many urbanites from the marked changes that a city can make in the physical environment in general and in the local atmospheric conditions in particular. These changes not only affect the sensory well-being of city dwellers, but the changes can also lead to conditions that threaten the health of city populations.

There are *three components* that are most influential in producing these changes: (1) the *mass of buildings* (concrete; brick; steel; tarred roofs), which absorb huge amounts of solar heat energy, (2) the *covering of soil,* which renders the surface impervious, combined with the removal of the vegetation, and (3) the *release of heat radiation* and of *pollutants,* which affects the atmosphere.

Building materials, especially those that have a low albedo (*reflective power*), absorb large amounts of heat. Long after sunset, after a warm summer day, the sides of a building and the surface of a parking lot still feet warm to the touch and radiate heat. In suburban areas (where usually only from 25 to 45 percent of the surface is occupied by buildings, streets, and driveways) a considerable amount of the incoming solar heat is consumed by the evaporation of water from soils and plants instead of heating up the mass of buildings in the inner city. The absorption of heat in the city can make the overall temperatures up to 4° F (2° C) warmer than those of the more open areas. Maximum temperature differences between cities and the open countryside may range as high as 10° to 15° F (5.5° to 8.3° C). This difference in temperature can be found from the surface to almost 1,000 feet (330 m) elevation (Neiburger, Edinger, Bonner; 1973, p. 494).

The buildup of excess heat in the *urban core area* is further aided by the fact that rainfall, which is normally absorbed by soils and vegetation, is not used for cooling through evapotranspiration (the combined loss of water by evaporation from the earth's surface and by transpiration from plants). In the *inner city,* where from 85 to 100 percent of the surface is impervious, the water flows into the storm sewers or is rapidly "boiled away" from the hot pavement and rooftops. The combination of these factors accounts for the reason that city air is both warmer and more humid (Schmid; 1974, p. 224).

In addition to the *reradiation* of heat by structures, additional energy from industrial processes, use of utilities, and from transportation systems is also released into the city air. Even the energy of human metabolism adds a little to the total heat output (Neiburger, Edinger, Bonner; 1973, p. 257).

The total heating produced by a city results in building up an area of markedly higher temperatures; this phenomenon was previously identified as a *heat island* in connection with aviation hazards (chapter 9). During calm atmospheric conditions, particularly when the skies are clear at night, the ground radiates surplus heat back to space. This process cools the lower air layer, but a cap of *residual* and much warmer air remains on top. This abnormal temperature stratification is called a *temperature inversion;* it is a very typical occurrence with large urban areas. The presence of a temperature inversion traps pollution that can be found over most cities, particularly on windless days (Fig. 14.1).

The combination of various types of *gaseous* and *particulate* emissions from many different sources, including automobiles, buses, smokestacks, and incinerators, accounts for the fact that city air is generally unhealthful and may, in areas of severe pollution, lead to a variety of respiratory ailments. Air in urban areas contains, on the average, 10 times more dust, 5 times more sulfur dioxide, 10 times more carbon dioxide, and 25 times more carbon monoxide than the air of suburban regions (Navarra; 1979, p. 487). Temperature inversions can trap such pollutants for several days.

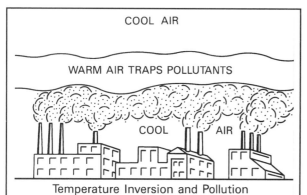

COOL AIR

WARM AIR TRAPS POLLUTANTS

COOL   AIR

Temperature Inversion and Pollution

Figure 14.1. An abnormal atmospheric temperature stratification (temperature inversion), with a layer of warm air overlying a layer of cooler ground air, prevents pollutants from rising and diffusing. This unhealthful situation often develops over industrial cities on windless days.

In the days before legislation in this country, and in some other nations, curbed dangerous emissions, such situations caused chronic health impairments, hardship, and even deaths. When a city is located in a *topographic depression,* such as in valleys or mountain basins, atmospheric pollution can be at its worst because the dirty air is virtually contained. Cities that have this particular setting include, among many others, Los Angeles—once called the "smog capital of the world"—Mexico City, and Guatemala City [Plate 81]. Although the 1970 Clean Air Act has helped to start the fight for clean air in this country, many problems remain: legal loopholes, lack of enforcement, and repeated extensions of compliance deadlines.

In addition to the problems associated with pollutants, the haze and particles suspended in the air may drastically reduce the amount of sunshine in urban areas. Climatologist H. Landsberg estimates that **ultraviolet radiation** in cities during the winter is as much as 30 percent lower than in nonurban areas (Neiburger, Edinger, Bonner; 1973, p. 493). Similar reports have come from Poland, especially from the industrial region of Upper Silesia, where air pollution cut sunshine by an estimated 20 percent.

A pollution disaster that struck London in December of 1952 was one of the worst in modern times. At that time Londoners still burned *high-sulfur coal,* both in power stations and in private fireplaces. During a persistent temperature inversion, and under almost windless conditions, a deep layer of pollutants built up over the city. The daily output of air pollutants was later estimated to have been: 1,000 tons of smoke particles; 2,000 tons of carbon dioxide; 140 tons of hydrochloric acid; 14 tons of fluorine compounds, and, probably the most damaging compound, 370 tons of sulfur dioxide (Whittow; 1979, p. 330). More than 4,000 people of all ages, in excess of the normal death rate, died during the pollution period.

Severe air pollution is much rarer in regions that experience adequate "ventilation." These are places that enjoy either well-developed **sea** or *lake **breezes*** (e.g., Seattle, Chicago, and Buffalo) or are exposed to higher average wind velocities typical of the midlatitudes under the **prevailing westerlies.** According to local climatological data from the National Oceanic and Atmospheric Administration (NOAA; U.S. Department of Commerce) several American cities have average wind velocities in excess of 10 mph (16 km/h). Some of these are Bismarck, North Dakota (10.3 mph or 16.5 km/h), Chicago, Illinois

Plate 81. Los Angeles' smog pollution is not restricted to the immediate urban area but drifts for many miles into neighboring valleys. (Photo by C.H.V. Ebert).

(10.3 mph or 16.5 km/h), Buffalo, New York (12.1 mph or 20 km/h), and Minneapolis, Minnesota (10.5 mph or 16.8 km/h). Wind can be annoying and may induce a strong chill factor (see chapter 8 on blizzards), but wind surely helps to clear the air!

Changes in the Air Around Us

There is no "*clean air*" or "*pure water*" on our planet, except when produced under sterile laboratory conditions. The "natural" atmosphere and "*natural*" precipitation, to use this highly overused and often misinterpreted word, contain many solid (particulate) and gaseous substances that are there without any influence from humankind's activities. Moreover, without these substances in the air and in precipitation, many biochemical functions and life-supporting processes could not take place. Air contains innumerable submicroscopic particles (e.g., sulfur, carbon, salt, and dust) that serve as necessary condensation nuclei (**hygroscopic particles**) needed to produce precipitation. Our "pure" rainwater is actually a mild acid with a normal **pH** around 5.6; it contains a number of acids, such as carbonic, sulfuric, and nitric acids. These acids, in great dilution, assist in breaking down minerals so that fresh *soil nutrients* become available. This overall cycling of inorganic and organic materials, active over millions of years, is a critical part of maintaining a viable base for life. But, as was already explained in the preceding sections, humans have entered the picture and now influence nature in a disproportionate way. Much of this can be observed in the atmosphere: "Scientists now expect that a buildup in the atmosphere of certain carbon, nitrogen, and chlorine compounds will change the earth's climate more over the next 50 to 75 years than it has changed over the last 15,000 years" (Postel; 1986, p. 5).

Within the scope of this book it is impossible to go into all aspects of human-induced changes in the atmosphere. Many of such changes are not even fully understood, and we can only speculate about their ultimate consequences. But within the array of potential threats two dominant dangers are the *increase in carbon dioxide* (mainly because of burning fossil fuels and agricultural activities) and the phenomenon *acid precipitation.*

Carbon Dioxide: A Growing Threat

The balance of carbon dioxide on Earth is achieved by storing and releasing this gas, involving the oceans, the biosphere, and the earth itself. The exchange is governed by *equilibrium conditions* within the total system. Humans' interference with this total system is of relatively recent origin; by comparison to the quantities involved in natural systems, the amount contributed by humans is relatively small, but growing. On an annual basis the two major activities that affect the carbon dioxide balance, *agriculture* and the *use of fossil fuels,* add about 2 billion and 6 billion tons to the atmosphere, respectively. This amount is about 8 percent of the carbon dioxide released by volcanoes and hot springs (Plass; 1959, pp. 6–8).

But this rate of CO_2 release has increased since 1959. Between 1959 and 1985, the annual average concentration in the atmosphere increased from 316 ppm (parts per million) to about 345 ppm, or by 9 percent. The carbon dioxide level before the year 1860 was estimated to be near 265 ppm. This indicates that humans' activities increased the concentration of carbon dioxide by about 30 percent in 125 years (Postel; 1986, p. 9).

Burning fossil fuels is the major human addition of carbon dioxide. One 1,000 megawatt coal-burning power plant emits about 600 pounds (270 kg) per second, or about 32,000 pounds (16 metric tons) each minute. As yet, there is no economically effective way to capture any significant fraction of this gas (Hayes; 1979, p. 10).

The Greenhouse Effect

The most important aspect associated with the world's overall increase in atmospheric carbon dioxide is the intensification of the ***greenhouse effect*** and with it the warming of the earth's atmosphere.

The greenhouse effect results from the trapping of *terrestrial infrared radiation* by carbon dioxide as well as by *water vapor, nitrous oxides, methane, chlorofluorocarbons,* and *ozone*. However, the main culprit appears to be CO_2. These gases allow most of the incoming solar radiation to pass through our atmosphere and reach the earth's surface where it is changed to heat; the heat, in turn, is radiated back in the form of infrared radiation.

The issue of the greenhouse effect is complex and involves many contributing factors; it is not surprising that specific data in the literature vary considerably. Many conferences have been held in recent years, and the debate continues. Conservative scientists used to believe that no significant climate changes would occur before the last part of the twenty-first century. Now they believe that the changes will happen sooner, and some believe they are already underway (Milbrath; 1989, p. 34).

All indications are that the earth's climates will warm up if the present trends in human-induced CO_2 generation continue. If fuel consumption proceeds at the present rate, more than a trillion tons of carbon dioxide will be added to the atmosphere by the year 2000. As early as the late 1950s it was calculated that this addition of CO_2 alone would increase the earth's average temperature by 3.6° F (2° C) (Plass; 1959, p. 8). More recent conclusions certainly do not instill optimism about the present trends and those projected for the future.

The Intergovernmental Panel on Climate Change (IPCC) issued a comprehensive report in 1990; this report confirmed that the greenhouse effect is with us now and is keeping our planet warmer than it would otherwise be. The report states that it is carbon dioxide that is responsible for more than half of the enhanced greenhouse effect in the past, and it is likely to remain the main factor in the future.

The expected rate of increase of the *earth's mean temperature* per decade during the next century is projected at about 0.54° F (0.3° C) and possibly as high as 0.9° F (0.5° C). Land surfaces are expected to warm more rapidly than the ocean, and high northern latitudes will warm more than the global mean in winter (IPCC; 1991, p. xi). Other computer-based predictive models reveal the possibility that the average global temperatures could increase by about 9° F (5° C) by the year 2100 due to the intensification of the greenhouse effect only (Eagleman; 1991, p. 184).

To what extent the other greenhouse gases will affect the future changes in the earth's temperatures still is somewhat uncertain; this concerns especially the role of water vapor, which has not yet been addressed as thoroughly as the role played by CO_2. The dynamics of the water vapor, constantly being evaporated and condensed, make a precise assessment of its role more difficult as well as more controversial; however, with the projected warming the average water vapor content of the atmosphere should increase: This increase would certainly intensify the greenhouse effect.

A warming climate would increase the air's capacity for holding moisture. This increases the potentials for (1) more precipitation, and (2) more intensive evaporation. Recently developed models anticipate a marked summertime drying of the mid-latitude regions of the Northern Hemisphere. This drying effect would severely influence the grain-producing areas of the United States and Canada; these regions, according to the Food and Agriculture Organization (U.N.), account for more than 50 percent of the world's grain exports. On the other hand, the warmer and wetter conditions expected for India, and for most of Southeast Asia, could increase rice production in that part of the world (Postel; 1986, p. 10).

In addition to warming the earth's atmosphere and land surfaces, the greenhouse effect will also result in the warming of our oceans. If, for example, the average temperature of the oceans would rise by only 1.8° F (1° C) the ocean level would increase by about 6.6 feet (2 m); this change in *sea level*—based on volume increase caused by higher temperatures—is called a *eustatic process* and would severely affect the world's shorelines (Thurman; 1990, p. 210).

A rise in world temperatures would accelerate the melting of glaciers and of the ice caps of Antarctica and Greenland. The maximum rise of the oceans, assuming that all ice would melt, is generally estimated at about 140 to 160 feet (42 to 50 m); this does not consider shoreline changes brought on by either tectonic or eustatic processes. This potential rise means sea level would cover a very large portion of the present coastal plains of the continents, and these lowlands tend to be densely populated and industrialized. New York, Amsterdam, Yokohama, and Calcutta—just to mention a few major cities—would become relics under the sea!

Another and equally important aspect of a warming ocean is the effect on its ability to absorb gases: A warmer ocean can dissolve less carbon dioxide; an excess amount of CO_2, would have to be released into the atmosphere and would further enhance the greenhouse effect. On the other hand, an increase of total ocean surface would enlarge the interface between atmosphere and water, which would increase the exchange capability with gases.

The foregoing underscores the delicate and ever-changing equilibrium that results from the interaction of (1) the volume of the ocean, (2) the temperature of the ocean, and (3) the amount of carbon dioxide in both ocean and the atmosphere. It is believed that at this time the oceans exchange an estimated total of 200 billion tons of carbon dioxide with the atmosphere each year (Plass; 1959, p. 3). If the two-way exchange involves the same amount, the ocean-atmosphere system is in equilibrium, but fluctuations lead to short-term imbalances. It is estimated that *ocean-atmosphere equilibria* exist for time periods longer than 1,000 years (Eagleman; 1991, p. 178).

An interesting question is whether any factors could counteract the influence of greenhouse gases. The IPCC reported the possibility that sulfur emissions into our atmosphere—leading to the formation of sulfur haze—could partially offset greenhouse effect warming. Richard Monastersky summarizes this possibility in Science News:

> Atmospheric experts have long suspected that sulfur haze—the same kind that obscures skylines in much of the industrialized world—could exert a cooling effect by reflecting sunlight. Yet they are only now recognizing sulfur's potential power.
>
> Formed during the combustion of fossil fuels, sulfur pollution pours from the same smokestacks and tailpipes that belch out 6 billion tons of carbon dioxide each year. The sulfur dioxide wafts into the atmosphere, where it eventually turns into tiny sulfuric acid 'aerosols' which can be either droplets or particles.
>
> Sulfur aerosols tug on the climate in two ways—one direct, the other more subtle. Aerosols exert their most obvious effect by reflecting incoming solar radiation back toward space. They wield indirect climatic power by serving as nuclei around which water vapor can condense to form sunlight-reflecting cloud particles. In both cases, aerosol pollution acts as a giant shade, reducing the amount of light reaching the Earth's surface.
>
> Over the last few years, Charlson [Robert J. Charlson is an atmospheric scientist at the University of Washington] and other atmospheric experts have gradually realized that the aerosol effect warranted consideration. They wondered: Could the sulfur shade actually block enough of the sun's energy to slow the greenhouse warming? To answer that question, researchers would need to estimate the power of aerosols and compare it with that of greenhouse gases.
>
> (Monastersky; 1992, p. 232) (From Science News, the weekly news magazine of science, copyright 1992 by Science Service, Inc.)

Whatever the future development will bring, the clearly documented overall increase of human-generated carbon dioxide will have serious and diverse impacts once certain critical thresholds are reached; the speed with which we are approaching these thresholds is on the increase. It is possible, if unchecked, that this route will take us to global disaster.

Air Pollution and Acid Precipitation

The overuse or careless application of a word, or of a slogan, can reduce the effectiveness of such expressions; this certainly is the case with the word *pollution.*

The mere presence of a substance, or of a condition, by itself does not have to have any serious consequence even though the substance may be toxic. To be a pollutant, a substance must occur in sufficient concentration, over a critical period of time, and must—above all—*degrade* the *quality* of something else to the point of making it useless or actually destroying it. The "something else" could be air, water, foodstuffs, or the health of an organism. Again, pollution has critical *threshold levels* that determine the severity of the condition.

The subject of atmospheric pollution is extremely complex and at times quite controversial. For example, carbon dioxide itself is not viewed as a "*pollutant.*" However, as we have seen in the preceding section, its rising concentration poses a potential threat to the earth system and to human activities. Another complicating factor is that substances emitted into the atmosphere will change the characteristics of the atmosphere itself; these changes, in turn, will affect the very substances that are involved. The third difficulty in assessing air pollution is that in many cases the source of emission is not a point, such as a factory or a smokestack, but a composite of many sources and of different materials. Moreover, the *long-range transport* of air pollutants, and the resultant deposition, is now well recognized as having serious environmental consequences (U.S.-Canada Research Group LRTAP, Report; 1980, p. 3). The term *long range* covers both *continental* and *intercontinental* transport as well as the *stratospheric diffusion* of pollutants.

A great variety of substances (such as sulfur dioxide, hydrocarbons, nitrogen oxides, carbon monoxide, ozone, metals, and acids) can reach concentrations to become hazardous pollutants. Basically, two *physical states* of such pollutants are recognized: (1) *particulate matter* (dust, smoke, aerosols, and fogs) and (2) *gases.* Particulate pollution represents **colloidal dispersion** of both solids and liquids of organic and inorganic substances. A colloidal dispersion is a suspension of finely divided particles that do not settle out.

In addition to corroding many valuable materials (e.g., metal surfaces, concrete structures, marble and limestone artworks and stone facings, textiles, and painted surfaces) the impact of air pollutants on living things can be devastating. Such effects include many *respiratory illnesses* such as emphysema, bronchitis, or bronchial asthma, and the death of plants and animals [Plate 82].

Moreover, a complex pollution phenomenon—mostly created over urban environments during clear weather conditions—involves the development of **photochemicals.** The danger of photochemicals, also referred to as *photochemical smog,* is a combination of physical and chemical problems: severe irritation of eyes and lungs with potential *carcinogenic* effects (*cancer-producing*) in cases of chronic exposures; generation of dangerous levels of ozone, which are capable of destroying vegetation; and a sharp reduction of visibility, which affects metropolitan air traffic. The latter is frequently encountered over the Los Angeles Basin and over Mexico City.

The composition of photochemicals is not fixed, but harmful concentrations of *nitrogen oxides* and of ozone are usually involved. Nitrogen oxide, a residue of gasoline and petroleum products combustion, is a *toxic irritant* of a brownish color. In the atmosphere, nitrogen dioxide (NO_2) may combine with water to form nitric acid (HNO_3). In the presence of strong sunshine, more specifically, exposure to ultraviolet radiation, the NO_2 splits into nitric oxide (NO) and oxygen (O). The oxygen then combines with molecular oxygen (O_2) to form *ozone* (O_3). Ozone has a bluish color, an irritating odor, and can damage plants; it has already killed thousands of pine trees in the San Bernadino Mountains near Los

Plate 82. Acidic air pollutants from this metal processing plant caused a variety of respiratory illnesses and considerable damage to vegetation. (Photo by C.H.V. Ebert).

Angeles (Postel; 1984, p. 12). It seems ironical that the same gas, in the form of the ozone layer, protects us against excessive ultraviolet radiation, but it becomes a nuisance, and even a destroyer, at lower elevations. Exhaust emissions control devices that are now required in the United States, and are also being introduced in other countries, will help to reduce the generation of hydrocarbons and nitrogen oxides at least as far as gasoline engines are concerned.

Although photochemicals are a serious problem in sufficient concentration, acid precipitation (*acid rain*) has become very prominent and controversial in recent years; it threatens the environment in many parts of the world. This threat is human-made. How do we know this?

Precipitation that fell before the Industrial Revolution is well preserved in glaciers and in the *continental ice sheets* that are still partially present on Greenland and on Antarctica. Studies by Swiss workers in Greenland revealed that the pH of the ice, deriving from snow that fell about 180 years ago, had a reading of 6 to 7.6. The lowest pH of precipitation measured in modern times was recorded at 2.4 during a rainstorm in Scotland in 1974. That acidity corresponds to that of vinegar (Likens et al.; 1979, p. 3)! The average pH of rainfall over most of the eastern United States is less than 4.5 (EPA, 6000/8-79-028; 1979, p. 4). From where does this acidity originate?

It was said before that normal pure rainwater averages a pH of 5.6. This mild acidity derives from a variety of acidic substances that normally find their way into the atmosphere. Of these the dominant ones include *sulfuric acid* (H_2SO_4), *nitric acid* (NHO_3), and *carbonic acid* (H_2CO_3). But these naturally occurring acids could not overcome the **buffering capacities** of the earth as offered by the oceans, alkaline rocks, soils, and by the atmosphere itself. But the massive continuous output of sulfur dioxide and nitrogen oxides from human-made sources overwhelms the system. According to various government sources, the annual sulfur dioxide emission in the United States and Canada was 21.1 and 4.77 million metric tons in 1980, respectively. Electric utilities accounted for 66 percent of this emission in the United States, and for 16 percent in Canada. At the same time, nitrogen oxide output amounted to 19.3 million metric tons for the United States and 1.83 million metric tons for Canada. In both countries transportation contributed the lion's share with 44 and 61 percent, respectively (Postel; 1984, pp. 15–16).

These oxides are converted to acids in the atmosphere. They return to Earth as liquid or frozen precipitation; this is called *wet deposition. Dry deposition* is the process by which particles, such as fly ash or dust, or gases, such as sulfur dioxide or nitric oxide, are absorbed onto surfaces. When these oxides come in contact with water (rain, fog, or heavy dew) they may form acids that can corrode materials and severely damage plants.

The *acidification of lakes* has been well documented in Europe, especially in Scandinavia, and in North America. In the Adirondack Mountains the fish population is endangered in more than half of all the lakes and ponds, and more than 200 of the lakes have become fishless. When the pH of a lake falls below 5, most fish cannot survive. In Ontario, Canada, 48 percent of 2, 600 lakes have been identified as very sensitive to acid rain, and in Sweden and Norway, fish life has been destroyed in more than 6,500 lakes, according to a special report by the Canadian Embassy.

Whether or not a lake becomes acid does not depend on the input of acidic materials alone, otherwise most lakes would show the same degree of acidification. A very important controlling factor is the makeup of the bedrock, and to some degree also the nature of soils, that underlie a particular region. Rocks and soils that contain few *alkalizing* agents (compounds of calcium, magnesium, and sodium) have a low buffering capacity to offset acid input. If acid precipitation falls on a watershed region that is underlain by rocks of this nature (gneisses, quartzites, or quartz-rich sandstone), the acids are not neutralized and the lakes acidify (Likens et al.; 1979, p. 9). *Bedrock geology* correlates, at least in general terms, with the distribution of acidified lakes both in North America and in Europe. The effects of acidification within geologically sensitive regions of the United States are now being studied through an interagency agreement between the Environmental Protection Agency (EPA) and the Brookhaven National Laboratory in Upton, New York (fig. 14.2).

Lakes are not the only victims of acid precipitation. EPA field studies revealed leaf damage when crops were exposed to precipitation with a pH lower than 3. Microscopic cross sections of leaves showed extensive damage to *chloroplasts,* the centers for *photosynthesis,* and in surrounding cells (EPA, 6000/8-79-028; 1979, p. 11). It is not only the immediate impact on lakes and plants that is of grave concern, but also the long-range decline in soil nutrients.

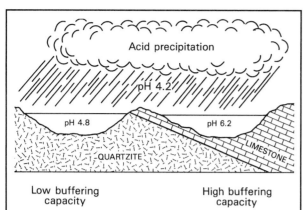

Figure 14.2 The level of lake acidification by acid precipitation depends to a large extent on the buffering capacity of the underlying bedrock. Acidic rock, high in silica and low in neutralizing materials, has a low buffering capacity; limestone, on the other hand, tends to prevent acidification. ─────────────

As rocks decompose in the normally mildly acid soil environment, they release *nutrient ions* (*electrically charged atoms*) such as Ca^{2+}, Mg^{2+}, and K^+. The *dissociated* hydrogens (H^+) of acids *displace* adsorbed nutrient **cations** (*positively charged ions*) from inorganic and organic colloids; this process makes the nutrient ions available to plants. For this reason a certain amount of active hydrogen, as it is typical for acid soils, is beneficial and essential. But if ion displacement and rock decomposition persistently exceed their absorption by plants, the ions are washed out of the soil. This results in an excessively acid and low-nutrient soil. Other constituents of acid deposition, such as *sulfates* and *nitrates,* initially act as fertilizers and boost plant growth; this was studied in the forests of West Germany and Scandinavia. However, the enhanced growth is of short duration. Once the input of sulfates and nitrates is greater than the trees' capacity to use them, very detrimental effects set in (Postel; 1984, p. 19).

Vast areas of forest in West Germany and in Central Europe have been severely affected by acid precipitation, and the number of damaged trees is growing ominously. According to the West German Ministry of Agriculture and Forestry, the area of damaged forest increased from 1.38 million acres (562,000 hectares) in 1982 to 6.28 million acres (2.5 million hectares) in 1983 (Postel; 1984, p. 9)! Similar conditions exist in Central Europe, especially in East Germany, Poland, and in the Czech Republic.

The degree of damage to forests is largely determined by the prevailing tree species. Conifers, in contrast to broadleaf deciduous trees, do not shed their needles each year; this means that their foliage is subjected to acid precipitation for extended time periods. This prolonged exposure to acidic precipitation destroys surface waxes and oily films which help the needles not only to retain water but also to shield them from insect damage (Coch; 1995, p. 287). The visual evidence is the browning of the needles which subsequently drop in ever-increasing numbers. The conifers' inability to quickly replace their foliage weakens the trees and invites diseases.

In 1995, the author noticed severe damage to the predominantly needleleaf forest on the upper slopes of Mt. Fuji in Japan. The most consistent destruction was observed at elevations of about 6,000 feet (1,800 m) in an area occupied by Japanese red pine and larch [Plate. 83]. Damage by acid precipitation in mountainous mid-latitude regions is greatest at elevations of around 4,000 to 6, 000 feet (1,200 to 1,800 m). At these altitudes, rising air usually reaches temperature levels which induce condensation and dense, acidic mist. This concentrated acidity not only corrodes the leaf surfaces but also destroys the important *mycorrhiza fungus* that thrives on the tree roots. This fungus aids the trees in obtaining soil nutrients. Without this beneficial relationship the trees, especially pines, will wither and die.

In addition to the damage to crops and to forests, many other potentially serious problems are associated with acid precipitation and with soil acidification. Many *metals,* some of which are highly toxic in excessive concentration, go into solution at high acidity levels. For instance, *aluminum* becomes quite soluble in acid soils. Subsequently the aluminum cation (Al^{3+}) will occupy *cationic exchange sites* on clays from which valuable nutrients are being displaced. When the occupation of these exchange sites exceeds 60 percent, aluminum appears in the soil solution. *Aluminum toxicity* for many plants occurs when the level of mobile aluminum exceeds 1 ppm in the soil solution (Foth; 1984, p. 201). Just as will other metallic ions in solution, aluminum will then concentrate in plants and in water supplies; in this fashion the metal could find its way into the food chain. There are many indications that aluminum may play a critical role in brain diseases and other degenerative illnesses such as Alzheimer's disease.

The preceding discussion outlined several of the dangerous aspects of atmospheric deterioration and

Plate 83. The author noted widespread effects from acid precipitation on needleleaf trees on the slopes of Mount Fuji, Japan. The greatest damage was observed at elevations of about 5,200 feet (1,584 m). (Photo by C.H.V. Ebert).

some of their disastrous consequences. Killer smogs, photochemicals, and acid precipitation play a key role in these threats to the environment and to human health. In view of existing world realities it seems utopian that these processes can be stopped altogether; but it is absolutely essential to slow down the present trends and to do this on a *worldwide* basis.

Legislation to control gaseous and particulate emissions has been passed in several countries; other nations are now giving the problem serious consideration. But one has to realize that emission control, especially the conversion of old plants, is very expensive; *cost* is a major obstacle to clearing up the earth's atmosphere. For the planet as a whole it seems that both atmospheric pollution and acid precipitation will continue to increase for decades to come. This belief, unfortunately, is reinforced by the fact that many *developing nations* are joining the ranks of heavy polluters. Moreover, just removing polluting substances from emission processes does not end the environmental threat. For example, up to

90 percent of sulfur dioxide in smokestack emission may be removed by so-called *scrubbers,* but the resulting sludge must also be removed in one way or another. This waste sludge could eventually become a serious source of water and ground pollution (Hayes; 1979, p. 6).

Upper Atmosphere Ozone Depletion

The importance of the *ozone layer* in our atmosphere, with the greatest concentration of O_3 at an elevation of about 15 miles (25 km), was mentioned briefly in the introduction to disasters involving the atmosphere. This layer is our shield against excessive *ultraviolet radiation* as it intercepts this radiation energy to form heat as well as the photochemical dissociation of molecular oxygen (O_2) and of existing ozone (O_3).

Single oxygens that split off molecular oxygen will then collide with other molecular oxygens to form new ozone. Similarly, the O_3 molecules split into atomic and molecular oxygens that, when conditions are right, may reform into new ozone molecules. This *photodissociation* obviously depends on the impact of ultraviolet radiation and thus occurs during the day.

The earth's ozone layer remained safe until an increasing amount of human-induced chemicals—released from industries, utility plants, and also from high-flying aircraft—began to attack it. The major culprits appear to be *chlorofluorocarbons, or CFCs,* and nitrogen oxides. The basic process leading to ozone destruction by chlorofluorocarbons takes place as follows: When a CFC molecule (CF_2Cl_2 + UV) is exposed to ultraviolet radiation, it splits off a chlorine atom (Cl + CF_2Cl). The chlorine combines with one oxygen from an ozone molecule and forms *chlorine monoxide* (ClO) leaving the O_2 behind. The unstable chlorine monoxide will lose its oxygen, which will then combine with another single oxygen to form molecular oxygen (O_2). The process described results in the net loss of one ozone molecule.

Studies by NASA revealed in 1985 that ozone levels over Antarctica were reduced to nearly 50 percent below normal; it was then when the expression "*ozone hole*" was coined, and the international community reacted.

The first significant step toward worldwide cooperation was undertaken in 1987 when many industrialized nations signed the *Montreal Protocol.* This agreement aimed at limiting and eventually reducing the amounts of CFCs injected into the atmosphere; the goal was to cut back the manufacturing of CFCs by 50 percent by the year 2000. Yet, as new data indicated a continued loss of atmospheric ozone, more meetings were called during which the urgency of the situation was fully realized.

Stern warnings reached the world's scientific community as revealed in the Proceedings of a Joint Symposium (Proceedings; National Research Council; 1989). Several facts stood out: (1) ozone depletion by chlorine is prominent in the upper atmosphere, but feedback effects lead to a smaller ozone increase in the lower stratosphere, (2) the ozone loss varies with latitude, with higher latitudes experiencing the greatest decline, (3) an annual growth rate of 3 percent of CFC release would produce a decline of about 10 percent of the average column ozone (upper and lower atmosphere combined) in less than 75 years, and (4) even with a freeze of the release of CFCs there would be 1-percent loss of ozone at the equator, a 4-percent decline at 40° N, and as much as

a 7 percent reduction at 60° N within the next 75 years.

Predictions are based on data, assumptions, and on models. Though some predictions are more optimistic than others, the basic threat of ozone depletion exists and must be taken very seriously. Anticipations of an ozone decline of 5 percent near the middle of the 21st century are overshadowed by reports published by J. C. Farman and coscientists which indicated a 50-percent decline in total column ozone for the Antarctic in 1985 (Solomon; 1990, p. 347).

That these fears are not unfounded is shown in widely publicized releases by NASA—based on data obtained by UARS (Upper Atmosphere Research Satellite) and by high-flying airplanes—which point to an ozone hole developing over northern United States, Canada, and over northern Europe.

It is difficult to portray the exact consequences that may derive from the destruction of ozone, but recent reports mention a decline in *phytoplankton* in high-latitude oceans; this plankton forms the base of the *marine food chain.* In 1987 the Environmental Protection Agency warned that for every 1-percent decrease in global ozone, there could be a 2- to 3-percent rise in skin cancer. In the past ten years, skin cancers have increased by 50 percent (Mathews; 1990, p. 95). The consequences deriving from global ozone depletion are complex and are in many instances still a question of speculation; however, the potential threats cannot be dismissed, but they should neither be exaggerated.

Much can be done by individuals to reduce the risks involved in the excessive exposure to ultraviolet radiation: (1) protect the eyes with quality sunglasses, (2) wear hats, and (3) avoid the hours of strongest sunshine between 10:00 A. M. and 3:00 P.M. Common sense and a basic understanding of the nature of the dangers of ultraviolet radiation can go a long way to mitigate risks. However, in the long run the world must win over the threat of ozone destruction to prevent the obliteration of vital components of the ecosphere.

Poisoning the Earth

There is not a week, and possibly not a single day, during which the public does not hear about a newly discovered toxic agent, or about an industrial accident, or of a waste disposal site that poisons the environment. These threats are not new; as early as 1962 Rachel Carson warned: "The most alarming of all man's assaults upon the environment is the contamination of air, earth, rivers, and sea with dangerous and even lethal materials. This pollution is for the most part irrecoverable; the chain of evil it initiates not only in the world that must support life but in living tissues is for the most part irreversible" (Carson; 1962, p. 6).

Mining operations, raw material processing, manufacturing, agricultural practices, waste disposal and dispersal, weapons testing, accidents, and daily life activities, all contribute to some degree to the altering, to the harm, or to the destruction of the natural environment. Some of this deterioration takes place over a relatively short time span; other events have a momentary, stupendous impact as witnessed in connection with industrial disasters. Many processes, however, work silently and remain undetected for decades until symptoms of illnesses or breakdown begin to surface. Many industrial accidents never make the headlines. But who can forget the following?

Industrial Accidents

1. The *Minamata incident* of 1953 to 1971 became a symbol of pollution that resulted in suffering and death through *mercury poisoning*. Minamata, a fishing town in Japan, was the home to a chemical plant that used mercuric oxide as a catalyst in its manufacturing process. Mercury waste was regularly dumped into the adjoining bay; here, marine organisms converted the mercury into mercuric chloride. This substance entered the food chain by the way of fish, and the people of Minamata eat a lot of fish. Neurological disorders emerged in 1953 and reached epidemic proportions by 1956. Although the cause was identified by 1959, the dumping of mercury waste continued until 1971 (Hayes; 1979, p. 15)! More than a thousand victims have been identified. Pictures of deformed children, some so seriously damaged that they lived in a vegetablelike state, shocked the world in many publications.

2. The *Seveso incident* of 1976 illustrated the danger of locating potentially lethal manufacturing plants close to communities. In this case, a chemical factory in the Italian town of Meda produced toxic defoliants and antibacterial substances. A short distance away is the small town of Seveso; several workers from Seveso had jobs in the plant at Meda.

On June 10, 1976, a tank ruptured at the factory and released a large cloud of toxic chemicals that contained traces of *dioxin.* No warning was given as the cloud slowly drifted into Seveso; excited chil-

dren playfully ran into the mysterious fog. By evening many inhabitants developed skin rashes; others vomited and showed signs of increasing illness. Seven hundred people were evacuated, and temporarily Seveso became a ghost town. Only long-range studies will reveal what the final outcome of this accident will be.

3. *Bhopal,* India, was the location of another industrial accident that took place in December 1984. This accident killed more than 2,700 people within a few days.

In the 1960s, in connection with the "*Green Revolution,*" India began to build up a chemical fertilizer industry. In 1975, a branch factory of Union Carbide located about 15 miles (24 km) away from Bhopal. But over the years many people moved toward the plant and settled in several shanty towns between the factory and the city. The plant manufactured a strong *pesticide (SEVIN),* which, as so many pesticides do, acts as a **nerve gas.** One early morning in December, 1984, a valve ruptured—the precise cause is still being debated—and a huge toxic cloud blew out from the plant and drifted into the still-sleeping town. People who tried to run away breathed more deeply; the searing gas destroyed their lungs. Death was mainly induced by *pulmonary edema.* Moreover, the chemical, as is true for nerve gas material in general, was easily absorbed through the skin. A recent book of this disaster written by Dan Kurzman, *A Killing Wind,* claims that as many as 8,000 persons lost their lives in this disaster. However, this number is denied by the Indian government.

4. At about 1:23 A.M. on April 26, 1986, an explosion ripped apart reactor No. 4 in the *Chernobyl nuclear plant* about 81 miles (130 km) north of the Ukrainian capital of Kiev in the former Soviet Union. The force of the explosion, and the subsequent intense fire, dissipated several tons of the *uranium dioxide* fuel and fission products, which included *cesium 137* and *iodine 131,* as well as tons of burning graphite. A plume of contaminants rose to a height of 3 miles (5 km) (Edwards; 1987, p. 634).

The Chernobyl nuclear accident was the first reactor disaster that came close to what in the professional literature has been described as the largest conceivable accident, an event in which a large reactor, with a mature inventory of fission products, breaches containment and releases an appreciable fraction of volatile fission products (Hohenemser; 1986, p. 2). The emission possibly ranged into the millions of *curies.* The release in connection with the Three Mile Island accident (1979) was only 15 curies.

Measurements of radiation tend to be confusing. This situation exists because some measurements relate to *radiation release,* whereas others refer to the dose of *radiation received.* The curie is a *rate of radioactive disintegration;* this means that atoms are splitting at a rate of 3,700 million disintegrations per second. The *r, or roentgen,* is the *measure of ionization* at 83 ergs of energy per cubic centimeter of air. The *rem (roentgen equivalent man)* is the *biological equivalent,* within living tissue, of roentgen (Calder; 1962, p. 29).

Over the years, as more knowledge about radiation emerged, the "safe" levels of radiation doses have been progressively scaled down, especially with the realization that radiation damage is *cumulative.* Using the limited data made available by the former Soviet Union, scientists of the West estimated that about 24,000 people of the 116,000 evacuees, or about one-fifth, received a radiation dose of 45 rem. The allowable "*once-in-a-lifetime*" dose is considered 25 rem (Edwards; 1987, p. 640).

In addition to placing 300 to 400 million people in 15 nations at risk, the Chernobyl accident also revealed major weaknesses in national and international *risk management:* (1) there is no adequate warning system in place; (2) there are as yet no internationally accepted standards for nuclear safety; (3) the density of sophisticated radiation monitoring stations is far too low to document local variability in a timely manner; and (4) given this policy vacuum, there is virtually limitless potential for poorly informed public officials to improvise, and thus to interfere with, a timely and effective response to disaster such as that at Chernobyl (Hohenemser; 1986, p. 2).

Hazardous Waste Disposal

The preceding chapter addressed the *waste avalanche* in general terms. However, the most threatening aspect of this problem is the discarding of hazardous waste that contains toxic substances.

The present ways of waste disposal are inadequate as far as their technologies are concerned. The U.S. Environmental Protection Agency estimates that 90 percent of the wastes produced may have been disposed of improperly. In the United States alone there are more than 15,000 *uncontrolled* sites. Many of these sites, in addition to landfills, have contaminated surface impoundments (*pits, ponds, lagoons*), with 90 percent of them posing a threat of groundwater contamination (Brodigan, et al.; 1986, p. 2).

One of the least-discussed techniques for discarding liquid waste is what is called *deep-well injection.* This process involves the forced injection of wastes into deep rock formations, often sandstone or limestone, that are sufficiently porous and are located below groundwater levels. The idea is to pump liquid waste into the rock formations with sufficient pressure to drive out existing fluids and to contain the waste materials. Under "ideal" conditions, injection wells should not affect water-bearing rocks, or *aquifers;* the injected layers should be sealed off by impervious rock strata; moreover, the area that contains the injection site must be free from *earthquakes.* In addition, intricate designs of pressurized steel casings are included to safeguard the installation.

Deep-well injection is cheap compared to other methods ($5 to $8 per barrel compared to $40 to $194 for incineration, and $30 to $40 for landfilling) and requires a minimal amount of maintenance and monitoring cost. The largest users of deep wells are petrochemical and pharmaceutical companies, which operate about half of the active wells; 92 percent of the deep wells are located on company property (Citizens Clearinghouse for Hazardous Wastes, Report; 1985, pp. 4–9). Although deep-well injection is considered "safe," many specific cases, including 39 incidents listed in the CCHW Report alone, involve leaks of highly toxic and corrosive materials and mandate shutdown orders.

One of the most serious problems modern technology is facing in the entire world is the disposal of *radioactive* waste. This waste is produced during all phases of the nuclear-fuel cycle, from the mining of ores, the making of the fuel, the reactor activities, the reprocessing of spent fuel, to the research activities involved. Certain amounts of *plutonium* and *uranium 235* will get into the waste during the chemical processing of fuel elements, and these radioactive materials have a *half-life* of 24,000 years and 4.5 billion years, respectively (Calder; 1962, pp. 205–206).

Probably the most controversial element in spent fuel is plutonium. It is toxic; it is carcinogenic; it can be chemically separated from the rest of the fuel and can be used to fuel reactors or to build atomic bombs. A 1,000-megawatt reactor typically produces about 826 pounds (375 kg) of plutonium per year of operation (Hayes; 1979, p. 20). The potentials for both accidental and deliberate harm are great.

Radioactive waste materials obviously vary in their levels of radioactivity, in their half-lives, and in their physical state. Some are liquids, others are in a gaseous state, such as krypton 83, and some are solids. Most of the nuclear wastes, except for low-level materials that can be diluted or diffused, are stored "temporarily" because there is no absolutely safe way to dispose of them securely on a permanent basis. Many techniques, aimed at *safe disposal*, are being used or are being considered. They range from dilution in the oceans, and in the atmosphere, to storage systems that allow the radioactive waste to *cool down* and decay. Other ways involve the concentration of low-level waste through *solidification*, conversion into glass (***vitrification***), and burial in old salt mines. Higher levels of safety probably can be achieved at much higher cost, and with improved technology, but absolute and timeless safety can never be assured (Hayes; 1979, p. 22).

◎ Love Canal: A Case Study

On August 7, 1978, the news media all over the United States announced that President Carter had declared that a state of emergency existed at Love Canal in the city of Niagara Falls, New York. The meteoric rise to notoriety had now placed this waste dump site into the limelight; it focused, at least temporarily, the nation's, and possibly most of the world's, attention on a place that had evolved from a dream into a nightmare.

As early as 1836 the site of Love Canal, north of the Niagara River and in the Village of LaSalle (now part of Niagara Falls), had been surveyed by the U.S. Government because of its favorable setting for the construction of a canal that would circumvent the Niagara Escarpment. Such a canal was also to become the site for an electrical power plant. However, it was not until 1892 that attention was given to this scheme. It was William T. Love, an industrial entrepreneur, who wanted to realize the bold dream of using the mighty Niagara River to supply a newly planned industrial center, Model City, with unlimited electricity. His newly founded Model-town Development Corporation was given a charter and enthusiastic support by the then governor of New York State, Rosswell Flower.

The construction of the water-diversion canal began near the Niagara River with a narrow drainage channel that connected the canal bed with the river. But, within a year after this promising start the plan collapsed; it left behind a water-filled

Plate 84. This aerial photograph of the late 1930s shows the 3,000-foot (915 m) water-filled section of the abandoned Love Canal before it was used as a dumping site for toxic waste materials. (Photograph credit: U.S.D.A.).

3,000-foot (915 m) section of the canal [Plate 84].

Two events mainly caused the downfall of the grandiose project: (1) the economic depression of 1894 induced the investors to pull their money out of the new company, and (2) physicist Nikola Tesla (1856–1943) had invented a new type of transformer—later known as the Tesla coil—which was capable of producing high voltage, high-frequency alternating currents that could be transmitted over long distances; this invention made it unnecessary to locate factories near the source of electric power generation (Ebert; 1980, pp. 2–3).

As time passed on, Mr. Love's canal faded into temporary oblivion, a monument to a dream within the pastoral setting of the Village of LaSalle. This community was not incorporated into Niagara Falls until May 1927. However, the development of electrical power attracted chemical plants to the nearby area. Among them was the Hooker Electrochemical Company that was established in Niagara Falls in 1905 (Levine; 1982, p. 9).

With growing industrialization came increasing amounts of waste materials. After the Love Canal site was studied as a potential dumping location, Hooker Company obtained permission to discard chemical waste into the canal; this operation started in 1942. Between 1942 and 1952 about 21,800 tons of various chemicals were placed into the old canal bed. A study by E. Zuesse cautioned: "It's also worth noting here that other wastes besides these 21,800 tons from Hooker have apparently been dumped into the Canal. According to New York State officials, federal agencies, especially the Army, disposed of toxic chemical wastes there during and after World War II. The city of Niagara Falls also regularly unloaded its municipal refuse into this Hooker-owned pit" (Zuesse; 1981, p. 20). Of course, no one was worried about the individual chemicals nor about the interactions of the chemicals. Hooker Company had bought the site shortly after 1947. By 1952 all dumping had stopped, and the canal was topped off with earth materials.

In 1953, the Niagara Falls Board of Education obtained the canal zone from Hooker Company for the price of one dollar. Hooker was quite aware of the potential dangers presented by the thousands of tons of chemical waste that was mostly contained in 55-gallon (208 liters) metal drums. It is not surprising that the company clearly stated in its sale's deed: "It is therefore understood and agreed that, as a part of the consideration for this conveyance and as a condition thereof, <u>no claim</u> (author's underline), suit, action or demand of any nature whatsoever shall ever be made by the grantee, its successors or assigns, for injury to a person or persons, including death resulting therefrom, or loss of or damage to property caused by, in connection with or by reason of the presence of said industrial waste." (Deed is recorded in County of Niagara, N.Y., July 6, 1953.)

The Board of Education, which now owned about 16 acres (6.5 hectares) of land (including the canal site itself), sold 6 acres (2.5 hectares) to the city of Niagara Falls and another 6 acres (2.5 hectares) to a land developer; it subsequently started to build a school (99th Street Elementary School) on the remaining area. During the initial construction the workers encountered some of the buried chemicals, and the school was moved some distance away from the original site. But, it still sat on the brink of Love Canal; its paved backyard and a baseball diamond were situated directly on the canal property. During the mid-1950s, and through the early 1970s, several hundred residences were built along new streets that paralleled the canal; two streets actually cut across it [Plate 85]. The areas north of the canal also saw a major buildup of homes.

Strong chemical odors were reported in the neighborhood around the canal as early as the mid-1950s. Many signs of a rapidly deteriorating situation began to develop as time passed on: Smelly, oily substances oozed out of low-lying spots; damp basement walls emitted a strong smell of chemicals, and a heavy odor hung over the area on warm windless days.

In January 1977, persistent urging from the New York State Department of Environmental Conservation convinced the city of Niagara Falls to hire Calspan Corporation to assess the situation with the result that, by mid-1977, it was established that additional testing was needed. In October of the same year, the Environmental Protection Agency was asked for assistance. Subsequently, a coordinated investigation, employing a number of subcontractors, tested surface water, soil samples, deepwell water quality, and basement contamination.

On August 2, 1978, Love Canal gained national attention when the New York Times devoted a front-page story to this problem. On the same day a public meeting was held in Albany, New York, when residents of the Love Canal neighborhood presented their concerns to the then State Health Commissioner, Dr. Whalen. Subsequently, the governor of New York, Hugh Carey, directed a task force to meet three basic objectives: (1) relocation of residents from the first two rows of houses adjoining the canal, (2) containment of canal leachates, and (3) immediate and continuing maintenance and monitoring of the area (Ebert; 1980, pp. 6–7).

Many local scientists, including the author, became involved in the Love Canal disaster. Dr. Beverly Paigen, of Roswell Park Memorial Institute, correlated the occurrence of a variety of illnesses and poor-health symptoms with suspected areas of chemical leachate migrations. Highly toxic materials (benzene, toluene, lindane, chloroform, trichloroethylene, and others) had been identified in the canal leachates [Plate 86].

This author cautioned, as early as the fall of 1978, that the physical conditions in and around Love Canal could induce the migration of leachates away from the canal. Although the soils in that vicinity are generally fine-textured (mostly silty clay loams and clay loams of glacial-lacustrine origin), there are numerous irregular sand layers and lenses that could facilitate the flow of water and leachates from the canal proper to surrounding areas. Furthermore, the chemicals could reach the underlying

Plate 85. This 1978 view shows the location of Love Canal between residential houses at the top and the 99th Street Elementary School at the bottom of the photograph. The baseball diamond and the paved area are located directly on top of the filled-in canal. (Photograph credit: Warren Philipson).

dolomitic bedrock and move either on top of it, or slowly through it, to the Niagara River; the dip, or slanting, of the bedrock formations is to the south.

An additional factor further complicated the

Plate 86. A test drainage pit at Love Canal, excavated before the start of the construction work of the abatement plan, quickly filled with water and chemical leachates. (Photo by C.H.V. Ebert).

situation. Several surface **swales,** defined as temporary low-lying drainage channels, existed in the vicinity of the canal; several swales actually cut gullies into the sides of the canal. This structure indicates that they had existed before canal excavations began. Subsequently, these swales were filled in with a variety of materials that ranged from soil to construction debris. The fill material is less compacted than the natural soil around it and contains, therefore, cracks and some cavities. These features would allow leachates to move more readily within such channels than through the undisturbed soils (Ebert; 1978, p. 3). Other conditions that affect the migration of leachates are dessication cracks; they develop in clayey soils during drying and facilitate the through-flow of liquids during the initial phase of wetting. Another concern was the potential movement of toxic waste in and along underground sewer lines and along utility channels. After heavy rains the hydrostatic pressure buildup in the canal many times forced water and leachates to spill over and into low-lying street sections [Plate 87].

Plate 87. Toxic leachates seeped out from Love Canal and spread into Read Avenue, which cut directly across the canal. (Photo by C.H.V. Ebert).

In addition to the evacuation of hundreds of families from homes surrounding Love Canal, the first priorities of the New York State Abatement Plan aimed at (1) containing the chemical wastes in the canal bed by placing a heavy clay cap on it, and (2) intercepting the leachates by placing drainage pipes around the sides of the canal.

The clay cap was extended to a considerable distance from the canal proper. Heavy bulldozers spread and built up the cap; they were also instrumental in tamping down the clay clods to achieve some degree of compaction [Plate 88]. Additional soil was then placed on top of the clay and seeded to grass. This action was to reduce subsequent soil erosion and would also lead to a more uniform absorption of precipitation and water from melting snow.

The drain pipes were placed in 16-foot (5 m) trenches that were partly filled with gravel to accept the leachates that continued to seep out of the canal bed. Catch wells were placed at intervals along the trenches, and the polluted liquids were pumped out to be processed by a treatment facility, using activated carbon, that was built within the fenced-in zone of Love Canal.

These efforts, at least initially, probably were the best under the given circumstances, but they certainly do not present a permanent solution. Many unanswered questions remain:

1. What will eventually be done with the mass of chemical waste still contained in Love Canal?
2. How far have toxic materials actually spread away from the areas tested up to now?
3. Have the leachates reached the Niagara River, and, if so, how much is seeping into it?
4. Will aquifers, dipping southward from Love Canal, be contaminated by the chemical pollutants?
5. Will additional homes that so far have escaped evacuation and demolition be found contaminated [Plate 89]?
6. What will be the long-range effects on people who have been exposed for years to these toxic substances?

These, and many other questions, illustrate that the environmental disaster of Love Canal is far from

Plate 88. Placing a clay cap on top of Love Canal was one measure taken in the attempt to prevent further spreading of toxic materials from the dump site. (Photo by C.H.V. Ebert).

Plate 89. Hundreds of homes around Love Canal had to be evacuated and were boarded up. Most of the homes, as well as the 99th Street School, were demolished later on. (Photo by C.H.V. Ebert).

over. Convincing correlations between birth defects, kidney diseases, leukemia, respiratory illnesses, and a higher-than-average occurrence of cancer in the vicinity of seepage areas, particularly along the swales, are still under study.

Love Canal is only one of hundreds, possibly thousands, of toxic dump sites, but it dramatically illustrates how past mistakes in handling chemical waste have already confronted us with severe environmental and societal disasters. It is hoped that we learn from this event so that future generations will not have to face other Love Canals (Ebert; 1980, pp. 9–10).

THE IMPACT OF WAR

Key Terms		
Paleolithic	ironstone	sorties
mace	laterite	napalm
assegai	jet stream	hyperthermia
bronze	defoliants	fission
"Iron Age"	Operation "Ranch Hand"	fusion
crossbow	Agent Orange	thermonuclear
glacis	teratogenic	ICBMs
Blitzkrieg	Operation "Pink Rose"	nuclear winter
plinthite		

The preceding chapters showed that humans have become a major force in changing our planet's environment, and in making these changes our race has taken an active role that once was the exclusive domain of nature itself.

The major steps in the modification of the natural environment were associated with "revolutions," such as the Agricultural and the Industrial Revolution; most of these changes, until recently, were viewed as unavoidable "side effects" of progress and development. In contrast, the *impact of war*—in its net effect and regardless of the reasons for which it was fought—is *destructive* by virtue of its clearly stated goals: to destroy, to kill, to make the area uninhabitable, so that one side may emerge as the victor.

We do not really know when and how the first war was fought, or how many casualties it caused. However, if one accepts the general definition that war is "*a conflict carried out by force of arms, or weapons,*" and if conflict is as old as the existence of humankind itself, then we may state that war is as old as humans' first appearance on our planet.

Human Conflict and Aggression

Competition between species, which in nature could be viewed as a milder form of conflict, appears to be a universal process that may have played a major part in the evolution of life-forms. Even plant succession could be considered a certain type of "*conflict,*" or competition. But humans' *struggle for existence,* to use Charles Darwin's classical expression, took on a very different slant in contrast to the *struggle for existence* in nature.

In the beginning, humans may have fought over the carcass of an animal; they may have defended a water hole in times of drought; or they may have driven off an intruder from a cave. But humankind quickly raised the intensity of the *struggle for existence* beyond the levels of necessity. This characteristic seems to be unique to the species; it eventually led to the most wanton destructiveness carried out on Earth: war. It seems that

Homo sapiens' two most outstanding abilities—*conceptual thought* and *verbal speech*—are not necessarily great blessings, or "are at least blessings that have to be paid for very dearly indeed." (Lorenz; 1966, p. 238).

The road from "*struggle for existence*" to competition and conflict, and finally to war is long and obscure. The introduction of tools, and later of weapons, accelerated the rate of "progress" along this road. In 1932 Sigmund Freud (1856–1939) expressed this theme in a letter to Albert Einstein as follows:

> To begin with, in a small human horde, it was superior muscular strength which decided who owned things or whose will should prevail. Muscular strength was soon supplemented and replaced by the use of tools: the winner was the one who had the better weapons and who used them more skillfully. From the moment at which weapons were introduced, intellectual superiority already began to replace brute muscular strength; but the final purpose of the fight remained the same—one side or the other was to be compelled to abandon his claim or his objection by the damage inflicted on him and by crippling of his strength.
>
> (Maple; 1973, p. 17)

When it comes to survival under different situations, it is difficult to draw a sharp line between "necessary force" and "wanton violence." Research shows that early peoples used stones not only to kill animals for food but also to kill other humans. Also, it seems that "Peking Man, the Prometheus who learned to preserve fire" used the fire to roast his companions (Lorenz; 1966, p. 239). It must also be conceded that a certain degree of conflict, in its milder form of competition, could have been beneficial in developing social progress. But after competition blended with open *aggression* and with *hostility,* the road to senseless and mutual violence opened up. Aggression involves an impulse to *attack* and to *destroy* (Stagner; 1965, p. 56).

Human aggression and violence are at times called animalistic or "*bestial*" behavior. This, as Dr. Edwin Hollander puts it, may be grossly unfair to the animals:

> They are usually disinclined to assault, kill, or maim their own kind, although they do engage in individual combat for dominance and predatory attacks against other species for food. If one had to choose, it would probably be preferable to meet a 'noble beast' than a 'bestial noble'.
>
> (Hollander; 1976, p. 432)

The Evolution of Weaponry

Conflict and violence between members of early human society were restricted to a few individuals and probably were carried out in virtual isolation. Heavy wooden branches, clubs, and stones served as weapons to which the use of fire was added at later times.

When people began to increase in number, and as they banded together in groups and tribes, the *scale* of rivalry and conflict grew accordingly. At some point the terms "*competition*" and "*conflict*" would no longer be accurate descriptions because of the change in scale, and the phenomenon of war evolved. War, of course, is a much more directed and organized endeavor; it requires a very different set of motives that exceed the "*struggle for existence*" discussed before. These motives extend far beyond the basic needs for survival.

As territories became defined (*territoriality*), as communities identified their places of residence, as agriculture made nomadic roaming unnecessary, it became essential not only to defend property but also to expand land holdings. Soon, such expansion was no longer based on the availability of more food and other resources to satisfy a growing tribe, or later, kingdoms and nations, but to acquire a new property: *power.* The struggle for power appears to be the basic *leitmotif* (repeated topic) of history, old or new. Hans J. Morgenthau, noted professor of political science, summarizes this concept in the following words:

> Whatever the ultimate aims of international politics, power is always the immediate aim. Statesmen and peoples may ultimately seek freedom, security, prosperity, or power itself. They may define their goals in terms of religious, philosophic, economic, or social ideal. They may hope that this ideal may materialize through its own inner force, through divine intervention, or through development of human affairs. But whenever they strive to realize their goal by means of international politics, they do so by striving for power.
>
> (Morgenthau; 1949, p. 13)

It is easy to see that if power exists, in the form of *arms* and *resources,* a barrier in the way to some goal may lead to frustration and subsequently to aggressive behavior. If the energies expended in the effort to reach such a goal prove continuously unsuccessful, they tend to flow over into generalized destructive behavior (Stagner; 1965, p. 53). War *is* destructive behavior, and weapons are the tools for achieving destruction.

It is difficult to follow with certainty the early evolution of weapons, and to separate hunting tools from arms of aggression, or their use in defense. For example, a wooden *spearlike pole* of **Paleolithic** age—which began hundreds of thousands of years ago—was found embedded in the rib cage of an ancient elephant in the region of Lower Saxony. A similar weapon was identified in Essex, England. Simple *bows and arrows* were used in southern Europe and in northern Africa about 15,000 B.C. (Reid; 1986, pp. 9–11). What can kill an animal could also kill a human, but organized warfare had to wait.

The *club* probably was the oldest instrument that served as a weapon; in its earliest form it was very likely a mere strong piece of wood, or just a heavy bone. From the club all of the cutting and thrusting weapons have been derived (Stone; 1961, p. 183). The effectiveness of the club was improved by attaching a stone to its end. This development not only increased the striking force but also allowed an aimed throw of this primitive weapon. Much later, following the development of metallurgy, the club evolved into the *mace*, a clublike device used in hand-to-hand fighting that was effective in smashing through simple armor. As the armor became stronger, the mace was supplemented by the use of the *battle axe* that was capable of cutting through the armor.

The development of daggers and other thrusting and cutting weapons followed a similar trend. The short, pointed stick set the model for the *dagger, sword,* and *spear.* However, the appearance of the dagger, and especially that of the sword, had to await the use of metal. Until that moment the thrust and throw of the wooden spear dominated. A sort of transition between spear and sword was the African **assegai,** a heavy-bladed weapon that could be used for thrusting as well as for short throws.

The first metallic daggers were fashioned of *copper.* This metal was found in its native state in several parts of the world, including the island of Cyprus—a major source of copper for the regions surrounding the Mediterranean Sea—and in Bolivia. Copper, with a melting point of 1981° F (1083° C)—compared to iron's 2795° F (1535° C)—could be readily smelted and shaped. The problem with copper was that it was too soft to hold a sharp cutting edge. Moreover, its softness restricted the length of the dagger and that of a spear tip. The advent of **bronze,** a copper alloy that included about 11 percent tin, changed the course of weaponry; it allowed the manufacturing of larger and stronger blades. Accordingly, the sword did not become a major weapon until after the year 2000 B.C. when bronze was replacing copper (Williams; 1976, p. 13). Just as bronze followed copper so *iron* displaced bronze. By 1500 B.C. iron was widely used by the Hittites, and by the late 13th century B.C. the *"Iron Age"* exerted its dominant influence on the making of tools and armaments. Five hundred years hence the first steel was produced in India (Bronowski; 1973, p. 131). The emergence of iron and *steel* made similar major impacts on the development of new weapons as well as that of defensive designs, such as stronger helmets and protective armor.

This short account, tracing the evolution from club to battle axe, from dagger to sword, illustrates the never-ending search for technology of the *"ultimate"* weapon; this search went on in the dark past and obviously continues in our times:

> Even before the terror of a new weapon wanes in the face of an effective defense, the quest is on for some more formidable means of attack. The resultant spiral makes war a struggle between the scientists, technologies and industrial complexes of great nations.
>
> (Reid; 1986, p. 8)

As we trace the evolution of weapons, we can detect three aspects that change over time:

1. The distance between combatants increases
2. The role of the individual becomes less significant, as the scale of war and the sophistication of weapons increases
3. The destructive power of weapons grows, and with it rises war's damage to the environment.

The development of weaponry is not restricted to means of attack only; it also gave rise to the rise of defensive designs. In the discussion of urbanization in chapter 14, it was pointed out that early cities served a variety of purposes; one was the protection of people and their property. This development was an important phase in the nature of warfare; it became evident when agriculturalists began to replace the hunter and concentrated in permanent settlements. With that arose the need for defense:

> A people capable of tilling the ground was liable to be attacked by less provident or less able

groups intent on seizing the fertile land and its produce. Here, then, were the beginnings of organized fighting on a larger scale than hitherto, with slings and stones, bows and arrows, and with a clear-cut economic motive—acquisition of food resources and living space.

(Williams; 1976, p. 8)

Permanent residences (villages, towns, and cities) had to be defended and had to be made defensible, or strong. This need resulted in the emergence of "strong places," or *fortresses* (Latin = *fort*[*is*], or *strong*). Strong places ranged from the placing of thornbush around a village, such as the African *boma,* to the use of massive stone walls. One of the oldest, if not the oldest, major fortification was the ancient town of Jericho, north of the Dead Sea; Jericho dates back to 7000 B.C. Its excavations clearly revealed the existence of a large wall built of unfinished stones and the presence of a *moat.*

Throughout subsequent history most major battles, in one way or another, were associated with the name of a fortress, with that of a fortified city, or just with cities that endured "*to the last of the defenders.*" As weapons became more powerful they found more and more weak spots in fortresses; eventually, the role of fortifications became dubious long before the atomic age. There simply is no such thing as an "impregnable" fortress! A prime example is the fall of Constantinople in 1453, which brought about the end of the East Roman Empire. For hundreds of years the Empire had repulsed the Ottoman Turks, and the symbol of this resistance was the walls of Constantinople. The Turks finally used very heavy guns—the barrels had to be cast on location!—that could hurl an 800-pound missile over a range of one mile (1.6 km); it had a rate of fire of seven rounds per day (Morgenthau; 1949, p. 294). In less than two months the massive walls of Constantinople were crumbling.

Gunpowder, and the associated evolution of *guns,* had a critical effect on world history. The dominance of guns made itself felt during the 15th and 16th centuries. Medieval castles no longer offered protection; cavalry—a major military weapon for more than a thousand years after Alexander the Great (356-323 B.C.)—rapidly declined in importance; battles became more *impersonal* and more *violent.* Because the bullets could pierce and strike strongly against armor, *chivalry* began to decline and was replaced by *mass destruction.* Guns were not the only means that made killing more efficient and

increased the distance between combatants. The flying dart, first propelled from the simple bow, then from the *longbow* and **crossbow,** was a lethal weapon for thousands of years.

The Welsh longbow, especially, contributed to the downfall of the heavily armored knights on horseback. The longbow was about 6 feet (2 m) in length. It could discharge several long heavy arrows, whereas the first mechanical weapon, the crossbow, could fire only one dart during the same time. The arrow from the longbow could easily penetrate *mail* (flexible armor composed of interlinked rings or scale-like platelets) and *plate armor* at close range.

The longbow could also be used as a *long-range weapon* (up to 400 yards or 366 m), which turned the tide in favor of the English in their battle against the French; thousands of heavy longbow arrows rained down on the massed French knights and their horses (Burke; 1978, pp. 60–61). The longbow did its share in dealing the final deathblow to feudal warfare. The dominance of the English longbow was driven home earlier at the battles of Crecy (1346), Poitier (1356), and at Agincourt (1415). It is ironical that the British also used a simple type of *artillery,* stones fired from *iron pots,* in the Battle of Crecy, but it was the longbow archers who proved decisive. Soon afterward, firearms of all sorts multiplied far faster than any weapons of earlier times (Williams; 1976, p. 47). Important modifications included a marked change in their sizes for specialization: Smaller firearms enabled cavalry to use them, and larger guns were designed to pulverize fortifications. Moreover, range and rate of fire increased dramatically.

Muzzle-loader musket fire, as late as the 1850s, still ranged up to only 300 yards (275 m); by 1866 *needle guns* reached 2,700 yards (2,470 m), and *repeating rifles,* by 1913, could fire a distance of almost 4,500 yards (4,115 m). Thus, between 1850 and 1913, the range had increased sixteen times (Morgenthau; 1949, p. 295).

The growth of artillery, both in range and caliber, was staggering. The *record range* of any artillery piece still stands at 75 miles (121 km). This distance was achieved during World War I by the so-called "*Paris Gun,*" which on March 23, 1918, fired into the French capital for the first time. Its barrel was set at 50 degrees elevation, and the range was controlled by the amount of powder charge used (*American Heritage;* 1964, p. 271). The record for the heaviest-caliber gun is held by a monstrous railroad cannon called "*Gustav,*" which was built by Germany during World War II. Its barrel had a length of 130 feet (40 m) with a caliber of 32 inches (80 cm). This gun could fire an armor-piercing projectile of 7 tons, which could pierce massive layers of concrete up to 11 feet (3.4 m) in thickness. Actual fire control and mechanical operation required a crew of 500 men (Hogg; 1970, p. 130). The gun was used to destroy the Soviet fortifications around Sevastopol (Crimean Peninsula) in 1942.

Fortresses, too, changed in response to the ever-increasing power of siege guns, but, as it was stated before, no fortress ever proved to be impregnable. The major changes in the design of fortifications included the lowering of their profile, protecting the low walls with a sloping *glacis,* and with *earthworks* that could absorb the impact of heavy projectiles more easily than walls of stone. Moreover, fortifications soon became *retaliatory* and were able to fire back at the attacker. Some of the fortresses, including Verdun and its satellite forts, attained ghastly, indelible fame in the history of war.

Probably the most costly and most ambitious scheme of modern fortification was the famous Maginot Line. It embraced a complex system of massive above-ground and below-ground forts and support facilities; it ran for a distance of about 200 miles (322 km) from Belgium to the Swiss border. Some of the underground facilities (munition and food storage areas, barracks, power plants) extended down to 200 feet (61 m) below the ground surface. The building of the Maginot Line was one of the most intensive construction-related impacts on the natural environment to that point in history. This huge fortress system, designed after World War I, aimed at protecting France *"forever"* against German aggression. In World War II, the invading German troops either bypassed the line of fortifications, coming across the unprotected border with Belgium, or they smashed through the line with the help of new weapons, including *paratroopers* and *dive bombers,* for which the Maginot Line was not prepared.

The *"Atlantic Wall,"* built by the Germans along the French coast to repel the Allied invasion in World War II, faced a similar fate. Massive steel-reinforced concrete strongpoints were reduced to rubble, or were put out of action, by direct heavy artillery fire from battleships, or by the use of new, deep-penetration bombs. The heaviest non-nuclear and most devastating aerial bombs ever used were the *"Grand Slam"* (22,000 lb) and *"Tallboy"* (12,000 lb) of British origin (Frankland; 1970, p. 111). The shockwaves of these bombs on the land and on structures (bridge foundations; viaducts) were similar to seismic disturbances, a fact that led to their nickname, *"earthquake bombs."*

The race for new weapons, as well as the creation of protective countermeasures, never stopped: The *sea* saw native war canoes, Viking ships, Greek triremes, the Spanish Armada, the *"dreadnought"* battleship of World War I, the super-battleship of World War II (the Japanese *Yamato* had a displacement of 72,000 tons), the aircraft carrier, and nuclear submarines with a cruiserlike displacement of 25,000 tons (Soviet Typhoon class) of contemporary times. The *air* witnessed observation kites and balloons, the dirigibles that carried bombs to England in World War I, the evolution of the fighter plane and the bomber, and the supersonic intercontinental missiles that travel through space to reach their targets within minutes. The *land* and *vegetation* endured fighting tribes, the first large-scale organized armies of the Assyrians, armored knights, massed infantry, the pulverizing force of artillery, earth-grinding tanks, underground tunneling and explosions, the killing gas clouds of chemical warfare, area bombing by massed bombers, napalm, phosphorous bombs, and finally the all-devouring searing heat and radiation from the nuclear bombs. As weapons grew in their arena of destruction, the earth seemed to shrink proportionally!

The Intensification of Warfare

The technological evolution of weaponry brought about a gradual intensification of war, but additional factors appeared, in part philosophical in nature, which fueled war and further increased its destructiveness. The new forces which propelled war into growing devastation of humans and of the environment included three major elements: (1) Emerging *nationalism,* (2) the *Industrial Revolution* coupled with new technologies, and (3) the drift toward what is referred to as *"Total War."* Quite clearly it is impossible to sharply divide these factors into narrow time segments of their emergence—their effects overlap in many ways—or into their individual influence.

Nationalism changed the motivation of people and introduced a more focused view of the state, its government, and of the cultural cohesiveness of nations within a defined space. "The French Revolution of 1789 marks the beginning of the new epoch of history which witnesses the gradual decline of the cosmopolitan aristocratic society and the restraining influence of its morality upon international politics."

(Morgenthau; 1949, p.189). Patriotism, including the concepts of "fatherland" or "motherland", generated new motives for the common soldier and for the population in general. This trend produced a corrosive and potentially destructive line of thought which centered on the "*they-and-us*" theme. This divisive concept, a necessary ingredient of both prejudice and conflict, coupled with the lure and illusion of war being a "*great adventure,*" changed war into becoming an acceptable extension of national policy; it could demand a patriotic orientation and total involvement by the entire nation.

The Prussian general Karl V. Clausewitz (1780–1831) was a professional soldier who participated in the historical battles of Jena, Borodino, and Waterloo which led to the eventual defeat of Napoleon. Clausewitz could be called a "military philosopher" who personally experienced war, yet he also recoiled from the brutality of combat. He propagated the concept that "war is a continuation of policy," referring, of course, to the policy set by a nation. War in the time of Clausewitz was butchery. Russian soldiers, at the battle of Borodino, stood fast in rows subjected to artillery fire for two hours; during that time the only movement was the stirring in the ranks by falling bodies.

It is interesting that Clausewitz, although he must have been hardened by his own war experiences, condemned the cruelty inflicted by the Cossacks in their pursuit of Napoleon's retreating army. Could it be, as a general phenomenon, that one rationalizes, or even justifies, cruelties practiced by "us" and our like, while displaying outrage and disgust concerning cruelties which, under the hands of strangers, take on a different form? (Keegan; 1993, p.9).

As outlined earlier in this chapter, the intensity of war increased with the evolution of weaponry. Over time new types of arms accounted for rising numbers of killed and wounded and also extended the range over which such weapons could be deployed. As cities became the foci of manpower and industrial production, the distinction between combatants at the front and civilians at home became blurred. With the advent of the 20th century, the concept of the so-called "*home-front*" became generally accepted. The long-range bombing of London by German dirigibles in World War I evolved into the massive bombing raids on cities in World War II which stabbed far into the heartland of nations. The *fire-death* of cities will be treated late in this chapter. The total territory of nations became exposed to the fury of war; civilian zones, centers of industry, as well as the people's "*will to fight*" became targets of war. This development is not to imply that the practice of total war is entirely a modern concept.

The goal of warfare is to defeat the enemy. Traditionally, the word "defeat" meant the destruction of hostile forces on the battlefield; however, history clearly reveals that many ancient cities—some being fortresses, such as Megiddo and Jericho—and their non-combatant inhabitants, were totally annihilated. In other words, even though the term "total war" was not really coined until World War II, it was *de facto* practiced throughout history. During the Punic wars between Rome and Carthage, which ended with the razing of Carthage in B.C. 146, Rome's strength was not only her military skills and tenacity but also her ability to create a society of sufficient resilience and determination to endure what could be called total war. The enormous losses of Roman and allied war dead amounted to about 400,000. This number was probably unprecedented in prior military history O'Connell; 1980, p.78)

Total War became an official concept after the disastrous defeat of the Germans at Stalingrad (now Volgograd) in 1943. At the same time, Allied bombers carried their deadly loads to one German city after another. The Nazi Minister of Propaganda, Joseph Goebbels, asked ten questions of the German people at a rally in Berlin on February 18, 1943. The fourth question asked: "Do you want total war? Do you want it—if necessary—to be more total and more radical than we can imagine at this moment?" (Zentner; 1963, p. 311). The crowd responded with stormy applause and fanatical affirmation. The same year brought the near-total destruction of some of the largest cities in Germany. Two and a half years later, total war spawned the use of nuclear weapons and erased the Japanese cities of Hiroshima and Nagasaki.

The Signatures of Battle

The 13th and 14th centuries witnessed the Mongol invasions, when these men from the Asian steppes moved westward to conquer. It is possible that as many as half a million deaths resulted from these wars. But, with the exception of a few ruins around ancient Moslem cities, such as Samarkand (today in the former Soviet Union) and Herat (Afghanistan), time has erased the traces of those wars. It was not until the machines of war turned up the earth, flattened hills and forests, gouged out craters, and pulverized cities that the marks of battle became more permanent. Time has softened the scars, but it is still possible to see the cratered landscapes of France, especially around Verdun and the infamous Vimy Ridge, which date back to World War I.

As the physical destructiveness of war increased, so did the number of war victims. Until modern times, accurate accounts of war deaths probably did not exist, and even recent figures can be very misleading. The tally becomes even more confusing when civilian casualties are included. For example, several sources give the number of war deaths of World War II as ranging from 15 to 17 million; but, the total deaths may be as high as 50 million. It is believed that more Russians died during World War II than the totals for all other combatants combined: "They died in numbers so vast the Soviet government for years would seek to conceal the totals—probably 20,000,000 of them at the front alone" (Salisbury; 1978, p. 219). It appears that the battle deaths of World Wars I and II account for somewhere between 50 and 60 percent of all soldiers killed since the beginning of the Egyptian conquest of lands in the Middle East in 1479 B.C.

World War I (1914–1918) can be viewed as the first *international mass conflict* that deployed huge numbers of *mechanized weapons* and sacrificed the lives of millions of soldiers. War at this stage had become a process of *attrition;* the number and destructiveness of weapons, not individual heroism, primarily determined the outcome of battles. New types of weapons were available at the outset, or they made their appearance during the four years of fighting.

When looking back at the kinds of weapons available during World War I (the modern machine gun, super-heavy and long-range artillery, poison gases, airplanes, tanks, and relatively advanced telegraphy) one would have thought that such an array of arms would have forced a quick decision. But, in contrast to the strategic plans and expectations of the times, the war, after initial mobility, ground to a stalemate and forced the attacking armies into the trenches. Concentrated firepower, from repeating rifles, machine guns, and rapid-fire field artillery, stopped the attacks involving massed infantry. Pierre Berton explained: "This was static warfare, the warfare of attrition, where the gain of a few yards of ravaged real estate was hailed as a victory. But the real purpose was to kill so many of the enemy that he could no longer function. It was confidently believed that the mounting pile of corpses would force him to sue for peace" (Berton; 1986, pp. 53–54).

The trenches evolved into an intricate network of bunkers, shelters, or sleeping quarters. The trenches also served as the starting point for reckless infantry attacks across a new type of territory which separated the enemies: the *no man's land,* "a barren, shell-holed, fire-raked waste where death was apt to strike the moment a man ventured into it" (Williams; 1976, p. 85).

It may be impossible to select one battle that could serve as an example to illustrate its impact on both land and soldier, but the *Battle of Verdun* (February to June 1916), one of the worst ever, could fulfill this role. This monstrous battle, with its code name *Gericht* (*Judgement*), was designed to terminate the stalemate of trench warfare and to permit a massive German break through the French lines.

The basic concept of this German offensive was based on (1) an unprecedented concentration of artillery fire on a relatively small front of about 6 miles (9.5 km), and (2) forcing, through attrition, France's army to bleed to death. Through the resulting breach, where no soul was expected to stay alive, the German attackers were believed to be able to storm to victory.

About 1,000 artillery pieces were brought into firing position; close to 3 million shells were stockpiled, or enough to supply the guns for six days of continuous fire (*American Heritage;* 1964, p. 170). At the peak of the battle about 100,000 rounds per hour, ranging from field guns to heavy mortars with calibers up to 16.5 inches (42 cm), crashed into the French frontlines.

The landscape disintegrated. Stands of forest, houses, bridges, hilltops, entrenchments, barbed wire entanglements—everything was battered into oblivion. A strange, new, and ugly landscape emerged: a moonlike scene revealed thousands of craters, debris, corpses, jagged metal fragments, and splintered wood; all was saturated with the stench of death, smoke, and gas. But contrary to the German expectations the Battle of Verdun was not decisive: "They expected to cross a passive field of mangled corpses and crazed derelicts. Instead, Frenchmen black as stokers, uniforms ripped off, looking more like scarecrows than soldiers, stirred amid the desolation and pumped away with their rifles" (*American Heritage;* 1964, p. 172). Even today, the remnants of this place of death and utter destruction (traceable trenches, pieces of rusty steel, crater depressions) can still be seen. The harshness of the *battle landscape* has softened under a cover of grass and new trees, but the memory of 900,000 French and German casualties lingers on.

Plate 90. In May of 1918 a desperate German offensive, preceded by a massive artillery barrage, turned the landscape near the Aisne River into a nightmarish scene of craters, mud, and shattered tree stumps. (Photograph used by permission of the Imperial War Museum, London).

Other locations, including the Somme sector, suffered similar intensities of almost stationary battles; here, British and French troops attacked to relieve the pressure on Verdun. The British lost 60,000 men in one day! Later, at the close of World War I, the Battle of the Aisne turned another landscape into a ghostly array of mud, shattered tree stumps, and craters [Plate 90]. Some, but not all, military strategists learned from the slaughter of Verdun, the Somme, and the Aisne, that massed artillery alone, despite its ability to hammer the landscape into an unrecognizable wasteland, could not force a decision.

Another way of dealing with trenches and barbed wire had to be found. As early as 1914, a tracked armored vehicle, designed to roll unscathed over barbed wire and to eliminate machine gun positions, was proposed by a British colonel, E. D. Swinton. But this vehicle, the *tank,* was not introduced until September 15, 1916, when the first model (Mark I) joined the battle (Reid; 1986, p. 243). To camouflage the real purpose of this weapon, the tank was initially referred to as a *"water carrier"* for use in the Middle East.

The first deployment of the tank was ineffective for both sides because, at any given time, it was used too sparingly and in insufficient numbers. The first use of *massed* tanks did not take place until November 1917, when the British sent 350 of them into a 6-

mile (9 km) narrow front sector at Cambrai; but this action resulted only in a momentary breakthrough. In contrast, the tank in World War II, in close coordination with airplanes and mechanized infantry, became the successful instrument of the highly mobile ***blitzkrieg,*** or *lightning war.*

The massive deployment of the tank reached its all-time high in the Battle of Kursk (July 12, 1943); this conflict became the greatest tank battle in history. The German aim of this incredibly destructive operation was to eliminate a Russian-held bulge that jutted deeply into the German front. The Russians expected the offensive and had created a defensive trench system 2 to 3 miles (3 to 5 km) in depth, with secondary defenses at depths of 6 to 8 miles (9.6–13 km) (Ziemke; 1968, p. 133). The total number of tanks and assault guns lined up for this battle amounted to a staggering 2,700 on the German side, and 3,306 for the Russians; the armored forces were backed up by 900,000 and 1,337,000 soldiers, respectively (Jukes; 1968, p. 79).

A wide variety of new, highly mechanized weapons of hitherto unknown dimensions of destructive power typified World War II. But this war, in contrast to the much more stationary nature of World War I, usually remained highly mobile. The scars on the land were not as deep and permanent as the ones that resulted from World War I. It was the massive use of *air power* that produced two new characteristics of war: (1) air power erased the old concepts of a rigid linear defense, as typified by the trench warfare of World War I, and (2) air power not only became the *flying artillery* of the front, but it reached far into the homeland of the combatants. There, it destroyed the production centers of the enemy, and bombing was also used to break the *morale* of the people; the city became the target of massive area bombing.

How utterly complete the destruction of towns and cities became in World War II is clearly shown in the ruins of the French town of *St. Lô.* [Plate 91]; It was caught in the path of the Allied invasion in 1944. Another town and its adjoining monastery, Cassino (Italy), was pulverized in February of 1944. On February 16, waves of bombers dropped 600 tons on the monastery that was suspected of having served the Germans as an observation post. Another 1,000 tons of bombs and 4,000 artillery shells completed the devastation of Cassino [Plate 92] (Wallace; 1978, pp. 142–145).

The Vietnam War offers another example of a devastating mutilation of the earth's surface [Plate 93]. The newly developed concept of *"denying the*

Plate 91. On July 18, 1944, after days of concentrated artillery and aerial bombardments, the town of St. Lô (France) fell to American troops. St. Lô was completely destroyed in this operation. (Photograph used by permission of the Imperial War Museum, London).

countryside to the enemy" led to highly destructive *area bombings* directed at forests and fields. To achieve the necessary bomb concentration, the B-52 D had been modified to increase its normal internal high-explosive bomb load from the original 27 bombs (750 lb each) to 66. An additional 24 bombs were carried on racks under the wings, adding up to a total of 90 bombs; this represented a weight of 67,500 lb., or more than three times the original design bomb load (Anderton; 1979, p. 361)!

From 1965 to 1971, the United States dropped more than 13 million tons of bombs on Southeast Asia; nearly 80 percent fell on South Vietnam. This resulted in about 26 million craters (based on the standard 500-lb. or 230-kg bomb used most frequently) in an area only slightly larger than Texas (Menard; 1974, p. 356).

The destructive *impact on the land* manifested itself in four ways: (1) cratering made the land nearly unusable; each crater occupied an area of about 685 square feet (64 m^2); (2) many craters filled up with water and became breeding grounds for malaria-bearing mosquitoes; (3) a great number of rural wells, including their water-bearing strata, were destroyed; and (4) plinthic subsoil material was thrown out of the bomb craters; this severely reduced the crop-producing capability of the soils [Plate 94].

Plinthite, a mixture of iron and aluminum hydroxides, clay, and some quartz, typically forms in the substratum of tropical wet soils. Once this material is exposed to sun and air, it will harden *irreversibly* into rocklike consistency that is sometimes referred to as ***ironstone,*** or ***laterite.*** As long as this subsoil layer remains moist and undisturbed below the root zone, the soil can be used for agriculture, but exposed plinthite can seriously affect plant growth (Foth; 1984, pp. 279–280). Depending on the condition of the surface materials, the average crater depth reached about

Plate 92. In February 1944 more than 1,000 tons of bombs and 4,000 artillery shells pulverized the town of Cassino, which blocked the advance of the Allied forces in Italy. (Photograph used by permission of the Imperial War Museum, London).

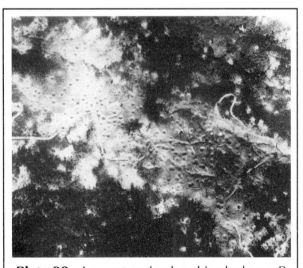

Plate 93. Area saturation bombing by heavy B-52 bombers turned the Mu Gia Pass of Vietnam into a devastated cratered landscape. (Photograph credit: U.S. Air Force Archives).

15 feet (4.5 m) and thus easily intercepted the plinthic soil layers [Plate 95]. The violent disturbance of this sensitive soil environment rendered the land unusable for growing crops and inefficient for grazing.

◎ The Gulf War

The environmental impact of war was dramatically demonstrated during the Gulf War (January 17 to February 24, 1991), and the effects will be felt for decades to come.

With the exception of the bombing and missile raids on Baghdad and on a few other urban areas, most of the direct war action took place in the sparsely populated desert areas of Kuwait and Iraq; however, the world watched with great concern and dismay the hundreds of burning oil wells and the mammoth oil spill that slowly crept into the Per-

Plate 94. Between 1965 and 1971 more than 10 million tons of bombs were dropped on South Vietnam alone. Each crater in this picture covers an area of about 685 square feet (64 m²). (Photograph credit: U.S. Air Force Archives).

sian Gulf. The 600 roaring oil-well fires, set ablaze by Iraq, generated a huge pall of black smoke that drifted as far as Pakistan, about 1,200 miles (1930 km) east of Kuwait.

The fires, as estimated by P. Hobbs of the University of Washington, carried about 20,000 tons of sulfur dioxide a day into the atmosphere; this amount is about 57 percent of the daily sulfur emissions from all coal-burning electric utility plants of the United States. Soot, comprising 30 percent of all particles emitted by the burning wells, amounted to 12,000 tons a day, while carbon dioxide generation reached 1.9 million tons daily, or about 2 percent of the per-day emission of carbon dioxide from the worldwide burning of fossil fuels and biomass.

In the initial stage of the disaster, no one could predict what the environmental impact would be in both short-term and particularly not in long-term effects. Days appeared like night and local temperatures dropped by as much as 36° F (20° C) below normal levels. Black and greasy rain fell in parts of Iran, and a black layer of oily soot discolored the normally light-colored deserts of Kuwait; the regions around the northern Persian Gulf seemed to be doomed. Fortunately, the worst predictions of the total environmental damage did not become true. Many scientists anticipated the fires to rage for two years, or more; yet, the last well was capped on November 6, 1991.

According to Bruce Hicks (director of NOAA's Air Resource Laboratory) the massive plume of smoke—in order to be transported around the globe by the **jet stream**—needed to reach an elevation of at least 45,000 feet (13,700 m); instead, the smoke cloud moved at an altitude between 10,000 and 16,000 feet (3,050 and 4,880 m) and only occasionally rose to 22,000 feet (6,700 m).

Smoke particles are hygroscopic; they attract water molecules from the atmosphere and foster condensation. The unusually heavy rainfall that occurred in many parts of the region cleansed large volumes of the polluted air. Although the early pollution levels of the air and land surfaces were appalling, the long-distance and long-term effects were less severe than first expected. The National Center for Atmospheric Research, and members of the University of Washington, sponsored by the National Geographic Society reported:

> While particle counts several thousand feet up in the plume were higher than U.S. air quality standards—even 80 miles downwind—concentrations of ozone and nitrous oxide were well within the limits. Sulfur dioxide levels were occasionally higher than those government standards. One thousand miles downwind—the greatest distance flown (by the research team) from the fires—pollution levels were no worse than those typically found in the U.S. urban areas.
>
> (Earle; 1992, p. 130)

Plate 95. This huge water-filled bomb crater in South Vietnam reveals the plinthic subsoil. Plinthite—a mixture of iron and aluminum hydroxides, clay, and silica—hardens irreversibly to a rocklike consistency once it is exposed to air. (Photograph credit: U.S. Air Force Archives).

The second disastrous impact deriving from the Gulf War was the vast oil slick that spread into the Persian Gulf from blown-up oil lines and ruptured tankers. The amount of oil was close to 126 million gallons; this is about ten times the quantity of oil released during the Exxon Valdez oil-spill disaster in the Prince William Sound of Alaska in 1989.

The greatest ecological damage done by the Persian Gulf spill occurred along the coast of Kuwait and that of northern Saudi Arabia. Thousands of seabirds were killed, and many other aquatic animals were destroyed or injured by the black mass of oil, which also devastated the Saudi shrimp industry. The cost of cleaning up the worst of the spill may rise as high as $5 billion, according to Saudi and Western estimates (Friedrich; 1991, p. 202). ── ◉

Chemicals at War

The records of *chemical warfare* go back to ancient times and considerably predate the age of modern chemistry. Poisonous and toxic materials (including the use of salt to sterilize the land) were well known to the ancient people of Greece and Rome; however, the systematic use of such substances in war was relatively rare; they were not very damaging to humans and the environment until World War I.

The early use of chemicals as weapons includes the mild poisoning of a river that served as drinking water to the enemy. This was carried out by the Athenian legislator, Solon, in 600 B.C. The enemy troops subsequently developed diarrhea, which effectively lowered their fighting ability (Clark; 1968, p. 13)! The use of *toxic smoke* was reported during the siege of Delium in 424 B.C., when sulfur was burned and led to the defeat of the Athenians. In 190 B.C., Roman soldiers, while attempting to tunnel underneath the walls of Ambracia, were driven back by the "nauseating and pungent smoke" of burning charcoal and feathers (Reid; 1986, pp. 237–238).

The aversion against the use of *nonchivalrous* weapons was reflected by the fact that *poisoned bullets* were outlawed by the Treaty of Strasbourg (1675). Peace conferences held at The Hague in 1899 and 1907 approved a resolution that outlawed "the use of projectiles the sole object of which is the diffusion of asphyxiating or deleterious gases" (Hersh; 1968, pp. 4–5). Unfortunately, the United States did not sign the agreement at that time. It was not until after the devastating experience of poison gases in World War I that the United States (with England, Japan, France, and Italy) agreed not to use

asphyxiating and poisonous substances (Clark; 1968, pp. 237–238). Finally, the Geneva Protocol of 1925 prohibited the use of poison gas and other chemical weapons. Some nations, including the U.S.A., the U.S.S.R., and China, viewed this accord only as a *no-first-use* agreement and reserve the right to retaliate in kind if the Protocol is violated (Meselson, Robinson; 1980, p. 38). In 1966, the General Assembly of the United Nations called on all nations to abide by the Geneva Protocol.

World War I saw the first systematic, large-scale use of *poison gas;* it started at the Ypres front on April 22, 1915. On that day the German forces released bottled *chlorine gas* that drifted toward the French lines. The French troops were utterly unprepared to deal with this new threat. Many choked to death as they attempted to outrun the greenish cloud that hovered near the ground and filled topographic depressions; chlorine gas is nearly 2 1/2 times heavier than air. The British soon followed the Germans in the use of chlorine gas (September, 1915); consequently, the troops of all combatants became equipped with gas masks and underwent antigas training (*American Heritage;* 1964, pp. 107–108).

More than 40 different gases emerged during World War I; the most effective ones included *phosgene, mustard gas,* and *chlorpicrin* (CCl_3NO_2). All of these were *antipersonnel* gases and had no notable effect on the environment. The total number of gas casualties reached 1.3 million soldiers, but only 91,000 deaths were attributed to gas warfare (Hersh; 1968, p. 5). Despite the availability of much more deadly gases to the combatants of World War II, poison gas, contrary to some rumors, was not used during that war. It is possible that stark fear, reinforced by the fact of not knowing the strength of the opponent in this area of warfare, kept both sides from using gas.

Chemical warfare took a new direction during the Vietnam War, when massive amounts of *herbicides* were used to (1) *destroy the foliage* of trees and undergrowth to deprive the enemy of cover, (2) *kill forests* to make them more *flammable* in preparation for their ultimate destruction by fire, and (3) *eradicate crops,* mostly rice, on which the Viet Cong depended.

The use of herbicides in warfare was not a new idea. During World War II, more than a thousand chemicals were tested in the United States in an effort to perfect militarily usable crop-destroying materials. The most important herbicidal compound developed was 2,4-D; it still is the most widely used herbicide in the world (Westing; 1984, p. 4); however, none of these herbicides were used in that war. In the 1950s, the British employed a limited amount of *defoliants* to deprive the communist insurgents of cover along the main roads in Malaya (Buckingham; 1982, p. iii); but some was also used to destroy crops in that conflict.

The herbicide operations in Vietnam were carried out on a totally different scale. They began after the government of South Vietnam asked the United States to start aerial applications of such chemicals. The major phase of this kind of war, dubbed *Operation "Ranch Hand,"* began in January of 1962. Nine years later, about 18 million gallons (68 million liters) of chemicals had been applied to about 20 percent of South Vietnam's tropical forests; this included 36 percent of its *mangrove forest* (Buckingham; 1982, p. iii). In addition to the severe biological damage done to the mangroves, which are primarily found in the coastal areas of South Vietnam, widespread erosion of the denuded river and estuary banks set in (Westing; 1984, pp. 12–13) [Plate 96].

The spraying of herbicides was carried out by modified UC-123s, equipped with a 1,000-gallon (3,780 l) tank. The tank could be emptied in four minutes to achieve a rate of application of 3 gallons (11 liters) per acre (0.4 hectare). The release was carried out at the lowest flight altitude possible; each plane covered a swath 260 feet (80 m) wide and about 9.5 miles (16 km) long [Plate 97] (Buckingham; 1982, p. 132).

In addition to the direct destruction of the forests, the spraying of defoliants had also a long-range effect on soils. According to a report by Hoang Van Huay (University of Hanoi), as reported by Westing, the chemicals interfered with the soil's microbiology by reducing the *rate of degradation* of organic matter; they also affected the root system of plants (Westing; 1984, p. 65). Moreover, the soils showed seriously depleted nitrogen levels and a low *humus* content. These symptoms, combined with the accelerated soil erosion, drastically reduced the land quality; this combination resulted in poor forest regeneration.

Plate 96. This photograph shows modified UC-123s in the process of releasing herbicides (defoliants) over a mangrove forest that borders a canal in South Vietnam. (Photograph credit: U.S. Air Force Archives).

Plate 97. Three modified UC-123s engaged in releasing liquid defoliant over a forested area in South Vietnam. Each plane covers a swath about 260 feet (80 m) wide and about 9.5 miles (16 km) long. (Photograph credit: U.S. Air Force Archives).

At this point it is difficult to say what the long-term effects of the various herbicides will be on humans. The deep concerns on the part of Vietnam veterans who were exposed to these chemicals, such as ***Agent Orange,*** were well publicized. About 60 percent, by volume, of all herbicides sprayed in South Vietnam was Agent Orange. This defoliant contains a high level of the toxic contaminant *dioxin;* dioxin has proven to be both *carcinogenic* and ***teratogenic,*** or causing birth defects (Westing; 1984, p. 20).

The use of defoliants in Vietnam also aimed at making it easier to burn out large forested areas that held enemy strong-points and military supplies. One of the largest efforts to burn forest tracts took place in 1967; the operation was coded ***Operation "Pink Rose."*** Each of three areas covered about 12,000 acres (4,850 hectares); each area was sprayed during the month of November of the preceding year, and again in January 1967. The planes flew 200 ***sorties*** (individual missions) and released a total of 255,000 gallons (964,900 liters) of defoliants. Subsequently, B-52s dropped incendiary bombs, and smaller planes released ***napalm*** to complete the destruction of the dying forest (Buckingham; 1982, p. 127). It is not possible to predict whether these devastated forest biomes will ever fully recover or whether a new low-level plant association will establish itself. Many more decades will have to pass before a valid assessment can be made.

Fire-Death of Cities: Firestorms

Earlier in this book (chapter 11) we saw that humankind used fire long before knowledge existed about how to make it. Fire, according to Greek mythology, was delivered to humankind by Prometheus; but this gift was given against the wishes of Zeus. Would fire become too powerful a tool in the hands of mortals? Fire, indeed, grew into a devastating force during the history of war: a burning branch, a flaming arrow, Greek fire, flamethrowers, incendiary bombs, phosphorous, napalm, and finally, the searing cosmic heat of the hydrogen bomb.

The most destructive fires associated with warfare occurred at the end of World War II as a result of massive *incendiary raids* against German and Japanese cities. These fire raids preceded the nuclear bombs of Hiroshima and Nagasaki and, initially, caused greater casualties than the atomic bombs!

Bombing of cities from the air began on a moderate scale in World War I. In the course of that war German airships (dirigibles), in addition to some bombing planes, dropped a total of about 200 tons on England and killed some 550 people (Frankland; 1970, p. 10). In World War II, within 14 hours, three air raids on Dresden (February 13–14, 1945) killed an estimated 135,000 people, while the massive attack on Tokyo (March 9, 1945) resulted in a death toll of nearly 84,000. By comparison, the initial number of people killed by the atomic bomb in Hiroshima was about 72,000.

Throughout the war the number of bombers used in raids increased, and so did the size and types of bombs. But, in retrospect, it was *fire* that killed, directly and indirectly, more people in bombing raids than explosive bombs; fire also did more permanent damage to buildings and industries. Especially the so-called *firestorm* achieved an unmatched intensity of destruction and death. Chapter 11 mentioned firestorms in connection with disastrous forest fires. But the overwhelming terror of human-induced firestorms exceeds by far that created by nature.

The history of mass fires in cities is long (London, 1666; Hamburg, 1846; Chicago, 1871; Tokyo, 1923), and each fire showed its particular characteristics. However, basically, there are two major types of fires: (1) the *conflagration,* and (2) the *firestorm.*

In a conflagration the fire front is driven forward by a preexistent ground wind. This mass fire, once it is fully developed, *slants forward* and *downwind;* therefore, it forms a distinct *fire front,* which is preceded by a turbid mass of preheated vapors. The fire continues to move forward until there is no further combustible material (Sanborn; 1946, p. 169). This type of fire developed in connection with the most damaging air raid flown against Tokyo on March 9, 1945. This violent conflagration destroyed about 16 square miles (41 km^2) of Tokyo and killed nearly 84,000 people.

The firestorm is different in three ways:

1. It forms under calm atmospheric conditions; that is, no ground wind exists at the outset.
2. It develops a stationary convectional column composed of burning gases and heated air.
3. It usually results in a complete burnout of all combustible materials within its boundary (Musgrove; 1981, p. 103).

The most devastating firestorm ever created by a single air raid was the one in connection with an attack by the British Royal Air Force on July 27, 1943. This attack aimed to destroy, once and for all, the industrially important port city of Hamburg, which is located about 90 miles (145 km) southeast from the mouth of the Elbe River in northwestern Germany. This bombing mission was a part of a series of raids planned against this target; the operation, envisioning ten attacks, was given the code name Operation "*Gomorrah*" (Caidin; 1960, p. 8).

The raid was carried out by more than 700 heavy bombers that carried a 2,400-ton bomb load (1,300 tons of high-explosives and 1,100 tons of mixed incendiaries). It was later estimated that an average of about 840 explosive bombs and 99,160 incendiaries per 0.4 square mile (1 km^2) fell into the target area (Klöss; 1963, p. 47) [Plate 98]. Immediately, two out of three buildings were afire within the concentrated target area of 4.5 square miles (12 km^2), while general fires were started over an area of 17 square miles (44 km^2). In less than half an hour the fires in the concentrated area merged and set in motion a gigantic firestorm. The *convectional column* rapidly extended upward; it formed a towering *cumulonimbus cloud* that extended to about 30,000 feet (9,100 m) over the burning city (Ebert; 1963, p. 256) (fig. 15.1).

In the absence of ground wind, and as the heated air rises, this type of fire draws in air from all directions. This could be likened to a *chimney effect*. As the intensity of the fire increases, so will the wind velocity of the inrushing air at ground level; this is the beginning of a *thermal firestorm* [Plates 99 and 100].

Plate 98. The firestorm of July 27, 1943—an event of unprecedented fury—left 4.5 square miles (12 km^2) of Hamburg (Germany) completely burnt-out. The devastated district was declared a "dead zone" by the German authorities. (Photograph credit: Hans Brunswig).

In the case of the Hamburg firestorm, special atmospheric conditions intensified the updraft and gave rise to a new concept: the *atmospheric firestorm*. This type was unknown until 1943 and attained storm effects that had been beyond imagina-

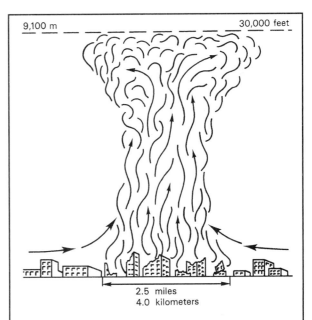

Central Firestorm Area

Figure 15.1. A towering convective cloud, fed by a raging firestorm, rose to about 30,000 feet (9,100 m) above the burning city of Hamburg on July 27, 1943. Concentric winds of hurricane force were drawn into the core of the firestorm area.

Plate 99. Sixty-four candles, of a total of 1,444, were ignited in this firestorm experiment conducted by the Hamburg Fire Department. Within less than 50 seconds the convectional air currents lengthened the flames and tilted them toward the center. The progress of this experiment is shown on Plate 100. (Photograph credit: Hans Brunswig).

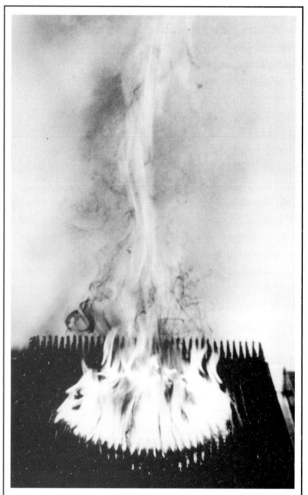

Plate 100. Within two minutes the flames of the candles had merged and sent smoke and burning gases to a height of about 5 feet (1.5 m). Note that the "firestorm" did not spread outward from the ring of candles originally ignited in this experiment. Compare this picture to Plate 99. (Photograph credit: Hans Brunswig).

tion up to this event (Brunswig, 1978, pp. 264–265). The *inrushing* air currents produced a *blast-furnace effect* that raised the temperatures at street level to an estimated 1,400° F (760° C) in some of the glowing street canyons; these temperatures caused a complete burnout within the core area of the firestorm [Plate 101].

The specific atmospheric conditions that contributed to the unprecedented fury of this firestorm were:

1. A strong lapse rate (decrease of air temperature with elevation), which encourages warm ground air to rise quickly.
2. A very low level of relative humidity that had prevailed for several days.
3. The absence of ground wind of any significant velocity at the beginning of the raid.

The strong vertical lapse rate was the result of intense solar heating of the city and the influx of cool maritime air aloft. The heating effect of the sun's rays had been reinforced by the greenhouse effect of smoke and dust, which derived from preceding air attacks on other sections of the city. The *normal lapse rate* is about 3.5° F/1000 feet (6.4° C/1000 m). The lapse rate at Hamburg at 5:00 P.M. on July 27 averaged 4.8° F (2.6° C), with a sharp temperature decrease of about 16° F (8.8° C) within the first 2,000 feet (610 m) (fig. 15.2)! As the heat of the fires

was added to the lower atmosphere, the air became totally unstable; this initiated a powerful vertical surge and the beginning of the firestorm. At maximum development the storm's peak wind velocities may have been as high as 120 to 150 mph (193 to 240 km/h). The howling winds picked up heavy debris and twisted off mature trees [Plate 102]. The low relative humidity of 30 percent, combined with the daily temperature highs ranging from 80° to 92° F (27° to 34° C), had dried out the woodwork of buildings and other combustible materials rendering them highly flammable.

The final casualty list probably will never be known, but many estimates go as high as 52,000. The lowest official figure was given as 31,647, and the

Plate 101. During the Hamburg firestorm on July 27, 1943, fire vortices and street-level temperatures estimated at 1,400° F (760° C) left this Civil Defense messenger in unrecognizable condition. (Photograph credit: Hans Brunswig).

Plate 102. Inblowing wind velocities during the 1943 Hamburg firestorm reached up to 150 mph (240 km/h) and twisted off mature trees. (Photograph credit: Hans Brunswig).

highest as 41,800 (Brunswig; 1978, pp. 400–401). Thousands of victims were found in unrecognizable condition in the streets where *heat radiation, fire whirls, smoke,* and *falling debris* had killed them

[Plate 103]. Equal numbers were discovered entombed in basement shelters where excessive heat, later estimated at ranges from 200° to 900° F (94° to 482° C), the lack of oxygen, and the accumulation of carbon monoxide ended their lives. **Hyperthermia** and *carbon monoxide poisoning* ranked highest as the cause of death in the shelters and cellars (Gräff; 1955, p. 205).

The high concentration of carbon monoxide in the basements was caused by the *incomplete combustion* of debris that had buried the shelters. The carbon monoxide combines with the hemoglobin and thereby reduces the oxygen-carrying capacity of the

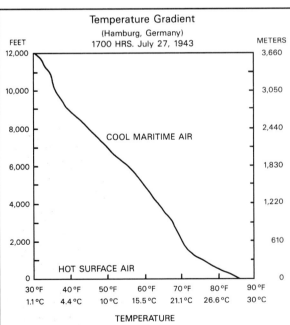

Figure 15.2. Strong ground heating, in combination with the introduction of cool maritime air aloft, resulted in a very steep temperature lapse rate over Hamburg on July 27, 1943. When the heat from the mass fire was added, the air became totally unstable and initiated a powerful vertical surge. This violent surge was the beginning of an atmospheric firestorm.

Plate 103. In their vain efforts to escape the Hamburg firestorm, thousands of victims were felled by the fire whirls and flames that roared through the open areas and street canyons. Similar scenes occurred during the firestorms that destroyed many Japanese cities. (Photograph credit: Hans Brunswig).

blood; this leads to the suffocation of the victim when about one-third of the hemoglobin has combined with carbon monoxide.

A firestorm and similar effects were observed during the triple air raid (three raids within 14 hours) against the hitherto unscathed city of Dresden (*February 13–15, 1945*). The raids killed 135,000 people, totally destroyed 35,470 buildings, and damaged another 27,939 (Irving, 1963, pp. 237–238).

Instantaneous Obliteration: The Nuclear Bomb

Earlier in this chapter it was stated that, as weapons evolved over time, the distance between combatants increased, the destructive power of the weapons increased, and the impact of weapons grew in severity. The first use of the *atomic bomb* in war, which resulted in the instantaneous obliteration of the Japanese cities of Hiroshima and Nagasaki on August 6 and 9, 1945, respectively, epitomizes these propositions [Plate 104]. Let the destruction of Hiroshima illustrate these points: (1) a single bomber (the B-29, *Enola Gay*) flew at an elevation of 30,000 feet (9,100 m) and carried a single bomb, (2) the bomb ("*Little Boy*") was about 10 feet (3 m) long, had a diameter of 28 inches (71 cm), weighed about 4 tons, and had a destructive power of about 13 kilotons TNT (13,000 tons of trinitrotoluene), and (3) the nuclear explosion, in combination with a resulting

Plate 104. A new era of total destruction began with the use of atom bombs in war. This picture shows the typical mushroom cloud of a nuclear explosion that devastated Nagasaki. (Photograph credit: U.S. Air Force Archives).

firestorm, totally destroyed an urban area of 5 square miles (13 km^2) and killed an estimated 151,900 to 165,900 people (including 32,900 military personnel) according to Japanese statistics (Hiroshima-Nagasaki Committee; 1985, p. 19).

The first *controlled* release of atomic energy at the University of Chicago on December 2, 1942, presented the world with new potentials and new problems (Calder; 1962, p. v). But the priority of those years was to create a new weapon: the atom bomb. President Roosevelt believed that Germany was leading in the effort to produce nuclear weapons, and he eagerly launched the atomic bomb project. The war was almost over before the United States uncovered "the pitiful state of the German nuclear art" (Wyden; 1985, p. 368).

The climax of the American development of the first nuclear bomb came on June 16, 1945, when a sunlike flash poured over the desert site of New Mexico where the bomb device was exploded atop a tower: "One observer, who lay flat on the ground 10 miles (16 km) from the explosion, with his head down and only his neck exposed, said that it felt as if a hot iron had been held next to his skin" (Hecht; 1955, p. 3). Three weeks later, the new weapon descended on Hiroshima.

To understand the three major impacts of the atomic bomb (*blast, heat, radiation*) it is helpful to describe briefly the basic working principles. There are two general types of nuclear bombs. The older one, used on Hiroshima and Nagasaki, is a *fission* device. When slow-traveling neutrons strike the isotope U-235 atom, the atom will split into two particles. This split is accompanied by an enormous release of energy as part of the *mass* is changed into *energy*. The mass of all fission products is not equal to the original mass. Moreover, during the fission of the U-235 atom additional neutrons are released. By striking other uranium atoms they initiate the *chain reaction*. The Hiroshima bomb used *uranium* as fissionable base, whereas the Nagasaki type employed *plutonium*.

The *fusion* process, leading to the development of the *thermonuclear* or *hydrogen bomb*, is based on the principle of smashing together hydrogen atoms to form heavier atoms. Extremely large amounts of energy are released during fusion. This process requires very high temperatures (millions of degrees) that normally are not available, except during the fission process. Thus, the hydrogen bomb requires the fission-bomb device to trigger fusion.

The atomic bomb that obliterated Hiroshima was released from a high-flying B-29 bomber and descended to its predetermined detonation altitude of about 2,000 feet (610 m) in a little over 40 seconds. The bomb exploded about 800 feet (240 m) off its aiming point, the Aioi Bridge, directly above the Shima Hospital (Wyden; 1985, p. 253) [Plate 105].

About 50 percent of the bomb's energy was used up in *blast* and *shockwaves,* 35 percent was expended in *thermal radiation,* and the remaining 15 percent produced *nuclear radiation* (U.S. Department of Defense; 1962, p. 8). The same source indicates that an estimated 50 percent of the deaths were caused by burns of one kind or another. The number of people killed in the inner zone (0 to 0. 6 miles, or 0 to 0.9 km) was 26,700 out of 31,200, or 86 percent. This number declined to 27 percent (39,600 out of 144,800) in the surrounding zone (0.6 to 1.6 miles, or 0.9 to 2.6 km).

In contrast to the more mountainous topography of Nagasaki, Hiroshima is located in the flat delta lands of the Ota River. This fact partially explains the uniform distribution of the bomb effects; there were no hilly ridges that provided a *shadow effect* as observed for some areas in Nagasaki. The highly flammable nature of the predominantly wooden structures in Hiroshima, the open and flat terrain, and the absence of ground wind led to the development of a massive firestorm [Plate 106]. Wind velocities reached about 40 mph (64 km/h) in Hiroshima, while the winds reached only 34 mph (55 km/h) in Nagasaki. The more open landscape, and greater uni-

Plate 105. This view shows the total devastation in Hiroshima near "ground zero" above which the atom bomb exploded. The stark remains of the Agricultural Exposition Hall—on the right side in the picture—now serve as a focal point in the war memorial park of rebuilt Hiroshima. (Photograph credit: U.S. Air Force Archives).

Plate 106. The highly flammable nature of Hiroshima's predominantly wooden buildings, the flat terrain, and the initial absence of ground wind led to the development of a massive firestorm. (Photograph credit: U.S. Air Force Archives).

Plate 107. Within seconds these municipal fire engines in central Hiroshima were destroyed by the impact of the nuclear explosion and radiation. (Photograph credit: U.S. Air Force Archives).

those who survived, in horrible conditions, for either hours, days, or several weeks. The few photographs selected for this book merely give a slight indication of the destructive power of the atom bomb [Plates 107 and 108]. It is even more incomprehensible that the nuclear weapons of today have a destructive power thousands of times greater than the atomic bombs used at the end of World War II. This overwhelming destructive power will not be restricted to the devastating physical effects but will extend—

formity in the urban pattern, may explain the difference in the intensity of the firestorm and in the totally burnt-out areas: 5 square miles (13 km^2) for Hiroshima and 2.6 square miles (6.7 km^2) for Nagasaki (Hiroshima-Nagasaki Committee, 1985, pp. 14–15). Nevertheless, the blast effect, that is, the scale of destruction, was much greater at Nagasaki; here, even large concrete buildings had their sides facing the explosion pushed in. The more severe destruction of structures in Nagasaki, although more restricted in area, must be attributed mainly to the greater force of the plutonium bomb "*Fat Man*" which, in its testing stage at Alamogordo (New Mexico), was equivalent to 20 kilotons of TNT (*Wyden;* 1985, p. 194).

Statistics and descriptive language cannot realistically portray the intensity of utter destruction left in the wake of the two nuclear bombs that were dropped on the two cities. Above all, no accounts, or pictures, can even approach the level of suffering of

Plate 108. The small strips of pavement on this bridge across the Ota River in Hiroshima were shielded by the concrete pillars from the atom bomb's heat radiation. (Photograph credit: U.S. Air Force Archives).

with many factors still unknown—into the dark realm of mutations, long-range carcinogenic threats, and the poisoning of the land, the waters, and the atmosphere of our planet.

The first American full-scale thermonuclear (*fusion*) test explosion at Eniwetok in 1952 generated *200 times* the power of the Hiroshima bomb (Hecht; 1955, p. 194). One of the largest, if not the largest, hydrogen bomb tests reported (Soviet Union; October 1961) was said to have been in the 58-megaton range (*58 million tons of TNT*), or about *3,000 times* more powerful than that which erased Hiroshima!

The use of the two nuclear bombs against Japan brought the war to an end. A full-scale invasion of Japan would have cost untold lives on both sides, but a *new era* had begun. John Williams, a British writer who specialized in the history of World Wars I and II, summed it up: "Nobody regretted the end of the war, but the manner in which it ended has weighed on the world's conscience ever since. And the advent of nuclear weapons has set problems that have ever since plagued the world in peace and war alike" (Williams; 1976, p. 111).

A Difficult Afterthought

It is, indeed, impossible not to be horrified and disturbed by the overwhelming events of destruction and human misery in connection with the fiery death of great cities during World War II. It may even be more difficult to accept the reasons for having used the nuclear bombs against Japan. However, it may be necessary, to achieve some degree of *historical balance,* to broaden one's focus as did Lt. General Ira Eaker, USAF (Ret.) in the introduction to David Irving's book *The Destruction of Dresden,* in which he said:

> I find it difficult to understand Englishmen and Americans who weep about enemy civilians who were killed but who have not shed a tear for our gallant crews lost in combat with a cruel enemy. I think it would have been well for Mr. Irving to have remembered, when he was drawing the frightful picture of civilians killed at Dresden, that V-1's and V-2's were at that time falling on England, killing civilian men, women, and children indiscriminately, as they were designed and launched to do. It might be well to remember Buchenwald and Coventry, too.
>
> (Irving; 1963, p. 7)

Many more places, names, and events could be added that would bring back memories of suffering and death: More civilians died in the siege of Leningrad (now St. Petersburg) than ever died in a modern city, more than ten times the number who died at Hiroshima (Salisbury, 1969, p. 513). Or, out of the original 75,000 prisoners taken by the Japanese after the surrender of Bataan (April 9, 1942), only 54,000 survived the *death march* to the prison camp. Of these, another 21,600 prisoners (40 percent) died of disease, hunger, or maltreatment during the next three months (Sulzberger; 1966, p. 169).

End of the J-Curve?

No part of the world would be able to escape the effects of a massive thermonuclear war. Furthermore, there would only be losers. These thoughts were expressed long before the peace demonstrations of the late 1960s and 1970s, and—of all people—by an elderly, highly conservative British military officer, Sir Robert Saundby, Air Marshal: "Since, under modern conditions, it would be impossible to win a full-scale war, in the sense that the victor could gain anything from it, the main object of armed forces nowadays is to prevent its occurrence" (Saundby; 1961, p. 224). This realization, in addition to the fact that every place on earth can easily be reached by contemporary weaponry, is probably the main barrier that is holding back World War III.

The advent of the airplane and of the submarine allowed war to move high above and deep below the surface modes of warfare. Adding *space* to the arena of weapons opens another dimension of conflict that erases all preceding scales of distance and time. It is no longer realistic to project the impact of a future large-scale nuclear war on the *environment* because its impact will be *global.*

Two weapon systems, the rocket (i.e., missiles) and the advanced versions of nuclear devices, are primarily responsible for creating this drastic situation that offers two basic alternatives: *stalemate,* or *annihilation.*

It is difficult to assess when rockets were first used in warfare. They were employed in ancient times in the Orient, and they were first mentioned in Europe when a tower of Chioggia was set ablaze by a rocket in 1379 (Reid; 1986, p. 184). By 1805, the English reintroduced the rocket to the army; this device was first developed by William Congreve (Grun; 1979, 1805 F). Rocketry based on a *liquid fuel* system owes its existence to three scientists, a Russian (Konstantin E. Tsiolkovsky), an American (Robert H. Goddard), and a German (Hermann

Oberth). Almost all modern military rockets are based on the basic design concepts of these men.

World War II saw the deployment of two devastating longrange rocket weapon systems; the two rockets were the German-designed V-1 and V-2 (V = *Vergeltungswaffen = revenge weapons*). The smaller aircraft, like V-1, was aimed at the Greater London area and killed more than 5,000 people while injuring another 40,000. The V-2, 46 feet (14 m) long and carrying about 1,650 lb. (750 kg) of explosives, was first sucessfully tested on October 3, 1942, at the rocket development center of Peenemünde on the Baltic Sea. It reached 60 miles (96 km) up into the atmosphere and thus more than doubled the record altitude of 25 miles (40 km) previously achieved by a projectile fired from the famous "*Paris Gun*" in World War I (Dornberger; 1954, p. 25). The supersonic rocket was used toward the end of World War II and struck England at a rate of 20 a day.

Since the days of World War II missile warfare has come a long way both in terms of physical impact and accuracy. This development was clearly demonstrated during the Persian Gulf War (January 17 to February 24, 1991) when *cruise missiles* were launched from naval units in the Persian Gulf at targets in Bagdad, Iraq.

The TLAMs (Tomahawk missiles) were directed to critical targets such as command-and-control facilities, surface-to-air missile sites, and nuclear and chemical weapons installations. It was estimated that about 85 percent of the missiles struck their targets.

Just because particular targets were struck did not mean that they were effectively destroyed; the results at times were less than what expected. One reason for this lack of assurance was that a modern 2,000 lb. bomb is not much more destructive that the 2,000 lb bomb used in World War II (Friedman; 1992,p.189).

This problem of uncertainty existed in World War II as well as in all bombing missions since that time as in the Korean and Vietnam wars and more recently in NATO operations in the air war against Serbian forces. In some incidents it could not even been asserted that targets had been legitimate or merely decoys. Fully valid assessment of bombing of bombing results simply cannot be achieved unless targets are examined on the ground.

The TLAMs were guided to their targets by way of an *inertial navigation system;* the system was updated by references to known elevations shown on maps as well as by direct visual reference scenes near the targets. The topographic reference maps contain a grid of preplotted elevation data. The missile computer analyzes the terrain and compares the profile with the stored data and instantaneously calculates the missile's precise position. This procedure allows the missile to update the inertial navigation system and correct its course to the target (Frogett; 1992, p. 72).

Before the Tomahawk missiles could be used in the Gulf War, a last-minute effort had to be made by the U.S. Defense Mapping Agency to map a topographic and other salient features of Iraq. This project was made possible because of the five-month preparation period needed for the military build-up of the Allied Forces. Today, new types of Tomahawk missiles take advantage of the GPS (*Global Positioning System*) technology which makes any detailed pre-programmed terrain survey unnecessary (Friedmman;1992,p.242).

Contemporary intercontinental missiles (*ICBMs*), and submarine-launched ballistic missiles (SLBMs), combine the unimaginable destructive power of thermonuclear bombs with the almost unlimited reach of modern rockets. With that, the anonymity of the opponents, the reach of their weapons, and the global effects of their combat have reached the realistic peak of the J-curve: A key is turned, a button is depressed, a fiery trail cuts through the sky, and *within minutes* a city 10,000 miles (16,800 km) away incinerates.

In December 1991, far-reaching political and geographical changes shifted the *power* foci of the world. One of the major events was the disintegration of the Soviet Union; it was the last stage of an empire which had been so bloodily assembled since the times of Peter the Great (Kissinger; 1994, p. 785). The disappearance of this superpower, as well as the fracturing of Yugoslavia, added another twenty nations to the long roster of independent countries, a fact which further complicated the *global geopolitical array* since the end of World War II.

Fear of mutual destruction led to several arms reduction and nuclear test ban treaties; the earliest of these was the *Nuclear Test Ban Treaty of 1963*. This agreement banned the testing of atomic weapons in space, above the ground, and in the oceans. Unfortunately underground testing was permitted, and since that time many other nations became nuclear powers.

In the former Soviet Union alone, the nuclear arsenal is now stored in Russia, in the Ukraine, in Kazakhstan, and in Belarus. Whether the last three nations will fully ban all nuclear weapons remains to be seen. Moreover, the existence of such weapons in other parts of the world should remind us of the fact that this specter of mass annihilation has not left our planet.

In addition to the direct impact of atomic warfare, there are several long-range effects, many are in the realm of speculation but nonetheless ominous. Such effects include the long-range consequences of various types of *radiation,* as well as *toxicity* of residual materials connected with nuclear explosions. Others deal with influences on weather and climate, and with destroyed or severely damaged ecosystems.

The extremely high temperatures that are generated in thermonuclear explosions, or *fireballs,* are capable of producing excessive amounts of nitrogen oxides deriving from oxygen and nitrogen in the air. After the atom bomb tests in the atmosphere in the early 1960s, a marked *depletion of ozone* was detected. A full-scale nuclear war could severely affect, or even destroy, the world's ozone layer. The vital role of this layer in protecting the biosphere against excessive amounts of ultraviolet radiation was stressed before in connection with disasters involving the atmosphere. The breakdown of the ozone layer would bring about a global environmental disaster that, in combination with the impact of nuclear war, could terminate life on earth as we know it today.

Another global effect of a thermonuclear exchange may bring about a dangerous lowering of the earth's temperatures, a phenomenon frequently referred to as **nuclear winter.** This ominous expression made its first appearance in an article by a group of scientists that included Carl Sagan. The concept of nuclear winter rests on the blockage of solar radiation by massive amounts of dust and smoke particles injected into the atmosphere; this blockage would result in lowering the earth's overall temperatures. The intensity and duration of such a temperature change obviously would depend on the number of nuclear weapons used and the amount of particles raised. It is believed that the latitudinal belt between 30 to 70 would see a temperature drop of about 56° F (31° C); the largest decline would occur 25 days after the nuclear explosions. It would require more than 150 days for the temperatures to rise back to near-normal levels (Eagleman; 1991, p. 198). While *volcanic dust* emissions would be totally outscaled by a nuclear war, the measurable effects of volcanic ejecta in the atmosphere give us a hint of this threat. During 1812 to 1815 Tambora, a volcano in Indonesia, ejected about 24 cubic miles (100 km³) of tephra

and finer dust. Unseasonably cool weather, excessive rainfall, and a short growing season resulted in calling 1816 *"the year without a summer."* J. M. Mitchell, a research climatologist for NOAA, pointed out that cooling trends in the lower atmosphere occurred after some large volcanic eruptions as those of Krakatoa (1883), Agung (1963), and several others that took place around the turn of the century (Simkin, Fiske; 1983, pp. 422–423). Lower global temperatures would severely affect agricultural potentials by delaying germination and by shortening the growing season. The effects would bring on extensive food shortages and famines all over the world, not just in the areas affected directly by a nuclear war.

Earth, at the end of the J-curve, would be indeed, a desolate, poisoned, and dark place. The words of a French priest, spoken after the Battle of Verdun (1916), would drift once more across the wasted earth: "Having despaired of living amid such horror, we begged God—to let us be dead" (*American Heritage;* 1964, p. 192).

The impact of war in general, or of a potential thermonuclear confrontation in particular, casts, indeed, a deep shadow over the future of humankind on our planet. It is sad that the human mind, so uniquely evolved and capable of logical and creative thought, also made possible the development of weapons capable of global destruction. But it is even more significant to realize that it is the educated mind, honed by the disciplines of natural and social sciences, that gives humankind the tool to turn the tide.

With the *historical* and *empirical* experience available, there simply is no excuse for denying what we have learned from war. Nor is there an excuse for blotting out the realistic assessment of damaging effects on the planet of powerful weapons not yet used in anger. The people of the world, once this information is known, must use their conscience and creativity to find a road to survival based on knowledge and reason.

The late Dr. Jacob Bronowski, writer, scientist, and humanist—a true Renaissance man—offered these thoughts:

> We are all afraid—for our confidence, for the future, for the world. That is the nature of human imagination. Yet every man, every civilization, has gone forward because of its engagement with what it has set itself to do. The personal commitment and the emotional commitment working together as one, has made the Ascent of Man.

(Bronowski; 1973, p. 438)

SOCIAL ASPECTS OF DISASTERS

16 C H A P T E R

Key Terms		
ecosystems	post-impact stage	overwarning
vulnerability	rescue phase	evacuation
self-help	action groups	relocation
irreversible degradation	ripple effect	human factor
megacities	pre-disturbance state	reconstruction
non-traditional behavior	geostationary satellites	Economic Miracle
initial impact stage		

In the beginning of this book we inquired into the role of humankind in creating hazards and disasters. It was stated that ancient humans were merely an integrated part of the natural environment; usually they did not exert a major influence on *ecosystems*. In contrast, modern society, after having developed science and technology, achieved powerful capabilities which now threaten not only our existence but also nature itself.

After having considered the scope of natural and human-induced hazards involving the lithosphere and earth materials, the hydrosphere, the atmosphere, and the biosphere—as well as the impact of war on people and environment—it is necessary to ask an important question: "*What are the social aspects of disasters?*"

The relationship of humans to hazards has been as complex and diversified as the nature of hazards. People in their early stages of existence probably viewed lightning, earthquakes, volcanic eruptions, storms, and floods—to mention some of the most damaging hazards—with great fear and awe which probably have not left entirely modern humans either. However, early humans did not have the understanding of the world around them. Their respect and fright shown to nature's fury must have been so immense that they viewed it as manifestations of the gods of fire and water, and of the earth. As people developed a broader perspective about such events a disturbing question possibly arose: Could these threats by nature be signs of punishment for human transgressions?

Taking Risks

In several of the preceding chapters the question was asked why people keep on farming the land on the lower slopes of volcanoes, build their homes, entire subdivisions, and industrial facilities in endangered floodplains, or locate cities near or even on faults as we can readily observe in California, in Japan, in Central America and in many other locations. The question could be answered by saying that fertile soils are found near volcanoes, that floodplains are easy to develop and may offer aesthetically attractive features, that most fault systems tend to be inactive for long periods of time, and that people simply "*take a chance.*"

Taking a chance and facing risks is a well-known human behavioral trait which can be triggered by a provoking "*I dare you.*" Such challenges may induce a youngster to venture across the thin layers of ice of a freshly frozen pond, to climb to the highest branches of a tree, or to jump off a high point.

Risks faced by most mature persons tend to remain more in the background of their lives. Such risks are more subdued, less acute, and tend not to be provoked by dares. Businessmen in Los Angeles think of sales, of losses, and contracts; very likely the San Andreas Fault is not at the front of their minds, yet they know it is there. Millions of people in California simply have learned to live with this ever-present danger and are more or less consciously taking a risk.

Many factors enable people to face danger. In some instances a *latent threat,* such as a dormant volcano or a geologic fault, allows fear to fade away slowly. Others cope with dangerous situations because they rely on the experience of having lived through a disaster and believe firmly that they do know how to survive. The great majority of people very likely are too busy making a living or are uninformed about the potential hazards in their vicinity; many others just do not care.

Living under risk-prone circumstances may have a negative connotation, and one could even question the wisdom of people who knowingly expose themselves to obvious dangers. But there is another side to this situation: Taking risks is a necessary part in exploring new lands and crossing the oceans, settling frontier territory, building new cities, creating new transportation routes, and in developing agricultural land. It appears that civilizations which are willing to face risks, and which are able to withstand hazards, have the strength and endurance to conquer, or at least to weather, the threats of nature. Most humans have the willingness to start anew: Managua, the capital of Nicaragua, suffered 10 earthquakes in the 20th century, of which 3 reached several levels. The city was rebuilt each time!

Obviously, there are many conditions beyond the understanding and control of populations which have to live in precarious locations and are thus highly vulnerable to hazards. In many cases poverty and also political, economic, or social circumstances make it impossible for millions of people to move

into a better and safer environment. The complex issue of *vulnerability* is discussed in the following section.

Vulnerability to Hazards

In the strictest sense, the word "*vulnerability*" means that a person is open to attack and susceptible to harm. Whether or not people are able to stand up to the impact of a disaster—an earthquake, a massive volcanic eruption, tsunamis, raging floods, or cruel famines—is strongly influenced by factors such as age, health, economics, access to transportation, experience and preparedness, community spirit, the ability to engage in *self-help,* and the effectiveness of local and national government.

It is impossible to assign a hierarchical order to these diverse factors because they vary far too much from individual to individual, from culture to culture, and from one social system to another; however, the factor of poverty seems to be the most critical one especially when it comes to the aftermath of a disaster.

During the devastating Managua earthquake of 1972 (see Chapter 1) the wealthy minority lost its homes, and so did the poor; however, the number killed and injured was considerably lower for the well-to-do-citizens who lived in more substantial buildings outside the inner city, while the poor population perished in great numbers in unsafe structures that were crowded together and that collapsed completely [Plates 109 and 110]. Furthermore, the destitute poor majority had to remain in "*temporary*" hot and dusty camps while the wealthier people moved, within a short time, to other cities to normalize their lives.

Rebuilding and recovery are easier when bank credits and insurance payments are readily available. This fact was clearly demonstrated in other cases including the recent earthquakes in California and in Kobe, Japan. In both areas most public facilities, critical highway systems, and commercial buildings were repaired or were replaced by new structures.

The impacts of disasters on developing and on underdeveloped countries are especially severe, and in these places the largest number of people are usually the most endangered by hazards. This is not surprising if we recall that almost 80 percent of the world's population live in developing areas (see Chapter 11); these regions also show the highest birth rates and experience an ever-increasing pressure on their resources, such as land, water, and housing.

Plate 109. The 1972 earthquake in Managua, Nicaragua, totally destroyed the poorly-built "taquezal" houses and caused a high number of casualities in such structures. The more substantial homes fared better as shown on Plate 110. (Photo by C.H.V. Ebert).

As people require more land, and as urban areas expand relentlessly, increasing amounts of cropland are being converted to other uses or undergo *irreversible degradation.* This trend, unfortunately, is not restricted to any one area but occurs in almost all parts of the world, as was discussed with reference to

Plate 110. Compared to the flimsy "taquezal" houses of the poorer sections of Managua, the more substantial structures, found in the more prosperous parts of the city, survived the earthquake better. (Photo by C.H.V. Ebert).

Japan (see Chapter 3). There, a growing economy and rising land prices led to an ongoing conversion of landslide-prone areas for urban and industrial use, thereby rendering even more people more vulnerable to disasters. Three countries stand out in this respect: over the last few decades Japan lost 52 percent of its grain-producing land, South Korea lost 42 percent, and Taiwan converted about 35 percent (Brown; 1995, p. 18).

Cities in many developing countries, as well as in highly industrialized nations, slowly reach mammoth size and therefore become increasingly vulnerable to disasters such as earthquakes, fire, floods, or diseases. The basic fact remains that too many people are in the wrong place at the wrong time when disasters strike. The impact of natural disasters on *megacities* is being investigated by the Institution of Civil Engineers in London (Natural Hazards Observer Report; 1994, p. 7). It is a sobering thought that by the year 2000, close to 30 cities of the world will have a population of over 8 million each, and two will exceed the 20-million mark!

Poverty and disasters converge in developing countries where the impact of major calamities quickly narrows the parameters within which survival and recovery are possible (Ebert; 1982, p. 99). The poorest people of the world, in the words of social scientists, have been "*marginalized*" when they move into marginal environments where both hazards and unhealthy conditions increasingly threaten their existence.

Disaster Health Studies

The majority of health studies focus on the more common problems (cancer; heart diseases; HIV infections) but spend less effort in studying the effects of disasters on human health. Yet, the number of people who were killed, injured, or displaced by disasters increased from about 100 million to 311 million between 1980 and 1991.

What could be the reasons for this relative neglect in dealing with *epidemiology* related to disasters? It has been suggested that the following factors are involved: (1) most disasters cannot be predicted, (2) controlling disasters in very difficult, (3) curative medicine is given priority, and (4) lack of useful, reliable data that should be available during or soon after the impact of disasters (Noji; 1997,p.v.).

While these factors certainly play an important role in the relative dirth of disaster-related health data, other elements also enter the picture. Modern health research equipment, such as *Magnetic Resonance Imaging* (MRI) and *CAT Scanning,* is not only extremely expensive but also very vulnerable to damage from disaster impacts. Also, many developing countries do not have the facilities to collect and to process health data, while other nations restrict such data entirely or make it available in unreliable and delayed format. For example, western aid and scientists were barred from reaching Tangshan, China, after the devastating earthquake of 1976, and comprehensive data was not made accessible for several years. Immediate international teamwork, both in assistance and research appear to be more useful than the general efforts dealing with prediction, risk assessment, and mitigation (Booth, Fitch; 1979, p.vii).

Human Reaction to Disasters

Disasters constitute a sudden and usually a totally unexpected disruption of normalcy. It may be possible to adjust emotionally to potential dangers and to develop a high degree of tolerance for threatening circumstances. This attitude, which in some cases may border on apathy, may also reflect an optimistic view, although it has often truthfully been said that optimism could be only a manifestation of inexperience.

When disasters strike, the tranquility of everyday life suddenly vanishes, points of reference disappear, and familiar surroundings disintegrate. People suddenly have to face their inability to control their fate in the presence of chaos. Their responses, in one way or another, produce **non-traditional behavior** in reactions to totally new scenarios for which they are not prepared. Thus it is not surprising that catastrophes provoke a wide array of strong emotions that may result in extraordinary actions by persons who would consider themselves—in other circumstances—quite ordinary (Nash; 1976, p. ix).

Unfortunately many reporters foster misleading images of humans in distress. Stories and pictures often are the products of competitive sensationalism and frequently stray away from reality. Individuals, as well as groups of people, are portrayed as "*dazed humanity*" after first going through throes of panic (Wijkman and Timberlake; 1984, p. 105).

Human behavior certainly shows a great variety of emotions and spontaneous reactions that may range from paralyzing fear to irrational actions, but convincing evidence exists that most people do not succumb to panic. In many cases they remain calm and even show leadership under the most trying circumstances. Such calmness does not necessarily mean that these persons are not afraid, but instead indicates that they have the discipline and the fortitude to face adversity and to assist others while disregarding dangers.

It is, however, very difficult to predict human behavior when under severe pressure, because stress represents an abnormal situation; however, there seems to be a reoccurring pattern which evolves in a step-like fashion:

1. The initial impact stage when people must deal with shock and with totally unfamiliar happenings.
2. The post-impact stage where individuals take quick actions to come to the aid of others who are usually members of the family.
3. The more organized rescue phase where groups form, leadership emerges, and where the first contacts with outside helpers are being established.

The formation of *action groups* in disasters serves as a well-known phenomenon. Social interactions and interdependencies typically surface among soldiers when they are resisting the stress of battle (Hollander; 1976, p. 213). Similar trends toward forming action groups readily appear under other stressful conditions as in the case of Love Canal (see Chapter 14). Residents around Love Canal, after their precarious situation became public information in 1978, formed a number of *citizens groups* which served as self-help units as well as political pressure groups (Levine; 1982, p. 175). Thus, contrary to reoccurring statements that disasters bring out anti-social behaviors, most calamities tend to foster *community spirit* and *altruism* (Bryant; 1991, p. 268). But stories of widespread lootings, and stripping of victims of jewelry, and even the theft of gold teeth—as after the storm-surge disaster at Galveston, Texas, in September 1900 (see Chapter 4)—are sad but true happenings (Nash; 1976, p. 208).

As more outside help arrives at the scene of disasters other social reactions tend to emerge. After the intensive phase of self-help, and after the initial rescue efforts are replaced by relief forces, the mood of the survivors often changes. These changes are partly brought on by sheer physical and mental exhaustion and also by the realization of severe, unavoidable consequences of the event: homes were lost, farms are destroyed, jobs have disappeared, and worst of all, death took its toll among family members and friends of the community. At this moment of realization people show an increasing state of numbness and quiet despair. It is not at all unusual that these feelings may turn into anger against survivors who suffered little or no loss. Such anger may also turn against the outside helpers who, as the victims know, will eventually leave the scene to return to their own normalcy after their job is done. Another moment that can trigger *acute animosity* occurs when the people recognize that their misery is exploited for short-term political gains, a situation that can conflict with assistance (Blaikie et al.; 1994, p. 209).

The preceding discussions probed into questions about how people perceive risks, how they vary in their vulnerability to hazards, and how humans react to the impact of disasters. The following section looks at post-disaster problems such as recovery, prediction of future hazards, and the necessity for planning.

Facing the Future

The spatial and temporal aspects of disasters vary with the severity, the size, and the duration of the

event. A lightning strike may kill a single person on a golf course, a tornado can destroy a narrow path of houses, a hurricane can devastate a vast area, and a massive flood is able to paralyze widespread regions for months as we witnessed during the 1993 flood disaster of the Mississippi. A good analogy is to liken the scope of a disaster to a *"ripple effect"* as produced by dropping a stone in a pond (Slovic; 1994, p. 161).

The response to a disaster travels outward from the site of occurrence to reach eventually the offices of state or national governments. In severe circumstances *"states of emergency"* or official *"disaster areas"* are declared. This outward motion of response can have detrimental effects: decision-making and aid delivery processes become centralized and then are determined in places and by authorities far away from the disaster scene. The result can be that local governmental offices—at the town, city, or county level—are bypassed although they usually know the local conditions best and are most familiar with the people of the affected area (Blaikie et al.; 1994, p. 208).

Whether or not the problems associated with disasters can be solved locally depends, of course, on the scale of destruction; however, the ability to cope also directly correlates to the attitude, experience, and the general characteristics of the population involved. The initiative to engage in self-help and in initial reconstruction is often ingrained in societies, such as in the United States, which has a tradition of *self-rule* (Geipel; 1982, p. 43).

The old saying *"forewarned is forearmed"* certainly pertains to preparedness, and preparedness is closely tied in with disaster *prediction*. Prediction of natural disasters had to await the development of accurate and reliable instrumentation as well as the basic understanding of the processes that permit or create hazards. Although much progress in knowledge and technology has occurred, many problems certainly remain.

One of the difficulties is that change is an integral aspect of any environmental system. Scientific observations so far have extended only over a relatively short time span when compared to geologic and climatological time. This *"snapshot"* image of the physical environment tends to deceive us by implying that Earth is in a *"steady state"* and that disasters are merely momentary deviations from the *norm*. The weakest point of this perception centers on the reality that we do not know what this norm is. For example, the concentration of greenhouse gases has fluctuated throughout geologic time; the present increase could be, in part, a phase of yet unknown cycles.

Another factor that makes environmental and natural hazard prediction so difficult is that human activities have already resulted in major changes on our planet (see Chapter 14). Some of these changes are so profound that it is becoming more and more difficult to determine what the environment was like in its *pre-disturbance state* (Roberts; 1994, p. 4). This latter problem pertains especially to ecological systems where prediction is probabilistic rather than deterministic:

> Nobody is sure how the world's ecology hangs together in its totality.
>
> (Simmons; 1989, p. 389)

It is not unusual even in advanced technological systems that built-in errors can aggravate dangerous situations. For example, *geostationary satellites*—of great assistance to meteorologists—may not be able to predict accurately the notoriously erratic tracks of hurricanes. An error of only 10 degrees in predicting a storm's course in a time span of 24 hours could result in a hurricane making its *landfall* more than 114 miles (185 km) from the projected spot (Rogers; 1982, p. 19).

Heeding a warning remains a key part of being prepared to face a disaster. Warning systems have been developed in many parts of the world, as was brought out in connection with tornadoes, tsunamis, and earthquakes; however, warnings of pending disasters are not always effective. The failure to respond to a warning is partially rooted in the psychological make-up of individual people. Some research claims that individuals of low socio-economic status, some ethnic minorities, and women respond the least to warnings, and older persons often ignore warnings in order to avoid the physical and mental stress associated with them (Bryant: 1991, p. 260). How accurate this categorization is one cannot readily say because there are so many exceptions to any statement dealing with human behavior. It is also important to realize that *over-warning,* sometimes the result of insufficient understanding of hazards, as well as the use of sketchy data, can lead to false alarms that unintentionally undermine future warnings.

One of the most drastic measures associated with warnings is the need for *evacuation,* which in many cases meets with considerable resistance. The underlying fears of abandoning one's home and leaving a familiar neighborhood can create traumatic experiences. Not knowing whether it will be possible to return to one's place and friends becomes a demoralizing factor. Moreover, stress, and in some cases overt antagonism among evacuees, can lead to severe social upheaval. Considering these points, it is not surprising that many people resist evacuation and above all *relocation.* Children who were evacuated from London during the bombing in World War II were reported to have experienced more *anxiety* than the ones who stayed behind with their families (Hollander; 1976, p. 213).

Regardless of any suggested or replicable differentiation in responding to warnings, it appears that the success or failure of any warning system depends upon the most unpredictable and weakest link, the *human factor:*

> An accurate forecast, a well-designed disaster prevention system, and all the aids that technology provides count for little if the human response is not geared to the realities of the occasion.
>
> (Rogers; 1982, p. 20)

The last and often most difficult phase in coping with the impact of disasters is to find the road to recovery. The trauma of losing members of one's family and friends, the sight of a devastated home, the uncertainty of economic survival, and the total emotional turmoil of the event must be overcome. People yearn to return to the *status quo* and think of the conditions that prevailed before the disaster changed or uprooted their existence; however, in most cases this retreat to the earlier situation is not possible. People must accept the fact that the past is gone, and they must find a way that leads out and away from the darkness of what they had to go through.

There are some positive aspects found in disasters even though these positives are difficult to appreciate in the immediate aftermath. For example, the destruction of entire cities during World War II at first looked like a permanent eradication of the bases for human and economic existence. Later it became clear that *reconstruction,* resulting in better housing and modernized industrial capacities, can be a major advantage and stimulus to further growth. This scenario applied especially to Germany's and to japan's war-ruined cities and to their subsequent rebuilding during the period referred to as *"Economic Miracle."* (Geipel; 1982, p. 16). Yet the crippling and the death of thousands of humans will remain a painful memory.

Concluding Thoughts

The discussions in this book ranged from disasters that involved the lithosphere and earth materials, the hydrosphere, the atmosphere, the biosphere, and those interlinked or even caused entirely by human activities. As the world's population increases, and as people move more and more into hazard-prone areas, it is not surprising that scientists and humankind in general show a growing concern for our total environmental future.

Every person living on our planet, whether he or she acknowledges this fact or not, is affected by environmental problems: all people should be involved in the effort to solve them. Whether humanity will be able to do this task successfully depends mostly on the understanding of such problems and particularly on the willingness to do something about them even if the action includes making sacrifices.

An important part of facing disasters is to study their effects in times past. The study of previous calamities will help us to assess their potentials and their characteristics; however, because the costs are often incalculable, people may make irrational decisions.

When it comes to disasters, history seems to repeat itself. There have always been, and there will always be in the future, both natural and human-induced hazards. Their natures must be recognized and must be investigated to minimize their threats to humans and to the environment. Armed with such awareness we may—perhaps—be able to plan effectively for the future and to avoid mistakes committed in the past. Without planning there can be no rational hope to escape the effects and human and financial costs of future disasters.

The ultimate success in the struggle against world-wide hazards and violent disasters depends not only on our knowledge and technical preparedness. We must also be *willing* and be *creative* to apply such knowledge for the sake of humanity and for the soundness of the only home we have: *Planet Earth.*

GLOSSARY

aa   A basaltic lava type whose surface is broken up into angular blocks.

acid precipitation   Any atmospheric precipitation that has an acid reaction through the absorption of acid-producing substances such as sulfur dioxide.

aeronautical meteorology   A subfield of meteorology that deals with weather phenomena that affect aviation.

aerosols   Droplets or solid particles that are dispersed and suspended in a gaseous medium.

Agent Orange   A toxic defoliant used during the Vietnam conflict.

air instability   This state exists in a body of air that is marked by a strong vertical temperature decrease and high moisture content. Unstable air tends to rise.

air mass   A large volume of air that has uniform physical properties and occupies a vast area.

air pocket   A phenomenon often encountered during bumpy flights that is produced by a sudden downdraft.

airspeed   The velocity of the air as it flows along an aircraft in flight.

albedo   The amount of light reflected by a given surface compared to the amount received.

alluvium   Various types of sediments (gravel, sand, silt, clay) deposited by rivers.

altimeter   An instrument that measures the altitude of an object with respect to a fixed level of reference.

altimetry   Scientific methods of measuring altitudes by means of various types of altimeters.

amplitude   The vertical distance that represents one-half of the wave height.

aneroid barometer   This instrument measures air pressure. It consists of a partially evacuated metal cell, a connecting mechanism, and an indicator scale.

anion   A negatively charged ion, or an atom which has more electrons (−) than protons (+).

anomalies   Deviations of values (rainfall, temperatures, etc.) from the expected and normal levels.

anticyclone   An area of high atmospheric pressure that has, when viewed from above, a clockwise circulation in the Northern Hemisphere and counterclockwise in the Southern Hemisphere.

arctic haze   A type of ice fog composed of very small suspended ice crystals forming in air layers that are in contact with the ground.

area fire   A conflagration that occurs simultaneously over a large area in connection with area bombing and massive forest fires.

arrhythmia   An erratic rhythm of heart beat in response to disturbances in electrical impulses to the heart.

assegai   A spearlike weapon of South Africa that can be used both for throwing and thrusting.

asteroid   A small planetary body which orbits around the Sun. A major concentration of asteroids lies between Jupiter and Mars.

asthenosphere   The upper segment of the earth's mantle. It has an approximate thickness of 155 miles (250 km) and is capable of plastic flow.

astrobleme   A scar or "blemish" on the Earth's surface resulting from an impact of an asteroid or meteorite.

atmosphere   The gaseous envelope that surrounds the earth. About 99 percent of its mass is found within about 18 miles (29 km) of the earth's surface.

aquifer   A rock layer capable of storing and conducting water.

avalanche   A large mass of either snow, rock debris, soil, or ice that detaches and slides down a mountain slope.

backfiring   Setting a fire intentionally to create a burnt-out (fuel-starved) zone designed to check the spread of forest fires.

baguio   Another name for a hurricane, or tropical cyclone, used in the Philippines.

ball-lightning   a type of lightning which appears as a moving luminous sphere composed of plasma (see glossary) or highly ionized gas. Its size ranges from a few inches to over six feet (2 m).

barometer   An instrument that measures atmospheric pressure. The first liquid barometer was designed by Torricelli in 1644.

basalt   A fine-grained, dark-colored igneous rock composed of ferromagnesian minerals.

bedding plane   A plane that separates or delineates layers of sedimentary rock.

biodegradation   A decomposition process carried out by microorganisms.

biosphere   That part of the earth system that supports life.

Black Death   (see **plague**)

black dusters   Huge dust storms that darkened the skies in southwestern United States during the "Dust Bowl" years in the 1930s.

blitzkrieg   A German word that evolved in the early part of World War II. It means "lightning-fast war."

blizzard   A fierce winter storm accompanied by very low temperatures, high wind velocities, and near zero visibility caused by drifting and blowing snow.

body waves   Earthquake waves that travel through and within the earth.

bronze   An alloy of copper and tin that has a lower melting temperature than copper. It was used to make weapons during the Bronze Age.

buffering capacity   The ability of a substance, or of a system, to resist changes in pH caused by the introduction of either acids or bases.

caldera   A large depression in a volcanic mass that was produced by either a violent volcanic explosion or by the collapse of a volcano.

canopy fire   A forest fire that involves the crowns of trees. It is also called a crown fire.

carbon dioxide (CO_2)   A gaseous product of combustion about 1.5 times as heavy as air. A rise in CO_2 in the atmosphere increases the greenhouse effect.

carbon monoxide (CO)   A product of incomplete combustion. CO is colorless and has no odor and combines with hemoglobin in the blood, leading to suffocation caused by oxygen deficiency.

catalytic reaction   A chemical reaction facilitated by some elements or compounds among other elements or compounds to initiate a chemical process.

cation   A positively charged ion, or an atom that has more protons (+) than electrons (−).

CFCs   Chlorofluorocarbons, which tend to destroy the upper-atmosphere ozone layer, mostly CFC-11 (trichlorofluoromethane) and CFC-12 (dichlorodifluoromethane).

chain reaction   A self-sustaining nuclear reaction that, once started, passes from one atom to another (see also **fission**).

chaparral   A drought-resistant vegetation association of low shrubs and trees. It is typically encountered in those parts of California that have a Mediterranean-type climate.

Circum-Pacific Ring   A zone that surrounds the Pacific Ocean and is known for its volcanic and seismic activity. It is sometimes called the "Ring of Fire."

climate   The statistical sum total of meteorological conditions (averages and extremes) for a given point or area over a long period of time.

cloud seeding   The artificial introduction of condensation nuclei (dry ice or silver iodide) into clouds to force precipitation.

cloud-to-ground lightning   One of the more common lightning discharges, which proceeds from the cloud toward the earth's surface.

coastal ocean   The shallow part of the oceans (usually less than 425 feet or 130 meters deep) that borders the continents and averages about 40 miles (65 km) in width.

cold front   The boundary on the earth's surface, or aloft, along which warm air is displaced by cold air.

colloidal dispersion   This is the process of extremely small particles (colloids) being dispersed and suspended in a medium of liquids or gases.

colluvium   An accumulation of soil and rock fragments at the foot of a cliff or slope under the direct influence of gravity.

compression wave   Shock waves produced by the sudden expansion and heating of the air along the path of a lightning bolt.

condensation nuclei   Submicroscopic particles (carbon, ice, dust) in the atmosphere that attract water molecules and thus become nuclei around which condensation will take place.

cone volcano   A steep-sided and cone-shaped volcano composed of both lava flows and layers of pyroclastic materials. This type is also called a stratovolcano.

conflagration   A destructive mass fire that is less stationary than a firestorm.

continental shelf   The part of the sea floor that extends from the low-water mark to the so-called continental shelf break. The water depth over the shelf usually does not exceed 600 feet (180 m).

conjunction (astron.)   The apparent close position of two or more celestial objects on the celestial sphere. The conjunction of the Sun and Moon result in the new moon phase as seen from Earth.

convection   Mass motion within gases and liquids caused by differences in density brought about by cooling or heating.

Coriolis effect   The apparent deflective force caused by the earth's rotation; it deflects wind to the right (facing downwind) in the Northern Hemisphere and to the left in the Southern Hemisphere. There is no Coriolis effect along the equator.

cornice   An elongated shelflike protruding slab of snow that forms on the downwind side of a mountain peak.

crater eruption   In this type of volcanic activity the eruption forces the ejecta out of the crater and the force of the eruption is primarily expended upward.

creep   As low movement of unconsolidated surface materials (soil, rockfragments) under the influence of water, strong wind, or gravity.

crossbow   An old weapon—though still used for hunting—that is composed of a bow mounted across a stock. The dart is released by a trigger, which makes the crossbow one of the earliest mechanical weapons.

crown fire (see **canopy fire**)

crust (geol.)   The outer layer of the earth's lithosphere. It is about 21 miles (35 km) thick under the continents but is much thinner under oceans. The crust is separated from the underlying mantle by the Moho discontinuity.

crustal plates   In the theory of plate tectonics it is stated that the earth's crust is not continuous but is composed of many large and small plate units that are in relative motion to one another.

cumulonimbus   A towering cumulus type cloud of great height (up to 70,000 feet or 21,000 m) that frequently generates thunderstorms, hail, and tornadoes. It represents the ultimate in the growth of cumulus clouds.

cumulus A cloud of vertical development with a bulging upper portion that may resemble a cauliflower. Upon continued development it may evolve into a cumulonimbus.

cyclonic depression A large area of below-normal atmospheric pressure which has a counterclockwise circulation in the northern hemisphere and a clockwise circulation in the southern hemisphere. It may strengthen to become a mid-latitude cyclone (see glossary).

debris slide A sudden downslope movement of unconsolidated earth materials or mine waste, particularly once it becomes water saturated.

deep-well injection A technique for disposal of liquid waste materials by pressurized infusion into porous bedrock formations or cavities.

defoliants Various types of herbicides such as those used in Vietnam to destroy the foliage of trees as well as leafy field crops.

delta An accumulation of sediments at the mouth of a river often assuming general triangular shapes.

demographic transition Cyclic changes in population that begin with balanced birth and death rates, followed by high birthrates and lowering death rates, and end with falling birthrates in balance with death rates.

dendochronology The study of annual tree rings and their growth rate to interpret events of the past, such as climatic changes and drought cycles.

desertification The creation of desertlike conditions, or the expansion of deserts as a result of humans' actions, which include overgrazing, excessive extraction of water, and deforestation.

desertization A relatively new term that denotes the natural growth of deserts in response to climatic change.

deserts Permanently arid regions of the world where annual evaporation by far exceeds annual precipitation. They cover about 16 percent of the earth.

Devonian period A time span of about 60 million years that began about 405 million years ago in the Paleozoic era.

diamond dust This phenomenon occurs in arctic haze and produces bright, shimmering lights intermixed with rainbow colors that result in confusing images and reduced visibility.

Doppler radar An instrument that emits a radar frequency that appears to be changing as the wave is bounced back from a moving object. The frequency lengthens when the distance between transmitter and object increases, and it shortens as the distance decreases.

downdrafts Downward and sometimes violent cold air currents frequently associated with cumulonimbus clouds and thunderstorms.

downwind effect Severe turbulence can develop on the downwind [leeward] side of large buildings and mountains. This turbulence could be called the "snow fence" effect; it can be dangerous to aircraft.

drainage basin The area drained by a river and its tributaries.

drought An extended period of below-normal precipitation, especially in regions of sparse precipitation. Prolonged droughts can lead to crop failures, famines, and sharply declining water resources.

duff Organic debris on forest floors in different states of decomposition.

dust devil A small, vigorous whirlwind, usually of short duration, which forms in response to intense heating of surface air. It becomes visible as it picks up dust and debris from the ground.

dust storm A severe weather system, usually in dry areas, characterized by high winds and dust-laden air. Major dust storms were observed during the 1930s in the Dust Bowl region of the United States.

earthquake A sudden movement and tremors within the earth's crust caused by fault slippage or subsurface volcanic activity.

easterly wave A temporary wavelike motion that develops within the trade winds and travels from east to west, which may grow into tropical storms and subsequently into hurricanes.

ecosystem A functional system based on the interaction between all living organisms and the physical components of a given area.

elastic rebound This concept implies that rocks, after breaking in response to prolonged strain, rebound back to their previous position or one similar to it. This sudden breaking and rebound may cause earthquakes.

El Niño An abnormal warming of oceans along many low-latitude west coasts when warm equatorial waters displace cold coastal waters. It typically occurs offshore Peru and Ecuador in December about every 3 to 7 years. It causes severe disruptions in weather patterns and in the forming of marine biomass.

entomologist A scientist who studies insects.

environmentalism A concept, also called environmental determinism, that proposes that the total environment is the most influential control factor in the development of individuals or cultures.

epidemiology The study of diseases as they affect a large number of people at the same time.

epicenter The point on the earth's surface located directly above the focus of an earthquake.

evapotranspiration The combined amount of water transpired by plants and evaporated from the earth's surface over a given time and area.

extension fault A branch rupture extending from a major fault line.

eye (of a hurricane) The mostly cloudless, calm center area of a hurricane surrounded by near-vertical cloud walls.

false relief images   Variable color shades that occur in ice fog (see **arctic haze**), which can produce false relief features on a level snow surface that can be very confusing to a pilot who wants to land a plane on snow fields.

fault   A fracture in a rock formation along which vertical and lateral movements may lead to earthquakes.

fetch   The distance over a water body in which waves are generated by a constant wind blowing from the same direction.

firebreak   A natural or human-made strip of land from which flammable vegetation and materials have been removed to control the advance of forest fires.

firestorm   A violent and nearly stationary mass fire that develops its own inblowing wind system, mostly in the absence of preexisting ground wind.

fire vortice   An intense whirl of hot gases and flames that may develop in mass fires.

fire wind   Wind created by fires in response to strong convectional updrafts over a burning area.

fishery   The commercial extraction of fish in a given region.

fission   The splitting of an atom into nuclei of lighter atoms through bombardment with neutrons. Enormous amounts of energy are released in this process, which is used in the development of nuclear power and weapons.

fissure eruption   A type of volcanic eruption that takes place along a ground fracture instead of through a crater.

flank eruption   A type of volcanic eruption that takes place on the side of a volcano instead of from the crater. This typically occurs when the crater is blocked by previous lava eruptions.

flash flood   A local and very sudden flood that typically occurs in usually dry river beds and narrow canyons because of heavy precipitation generated by mountain thunderstorms.

flood crest   The peak of a flood event, also called a flood wave, that moves downstream and shows as a curve crest on a hydrograph.

floodplain   A stretch of relatively level land bordering a stream. This plain is composed of river sediments and is subject to flooding.

focus   The point of earthquake origin in the earth's crust from where earthquake waves travel in all directions.

fossil fuel   Fuels such as natural gas, petroleum, and coal that developed from ancient deposits of organic deposition and subsequent decomposition.

fuelbreaks   Firebreak lanes containing fire-resistant vegetation designed to retard forest fires.

fuel load   The total mass of combustible materials available to a fire.

Fujita scale   A scale, developed by Theodore Fujita (University of Chicago), which represents tornado wind velocities and associated levels of damage.

fumarole   A small vent which issues smoke and hot gases in areas of volcanic activities.

fusion   The combining of two lighter nuclei into a heavier one with attendant release of energy. This is the working principle of the hydrogen bomb.

geothermal energy   The extraction and conversion of the interior heat of the earth into useful energy.

glacis   A protective earthen bank that slopes away from the outer walls of a fortification.

glaze   A clear-ice deposit that forms on impact of water droplets on a surface at below-freezing temperatures.

glowing cloud (see **nuée ardente**)

GOES   Geostationary Operational Environmental Satellite. It receives and transmits electronic data emitted from various types of sensors. It is a warning device.

granite   A light-colored, or reddish, coarse-grained intrusive igneous rock that forms the typical base rock of continental shields.

great circle   A circle formed by passing a plane through two points at the earth's surface and its center. The arc passing through the two surface points establishes a great circle route, i.e., the shortest distance between the points.

greenhouse effect   The trapping and reradiation of the earth's infrared radiation by atmospheric water vapor, carbon dioxide, and ozone. The atmosphere acts like the glass cover of a greenhouse.

ground accelerometer   An instrument that measures the motion of the ground occurring during earthquakes based on values of the gravitational acceleration of 9.8 m/s^2.

ground avalanche   An avalanche type that involves the entire thickness of the snowpack and usually includes soil and rock fragments.

ground fire   A type of fire that occurs beneath the surface and burns rootwork and peaty materials.

groundspeed (aviation)   The forward speed of an airplane relative to the earth's surface.

ground-to-cloud lightning   A less common type of lightning, also called triggered lightning, that typically emanates from isolated high points, such as towers and mountain peaks.

gully   A narrow V-shaped small ravine caused by the erosional force of surface water.

haboob   An intense dust or sand storm that typically develops along the squall front of thunderstorms. The haboob occurs frequently in the low-latitude deserts of the Sudan and Egypt.

half-life   The time span that is needed for the disintegration of half the mass of a radioactive element.

heat island   Typically encountered over the core area of large urban areas with markedly higher temperatures compared to outlying zones.

hoar frost   Ice crystals that form on the earth's surface by sublimation of atmospheric water vapor that pass directly into the solid ice form.

homosphere The well-mixed lower layer of the earth's atmosphere that extends to a height of about 50 miles (80 km) above the surface.

hot spot (geol.) Excessively hot magma centers in the asthenosphere that usually lead to the formation of volcanoes.

humus The partially or fully decomposed organic matter in soils. It is generally dark in color and partly of colloidal size.

hurricane A tropical low-pressure storm (also called baguio, tropical cyclone, typhoon, willy willy). Hurricanes may have a diameter of up to 400 miles (640 km), a calm center (the eye), and must have wind velocities higher than 75 mph (120 km/h). Some storms attain wind velocities of 200 mph (320 km/h).

hydraulic action The pressure, friction, and suction imparted by moving water.

hydrograph A graph that shows the rate of river discharge over a given time period.

hydrology The branch of science which deals with the properties, behavior, and the distribution of water above and below the earth's surface.

hydrophobic The property of having no affinity for water, i.e., water repellent.

hydroponics The growing of plants in nutrient solutions.

hydrosphere The total water portion of the earth, in all its states, except the water that is chemically bonded in the lithosphere.

hygroscopic particles Condensation nuclei in the atmosphere that attract water molecules (carbon, sulfur, salt, dust, ice particles).

hyperthermia Abnormally high body temperatures that may exceed the level of heat tolerance.

hypothermia A condition when the body temperature falls more than 4° F (about 2° C) below the normal temperature of 98.6° F (37° C).

ICBM Intercontinental (long-range) ballistic missile.

ice core data Data on chemical and physical properties obtained from fossil ice that reveals paleoclimatic conditions.

ice fog Forms in very cold weather, and frequently in arctic areas, when water vapor sublimates directly into ice crystals. It can have a blinding effect to a pilot flying into the sun.

IFR (Instrument Flight Rules) Mandatory use of instrument flying under adverse atmospheric conditions.

ignition temperature The critical temperature at which materials will ignite and burst into flame. Also called the kindling temperature.

induction The creation of an electric charge in a body by a neighboring body without having physical contact.

intensity (earthquake) A measurement of the effects of an earthquake on the environment expressed by the Mercalli scale in stages from I to XII.

intercloud lightning A type of lightning that occurs between clouds and can be dangerous to airplanes.

internal waves The waves occur at interfaces (planes of discontinuity) of fluids having different densities.

Intertropical Convergence A seasonally shifting zone of interaction between the northeasterly trade winds of the Northern Hemisphere and the southeasterly trade winds of the Southern Hemisphere near or at the equator.

intracloud lightning A type of lightning that occurs between different charge fields within a cloud. It is also called sheet lightning.

ion An electrically charged molecule or atom that lost or gained electrons and therefore has a smaller or greater number of electrons than the originally neutral molecule or atom.

ionization The process of creating ions (see **ion**).

Iron Age The period that followed the Bronze Age when humankind began the use of iron for making implements and weapons around 800 b.c. The earliest use of iron may go back to 2500 b.c.

ironstone A term sometimes used to describe a hardened plinthite layer in tropical soils. It is primarily composed of iron oxides bonded to kaolinitic clays.

island arc A chain of volcanic islands which usually form along an oceanic trench or a plate subduction zone. Examples are the Aleutian and Kurile Islands.

isobar A line on a map along which there is the same atmospheric pressure level.

isobaric surface An imaginary surface along which there is the same atmospheric pressure.

isotherm A line on a map that connects points of the same temperature values.

isotopes Atoms of a given element having the same atomic number but differing in atomic weight because of variations in the number of neutrons.

J-curve A graphic curve, shaped like the letter "J," which indicates a geometric growth (as population) compared to an arithmetic increase.

jet stream A high-velocity, high-altitude (25,000 to 40,000 feet or 7,700 to 12,200 m) wind that moves within a relatively narrow oscillating band within the upper westerly winds.

joint (geol.) A natural fissure in a rock formation along which no movement has taken place.

karst An area underlain by limestone that shows sinkholes, caves, and depressions formed by solution. The drainage of a karst region is mostly underground.

khamsin A severe dust storm that is typically found in Egypt. It blows from the southern desert regions northward toward the Mediterranean and may darken the skies for several days.

kiloton A term used to express the force of an explosive device such as a nuclear bomb. One kiloton is equal to the force of 1,000 tons of TNT.

kindling temperature (see **ignition temperature**)

La Niña   The normal and opposite conditions in low-latitude west coast areas that are affected by El Niño (see glossary).

land breeze   A night wind, typically found in tropical areas, that blows from the land toward the sea when the land surface is cooler than that of the ocean. The atmospheric pressure thus is lower over the water compared to the higher pressure over land. This is the opposite of a sea breeze.

landslide   A general term that denotes a rapid downslope movement of soil or rock masses.

land subsidence   A gradual or sudden lowering of the land surface caused by natural or human-induced factors such as solution (see **karst**) or the extraction of water or oil.

lapilli   Small pieces of lava fragments ejected from a volcano. They are also referred to as cinders.

lapse rate   Expresses the rate of change (temperature or pressure) of atmospheric values with a change in elevation.

latent energy   Heat energy that produces changes of state in a substance without increasing the temperature of such substance. An example would be the melting of ice into liquid water and the subsequent evaporation to vapor. The latent energy is released when the processes are reversed.

laterite   A reddish tropical soil that is high in iron and aluminum sesquioxides. Upon drying the soil takes on bricklike (*later* = brick) characteristics. This material is now referred to as plinthite.

leachate   A liquid that carries materials in solution or in suspended state.

LIDAR   Light Detection and Ranging system. Instruments using infrared or laser beams to measure wind speed and direction from the movement of windborne aerosols.

lift deficiency (aviation)   A situation experienced by an aircraft that finds itself below a planned glide path in response to a series of downdrafts.

lift surplus (aviation)   A situation experienced by an aircraft that finds itself above a planned glide path in response to a series of thermal updrafts.

lithosphere   The outer solid layer of the earth that rests on the nonsolid asthenosphere. The lithosphere averages about 60 miles (100 km) in thickness.

loess   Fine siltlike soil particles transported and deposited by wind action. Some loess deposits may be hundreds of feet thick.

Love wave (geol.)   A seismic shear wave with sideway movements at right angles to the direction of wave travel. It is an earthquake surface wave.

mace   A clublike weapon made of metal or with a metal head used in hand-to-hand combat.

magma   Naturally occurring molten rock, which may also contain variable amounts of volcanic gases. It issues at the earth's surface as lava.

magma chambers   Underground reservoirs of molten rock (magma) that are usually found beneath volcanic areas.

mantle (geol.)   The intermediate zone of the earth found beneath the crust and resting on the core. The mantle is believed to be about 1,800 miles (2,900 km) thick.

mariculture   The commercial growing of food fish, shellfish, and shrimp in various types of ponds and confinements. Also referred to as aquaculture.

mass wasting   The mass movement of soil and rock fragments under the influence of gravity.

megaton   A term used to express the force of an explosive device such as a nuclear bomb. One megaton is equal to the force of 1,000,000 tons of TNT.

Mercalli scale   Used to describe the effects of an earthquake's intensity on a scale of I to XII ranging from "imperceptible" to "major catastrophe." It is not a quantified scale.

meteorology   The scientific study of weather and atmospheric physics.

mycorrhiza   A fungus in close association with plant roots where both fungus and plant benefit as far as nutrient uptake is concerned.

microbursts   Violent cold-air downdrafts from thunderstorms often resulting in dangerous wind shear.

microwave sounder unit   A satellite-based remote sensor capable of measuring temperatures in the lower stratosphere.

Mid-Atlantic Ridge   A major submarine mountain ridge that runs through the center of the North and South Atlantic Ocean. It has a rift valley in its center with a maximum depth of about 1.2 miles (2 km). Several volcanic islands have formed along this ridge system.

midlatitude cyclone   An area of low atmospheric pressure (a "Low" on a weather map) that is marked by cold and warm fronts and moves generally from west to east. This storm system has a counterclockwise wind circulation in the Northern Hemisphere and a clockwise circulation in the Southern Hemisphere.

Mississippian period   A time span of about 35 million years that began about 345 million years ago in the Paleozoic era.

Moho discontinuity   A zone between the earth's crust and mantle that shows a marked change in the travel velocity of seismic waves caused by density changes between these layers. Named after the seismologist Mohorovicic who discovered this discontinuity in 1909.

molting   The shedding of skin layers or feathers (reptiles, birds, insects) in the process of development.

monoculture   The growing of a single crop or tree species while excluding all others.

monolith   A single rock mass that stands above the surrounding land. Ayers Rock in central Australia is a typical example of this feature.

monsoon   A seasonal wind that blows from a cold continent toward the sea in winter and from the sea into a continent in the summer. In principle it acts like a large-scale and seasonal land and sea breeze.

Montreal Protocol   An international agreement to limit and eventually reduce the amount of chlorofluorocarbons injected into the atmosphere (September 16, 1987).

mountain breeze   A wind caused by the gravity flow of cold dense air downslope into the valley. It is very pronounced at night.

mountain wave   A severe air turbulence that forms on the downwind (lee) side of mountains. Both downdrafts and updrafts are created in this turbulence, which may extend more than 100 miles (160 km) downwind from a mountain range.

mudflow   A downslope movement of water-saturated earth materials such as soil, rock fragments, or volcanic ash.

mutagenic   Any substance or radiation type that can produce changes in an organism's genetic structure and in the rate of normal mutation.

napalm   A jellied and highly flammable substance of variable composition (naphtene and palmitates) used in incendiary bombs and flamethrowers.

nerve gases   Highly toxic chemicals (similar to organophosphate insecticides) that even in minute amounts become lethal to living organisms. Nerve gases are part of the chemical warfare arsenal.

non-native species   Nonindigenous (non-native) animals and plants to a given region. Their introduction can result in serious ecological disturbances and in the extinction of native species.

nonrenewable resources   Resources (coal, oil, ores, etc.) that cannot be renewed once they have been used up. In contrast, wood, air, and water are renewable resources.

northeasterly trade winds   The trade winds of the Northern Hemisphere that blow from the northeast toward the equator. They emanate from the Subtropical High, which centers on 30° N latitude. (Note: The equivalent winds of the Southern Hemisphere are the southeasterly trade winds.)

nuclear winter   The theory that a major nuclear war will saturate the atmosphere with dust— similar to volcanic dust—which would block sunlight. This blocking could result in a disastrous lowering of global temperatures.

nuée ardente   The French name of a glowing cloud of superheated gases and volcanic ash descending the flank of a volcano. A massive nuée ardente destroyed the town of St. Pierre on Martinique in 1902.

nymph   A developing insect in an incomplete state of metamorphosis.

Operation "Gomorrah"   Code name given to a series of massive air raids by the Royal Air Force aimed at the destruction of the German port city of Hamburg in July 1943.

Operation "Pink Rose"   Code name for one of the major American efforts to destroy large tracts of forests in Vietnam between November 1966 and January 1967. The operation involved the application of defoliants followed by fire raids using napalm incendiaries.

Operation "Ranch Hand"   Code name for a major phase in the application of defoliants to large forest areas in Vietnam. This effort started in 1962.

Operation "Stormfury Americas"   A series of cloud-seeding experiments aimed at modifying hurricanes conducted by NOAA's National Hurricane and Experimental Meteorology Laboratory between 1961 and 1971.

opposition (astron.)   The situation when the sun and a given planet, or the moon, are positioned on opposite sides (180° apart) of the earth.

orographic precipitation   Precipitation as a result of forced uplift of a moist air mass along the slope of a mountain range, or orographic barrier. The uplift causes cooling and condensation.

overgrazing   The grazing by animals in excess of the support capacity of pasture land. This activity leads to severe soil erosion.

ozone layer   A layer of ozone concentration in the earth's atmosphere that lies about 18 miles (30 km) above the earth's surface. This layer absorbs large amounts of potentially harmful ultraviolet radiation.

ozone self-healing   The hypothesis that a reduction of ozone in the higher stratosphere would allow more ultraviolet radiation to penetrate the lower stratosphere and create more ozone there, thus limiting total air column ozone depletion.

pahoehoe   A Hawaiian word used for a lava flow characterized by a smooth surface that may also have rope-like or billowy features.

Paleolithic   Referring to ancient times when early humans used stones for the making of tools and weapons.

paleontology   The branch of geology that investigates life of past geological times as indicated by fossils.

perigee (astron.)   The position of a planet, the moon, or a satellite when it is closest to the earth during orbital motion.

perihelion (astron).   The closest position of the earth in its orbit around the sun.

pH   A numerical expression (negative logarithm) that indicates the relative concentrations of H^+ ions and OH^- ions in solutions. A solution in which H^+ dominates is acidic. pH values range from 1 to 14 with 7 indicating neutrality.

photochemicals   Noxious chemicals that develop in the atmosphere as a result of the action of sunlight on oxides of nitrogen and hydrocarbons. Photochemical pollution results from these processes.

photolysis   The chemical decomposition induced by light.

phytoplankton   Microscopic plant life found in oceans as well as in freshwater bodies; it represents the base of the oceanic food chain.

plague (Black Death)   An epidemic disease spread by rats and bites of fleas. The bubonic plague spread from Asia to Europe in the fourteenth century and probably killed about 30 percent of the known population.

plankton   Microscopic aquatic organisms not capable of directional locomotion. They are divided into plants (phytoplankton) and animals (zoo-plankton).

plasma   A gas believed to be involved in ball-lightning (see glossary) which is made up of almost equal numbers of positive and negative ions. It displays unusual characteristics not found in common gases.

plate tectonics   A model of global tectonics based on the movements and interaction between segments (plates) of the earth's crust.

plinthite   A tropical soil material composed primarily of sesquioxides of iron and aluminum that hardens irreversibly to ironstone upon exposure to the air and drying. This material was formerly called laterite.

polar vortex   A strong wind system that encircles the South Pole at mean latitudes of 60° S to 70° S. A less consistent vortex exists around the North Pole during the colder months.

polder   Land areas, typically found in the Netherlands, reclaimed from the sea that lie below sea level and must be protected by dikes.

powder avalanche   A type of snow avalanche composed of billowing dry snow.

prefrontal squalls   Unstable weather conditions (gusty winds, thunderstorms, wind shear) typically found along a vigorously moving cold front of a midlatitude cyclone.

pressure altitude   Indicated by an aneroid altimeter. It may differ from the standard-pressure atmosphere and may give incorrect altitude readings unless corrected for local pressure conditions.

prevailing westerlies   A major wind belt of the midlatitudes in both hemispheres that blows predominantly from a westerly direction and coincides with the overall movements of midlatitude cyclones and other weather systems.

primary wave (P)   The fastest-moving earthquake wave, which is also called a "push-and-pull" wave (compression and rarefaction), in which the rock moves back and forth in the direction of wave travel.

pyroclastic materials   Collective terms of fragmentary volcanic ejecta such as bombs, tephra, or cinders.

pyrolysis   The exposure of flammable material to high temperatures leading to combustion and decomposition.

radar   A contraction of "radio-detection and ranging." An electronic device used for the detection and ranging of distant objects capable of receiving and returning radar waves.

rain shaft   A column of rain descending from a cloud that could be mistaken for a tornado when viewed from a distance.

RAWIN   A contraction of "radar wind." A system based on the reflection of radar waves from metallic surfaces carried aloft by weather balloons to track air currents.

Rayleigh waves (R)   An earthquake surface wave that behaves in the way ocean waves move at the water surface.

relative humidity   The ratio between the amount of moisture present in a given air mass compared to the total amount that the air could hold at saturation.

remote sensing   The detection, identification, and analysis of features using devices (cameras, electronic sensors) that are located at some distance from the object to be investigated.

renewable resources   Resources capable of being renewed such as forests, crops, water power, and materials that can be recovered by recycling.

response time   The time that is needed for a system to establish a new state of equilibrium after it has been disturbed by the input of matter or energy.

return stroke   An upward electrical discharge from the ground that meets the stepped leader of a lightning bolt coming from a cloud over a distance of about 150 feet (45 m). It follows the previously established path of the stepped leader back to the cloud.

Richter scale   A scale developed by Charles R. Richter to measure the magnitude of an earthquake. It is a logarithmic scale so that an earthquake's energy release, expressed by whole numbers, is about 31 times greater than that expressed by the next lower number.

rift valley   A structural valley, or graben, produced by the downfaulting of the earth's surface along roughly parallel fractures.

rime   A white and opaque deposit of ice formed by the freezing of supercooled water droplets upon impact of a surface at below-freezing temperatures.

rockfall   The detachment and downslope fall of rock fragments. It is a form of mass wasting and usually forms a debris slope (talus) at the foot of a cliff.

rockslide   A mass of bedrock that detaches and slides down a slope.

rotor cloud   A turbulent cloud that forms on the downwind (lee) side of a mountain. See also **mountain wave.**

running crown fire   A rapidly moving massive crown fire that develops if strong ground winds fan the flames of a forest fire. It is very difficult to control.

runoff   That portion of precipitation that is not absorbed by the soil and runs downslope on the surface. It is a major factor in flooding.

St. Elmo's fire   A bluish glow seen on high points (a ship's mast, church spires) in response to a buildup of positive electrical charges. It poses no danger.

salinization   The excessive buildup of soluble salts in soils or in water. This often is a serious problem in crop irrigation systems.

saltation   A form of wind erosion where small particles are picked up by wind and fall back to the surface in a "leap and bound" fashion. The impact of the particles loosens other soil particles, rendering them prone to further erosion.

sanitary landfill   A managed disposal site in which refuse and soil layers alternate to promote biodegradation.

savanna   A tropical tall-grass region with scattered trees, located between the tropical rainforest and semiarid regions.

sea breeze   A daytime wind, typically found in tropical areas, that blows from the sea toward the land when the sea surface is cooler than the land surface. The atmospheric pressure thus is lower over the land compared to the higher pressure over the sea. This is the opposite of a land breeze.

sea level   An imaginary average level of the ocean as it exists over a long period of time. It is also used to establish a common reference for standard atmospheric pressure at this level.

sea smoke   A form of steam fog that forms when cold air moves over a relatively warmer water surface.

secondary wave (S)   A body earthquake wave that travels more slowly than a primary wave (P). The wave energy moves earth materials at a right angle to the direction of wave travel. This type of shear wave cannot pass through liquids.

seismic sea wave (see **tsunami**)

seismograph   A device that measures and records the magnitude of earthquakes and other shock waves such as underground nuclear explosions.

seismology   The science that is concerned with earthquake phenomena.

seismometer   An instrument, often portable, designed to detect earthquakes and other types of shock waves.

semiarid regions   Transition zones with very unreliable precipitation that are located between true deserts and subhumid climates. The vegetation consists usually of scattered short grasses and drought-resistant shrubs.

sharav   A hot, humid wind encountered in the Middle East, especially in southern Israel, that provides oppressive weather conditions along coastal regions.

shear   The movement of one part of a mass relative to another leading to lateral deformation without resulting in a change in volume.

shear strength   The internal resistance of a mass to lateral deformation (see **shear**). Shear strength is mostly determined by internal friction and the cohesive forces between particles.

sheet lightning (see **intracloud lightning**)

shield volcano   A dome-shaped volcano formed by successive lava flows without interbedded pyroclastic materials.

SKYWARN   An organization of trained tornado spotters who work closely with the National Weather Service in preparation of tornado warnings.

slab avalanche   A type of snow avalanche. It consists of a thick slab of compacted snow that detaches from a snowpack and moves rapidly downslope.

slash-and-burn agriculture   A technique used in tropical agriculture where forests are cleared, the tree remains are burned, and the field is cultivated for a number of years until the soil is exhausted and a new site must be prepared.

soil failure   Slippage or shearing within a soil mass because of some stress force that exceeds the shear strength of the soil.

soil liquefaction   The liquefying of clayey soils that lose their cohesion when they become saturated with water and are subjected to stress or vibrations.

soil salinization   The process of accumulation of soluble salts (mostly chlorides and sulfates) in soils, caused by the rise of mineralized groundwater or the lack of adequate drainage when irrigation is practiced.

soil structure   The arrangement of soil particles into aggregates that can be classified according to their shapes and sizes.

soil texture   The relative proportions of various particle sizes (clay, silt, sand) in soils.

sortie   The individual military mission of an aircraft or of a group of soldiers.

spontaneous combustion   This type of fire is started by the accumulation of the heat of oxidation until the kindling temperature of the material is reached.

spring tide   An above-average high tide that occurs about every two weeks during full or new moon phases when the tidal pull of the sun and moon is combined.

squall line (see **prefrontal squalls**)

standard atmosphere   A hypothetical atmosphere of standard average sea-level pressure and temperature gradients with altitude. All atmospheric variations are measured against the standard conditions.

standing wave   An oscillating type of wave on the surface of an enclosed body of water. The wave acts similar to water sloshing back and forth in an open dish.

stallspeed (aviation)   The critical speed below which an airplane cannot be controlled because of the lack of aerodynamic lift and the ineffectiveness of steering surfaces.

static electricity   A nonflowing (stationary) electrical charge typically generated by friction.

stepped leader   The leading part of a lightning bolt as it descends from a cloud toward the ground.

storm surge   The massing and often dangerous rise of ocean waters in response to strong wind action and the low atmospheric pressure encountered with hurricanes. These surges may cause extensive coastal flooding.

stratosphere   The part of the upper atmosphere that shows little change in temperature with altitude. Its base begins at about 7 miles (11 km) and its upper limits reach to about 22 miles (35 km).

stratovolcano (see **cone volcano**)

stream terraces   Elevated remainders of previous floodplains; they generally parallel the stream channel.

subduction (plate tectonics)   The process of forcing one crustal plate beneath another plate in zones of plate collisions or convergence.

sublimation   The change of ice to water vapor, or the change of water vapor to ice, without going into the liquid state.

submarine canyons   Steep-sided valleys that extend across the continental shelf from the land toward the deep ocean.

subtropical highs   Almost stationary high-pressure (anticyclonic) atmospheric cells that are located in the subtropical latitudes of the North and South Atlantic and North and South Pacific. They have a clockwise circulation pattern in the northern hemisphere and a counterclockwise circulation in the southern hemisphere.

summit aridity   Dry conditions that may develop on convex hills because of excessive drainage and thin soil layers.

supercell   An extraordinarily large cumulonimbus storm cloud capable of spawning severe thunderstorms and tornadoes.

supercooled droplets   Water droplets that can exist in the free atmosphere at temperatures below the freezing point.

surface creep (see **creep**)

surface fire   Rapidly moving fires that consume dry grass, low shrubs, and litter on forest floors. They are relatively easy to control because of their low fuel load.

surface waves (geol.)   Earthquake waves, such as Love waves (L) and Rayleigh waves (R), that travel at the earth's surface.

swales   Low-lying shallow tracts of land that serve as temporary drainage channels after snowmelts and heavy precipitation.

synoptic   Pertaining to data obtained simultaneously over a large area typically used in meteorological, climatological, and oceanographic research.

system dynamics   A technique based on the concept that the total structure of any system often is as important in explaining its behavior as is that of the individual component.

tektite   Silica-rich glassy fragment of fused rock material which apparently forms by asteroid impacts.

temperature inversion   An increase in air temperature with increasing altitude. This pattern is a reversal of the normal temperature distribution.

tephra   Cinder and ash material ejected from a volcano.

teratogenic   The property of a chemical, or of radiation, of causing birth defects.

thermonuclear bomb   A nuclear bomb, such as the hydrogen bomb, that is based on fusion.

threshold event   Also called "jump event." A sudden change in a chemical or physical system in response to energy input.

threshold factor   The critical, minimum, and sometimes unpredictable amount of a substance or physical force that will result in harm.

TOMS   Total Ozone Mapping Spectrometer. A remote sensor employed on the NIMBUS-7 satellite to measure total air column ozone.

topsoil   The surface layer of a soil that is rich in organic materials.

tornado   A highly destructive and violently rotating vortex storm that frequently forms from cumulonimbus clouds. It is also referred to as a twister.

total war   The concept of war that does not make any distinction between warfare at the front and warfare against the civilian population. The development of long-range air power and missiles made this concept a reality.

Trade winds   Steady winds which blow from the northeast (NE Trades) in the northern hemispherer and from the southeast (SE Trades) in the Southern hemisphere from the subtropical highs (see glossary).

transform faults   Lateral faults that run between offset portions of midoceanic ridges. Plate segments move along such faults either in opposite directions between the offset parts of the ridges, or in the same direction when they extend beyond the offset zone.

triangulation   A survey technique used to determine the location of the third point of a triangle by measuring the angles from the known end points of a base line to the third point.

triggered lightning (see **ground-to-cloud lightning**)

tropical cyclone (see **hurricane**)

troposphere   The lower portion of the earth's atmosphere that shows a decrease of temperature with increasing altitude and contains most of our weather phenomena. It extends for about 6 miles (10 km) above the poles and about 12 miles (20 km) above the equator.

tsunami   A Japanese term that refers to a seismic sea wave that can be generated by severe submarine fault slippages or volcanic eruptions. The tsunami reaches great heights when it enters shallow waters, but it is unnoticeable on the high seas.

turbulence (meteorol.)   Any irregular or disturbed wind motion in the air.

twister   An American term used for a tornado.

ultraviolet radiation   The electromagnetic radiation just beyond the wavelength of visible light. This potentially harmful radiation is partially absorbed by the ozone layer.

urbanization   The transformation of rural areas into urban areas. Also referred to as urban sprawl.

valley breeze   A convectional wind caused by the warming of the valley floor and mountain slopes resulting in a gentle upslope wind. It is best developed during the hottest part of the day.

vapor pressure   That part of the total atmospheric pressure contributed by water vapor. It is usually expressed in inches of mercury or in millibars.

vent (volcanic)   The relatively narrow passageway that leads from the magma reservoir to the crater of a volcano.

vitrification   The conversion of a substance, such as radioactive waste, into a glasslike state.

volcanic bomb   A type of volcanic ejecta that consists of blocks of lava that cool to a solid consistency in the air before falling back to the ground.

volcanism (vulcanism)   All phenomena associated with volcanic activities.

vorticity (meteorol.)   Any rotary flow of air such as in tornadoes, midlatitude cyclones, and hurricanes.

wake turbulence (aviation)   The turbulence found to the rear of an aircraft in motion.

wall cloud   Vertically developed cloud borders in association with cumulonimbus clouds where tornadoes may form. The term is also used to describe the steep-sided inner boundary of a hurricane's eye.

warm front   The boundary on the earth's surface, or aloft, along which warm air overrides and displaces cold air.

wave crest   The upper rise of a wave that in connection with the low portion, or trough, constitutes a wave.

wavelength   The horizontal distance measured between two successive wave crests.

wave period   The time required for two successive wave crests to pass a given point.

wave reflection   The return of wave energy (radiation, heat, sound, water) after striking a surface.

wave refraction   The change in the direction of wave travel.

wave trough   The low portion of a wave that in connection with the high part, or crest, constitutes a wave.

weather   The physical state of the atmosphere (wind, precipitation, temperature, pressure, cloudiness, etc.) at a given time and location.

whiteout   The elimination of visibility caused by arctic haze or during blizzard conditions.

willy willy   A short name for whirlwind used in Australia. It may refer to severe tornadolike storms over a desert as well as to strong tropical storms.

windbreak   Natural or planted groups or rows of trees that slow down the wind velocity and protect against soil erosion.

wind chill factor   The combination of air temperature, wind velocity, and humidity that affects the rate of heat loss from the body.

wing vortices (aviation)   A rotary type of turbulent air flow that develops from the wing tips of an aircraft in flight. The strength of the vortices is proportional to the weight of the plane.

xerophytic vegetation   Plants capable of growing under arid conditions, i.e., xerophytes.

REFERENCES

Chapter 1

Bath, Markus, 1973, *Introduction to Seismology:* John Wiley & Sons, Inc., New York.

Bolt, Bruce A., 1980, *Earthquakes and Volcanoes,* Readings from Scientific American: W.H. Freeman and Company, San Francisco.

Bolt, Bruce A., 1993, *Earthquakes:* W.H. Freeman and Company, New York.

Bruneau, Michel, 1999, "Kocaeli (Izmit) Earthquake of August 17, 1999, "Multidisciplinary Center for Earthquake Engineering Research, *Bulletin,* Volume 13, NO. 3, SUNY/Buffalo.

Bruneau, Michel, 1999, "Structural Damage (Izmit Earthquake)," MCEER, *Response,* SUNY/Buffalo.

Butler, John E., 1976, *Natural Disasters:* Heineman Educational Australia, Richmond.

Ebert, Charles H. V., 1981, "The Seismic Geography of Managua, Nicaragua," *ECUMENE,* Vol. XIII.

Foster, Robert J., 1971, *Geology:* Charles E. Merrill Publishing Company, Columbus, Ohio.

Hays, Walter W., 1995, "Understanding Kobe," *Natural Hazards Observer,* Vol. XIX, No. 4.

Leeds, David J., 1973, "Destructive Earthquakes of Nicaragua," *Proceedings,* San Francisco Managua Earthquake Conference, Vol. I.

Longwell, Chester R., Knopf, Adolph, and Flint, Richard F., 1948, *Physical Geology:* John Wiley & Sons, Inc., New York.

Menard, H. W., 1974, *Geology, Resources and Society:* W.H. Freeman and Company, San Francisco.

Mitchell, William A., 1999, "Social, Political and Emergency Response (Izmit Earthquake), "MCEER. *Response,* SUNY/Buffalo.

Papageorgiou, Appostolos, 1999, "Seismological Observations, (Ismit Earthquake), MCEER, *Response,* SUNY/Buffalo

Rikitake, Tsuneji, 1982, "Interview with Gu Gongxu," *Impact of Science on Society,* Vol. 32, No. 1, UNESCO, Paris.

Scholz, Christopher, 1973, "Earthquake Prediction; A Physical Basis," *SCIENCE,* Vol. 181, No. 4102.

Seible, Frieder and Priestly, M. J. Nigel, 1995, "The Kobe Earthquake of January 17, 1995," Structural Systems Research Project, Report SSRP-95/03, University of California, San Diego.

Tank, Ronald W., 1983, *Environmental Geology:* Oxford University Press, New York.

Thurman, Harold V., 1985, *Introductory Oceanography:* Charles E. Merrill Publishing Company, Columbus, Ohio.

Wijkman, Anders and Timberlake, Lloyd, 1984, *Natural Disasters; "Acts of God or Acts of Man?":* International Institute for Environment and Development, London and Washington, D.C.

Wyllie, Peter J., 1976, *The Way the Earth Works,* An Introduction to the New Global Geology and its Revolutionary Development: John Wiley & Sons, Inc., New York.

Other Sources

Bulletin, Vol. 5, No. 1 (January 1, 1991); National Center for Engineering Research, State University of New York at Buffalo, Buffalo, New York.

Special Publication 104, "The Loma Prieta (Santa Cruz Mountains) California Earthquake of 17 October 1989." Department of Conservation, Division of Mines and Geology, Sacramento, California, 1990.

Special Report, "Preliminary Reports from the Hyogo-ken Nanbu (Kobe) Earthquake of January 17, 1995." *National Center for Earthquake Engineering Research Bulletin,* Vol. 9, No. 1, State University of New York at Buffalo.

Chapter 2

Bolt, Bruce A., 1980, *Earthquakes and Volcanoes,* Readings from Scientific American: W.H. Freeman and Company, San Francisco.

Costa, John E. and Baker, Victor R., 1981, *Surficial Geology,* Building with the Earth: John Wiley & Sons, Inc., New York.

Gross, Grant, 1982, *Oceanography;* A View of the Earth: Prentice-Hall, Inc., Englewood Cliffs, New Jersey.

Hoppe, K., 1992, "Mt. Pinatubo's Cloud Shades Global Climate," *Science News,* Vol. 142, No. 17 (October 24, 1992).

Krafft, Maurice and Krafft, Katia, 1975, *Volcano:* Harry N. Abrams Inc., Publishers, New York.

Macdonald, Gordon A., 1972, *Volcanoes:* Prentice-Hall, Inc., Englewood Cliffs, New Jersey.

Murton, Brian J. and Shimabukuro, S., 1974, "Human

Adjustment to Volcanic Hazard in Puna District, Hawaii," *Natural Hazards:* ed. by Gilbert F. White: Oxford University Press, New York.

Rittman, A., 1962, *Volcanoes and Their Activities:* John Wiley & Sons, Inc., New York.

Rondal, Jose D. and Yoshida, Masao, 1994, "Disaster Planning and Management," *Symposium Proceedings,* Bureau of Soils and Water Management, Manila, 1994.

Simkin, Tom et al., 1981, *Volcanoes of the World,* The Smithsonian Institution: Hutchinson Ross Publishing Company, Stroudsburg, Pennsylvania.

Other Sources

NOAA, Department of Commerce, "Mount Pinatubo—The June 1991 Eruptions," Product No. 739-A11-007, National Geophysical Data Center, Boulder, Colorado, 1993.

Chapter 3

Armstrong, Betsy and Williams, Knox, 1986, *The Avalanche Book:* Fulcrum, Inc., Golden, Colorado.

Butler, John E., 1976, *Natural Disasters:* Heineman Educational Australia, Richmond.

Coates, Donald R., 1981, *Environmental Geology:* John Wiley & Sons, Inc., New York.

Fraser, Colin, 1966, *The Avalanche Enigma:* Rand McNally and Company, New York.

Hausenbuiller, R. L., 1985, *Soil Science,* Principles and Practices: Wm. C. Brown Company Publishers, Dubuque, Iowa.

D.K.C., 1992, "Landslide Hazard Assessment in the Context of Development," *Geohazards,* Natural and Man-Made, ed. by G. J. H. McCall and S. C. Scott, Chapman & Hall, London.

Johnson, M, Douglas L. and Lewis, L.A., 1995, *Land Degradation; Creation and Destruction:* Blackwell Publishers, Massachusetts.

Keller, Edward A., 1979, *Environmental Geology:* Charles E. Merrill Publishing Company, Columbus, Ohio.

McClung, David and Schaerer, Peter, 1993, *The Avalanche Handbook,* The Mountaineers, Seattle, Washington.

Miller, Albert, 1971, *Meteorology:* Charles E. Merrill Publishing Company, Columbus, Ohio.

Miller, Raymond W. and Donahue, Roy L., 1990; *Soils:* An Introduction to Soils and Plant Growth: Prentice Hall, Englewood Cliffs, New Jersey.

Tank, Ronald W., 1983, *Environmental Geology:* Oxford University Press, New York.

Other Sources

Japan Society of Landslide, "Landslides in Japan," *Report,* National Conferences on Landslide Control, Tokyo, 1988.

Chapter 4

Barnes-Svarney, Patricia, 1983, "Tsunami: Following the Deadly Wave," *Sea Frontiers,* Vol. 34, No. 5.

Bernstein, Joseph, 1971, "Tsunamis," *Oceanography,* Readings from Scientific American: W.H. Freeman and Company, San Francisco.

Cazeau, Charles H. and Scott, Stuart D., 1979, *Exploring the Unknown:* Plenum Press, New York.

Costa, John E., 1981, *Surficial Geology,* Building with the Earth: John Wiley & Sons, Inc., New York.

Duxbury, Alyn C. and Duxbury, Alison, 1984, *The World Oceans,* Addison-Wesley Publishing Company, Menlo Park, California.

Ebert, Charles H. V., 1986, "Thera: Sleeping Giant of the Cyclades," *Sea Frontiers,* Vol. 32, No. 6.

Gross, Grant M., 1971, *Oceanography:* Charles E. Merrill Publishing Company, Columbus, Ohio.

Ingmanson, Dale E., 1985, *Oceanography:* Wadsworth Publishing Company, Belmont, California.

Islam, M. Aminul, 1974, "Coastal Bangladesh," in *Natural Hazards* (ed. Gilbert White): Oxford University Press, New York.

Khalequzzaman, Md., 1992, "The Flood Action Plan in Bangladesh," in *Natural Hazards Observer,* Vol. XVI, No. 4, National Research and Applications Information Center, University of Colorado, Boulder, Colorado.

Knauss, John A., 1978, *Introduction to Physical Oceanography:* Prentice-Hall, Inc., Englewood Cliffs, New Jersey.

Melosh, H.J., 1989, *Impact Cratering:* A Geologic Process: Oxford University Press, New York.

Menard, H. W., 1977, "The Sea and Its Motions," *Ocean Science,* Readings from Scientific American: W.H. Freeman and Company, San Francisco.

Menard, H. W., 1974, *Geology, Resources, and Society:* W.H. Freeman and Company, San Francisco.

Murck, Barbara, Skinner, Brian J., Porter, Stephen C., *Dangerous Earth:* John Wiley & Sons, Inc. New York.

Nash, Jay Robert, 1976, *Darkest Hour:* Nelson-Hall, Inc., Chicago.

Parker, Dennis J., 1992, "The Flood Action Plan: Social Impacts in Bangladesh," in *Natural Hazards Observer,* Vol. XVI, No. 4, National Research and Applications Information Center, University of Colorado, Boulder, Colorado.

Pond, Stephen and Pickard, George L., 1978, *Introductory Dynamic Oceanography:* Pergamon Press, New York.

Robinson, H. W., 1953, "The Storm Surge of 31st January–1st February, 1953," *Geography,* Vol. 38.

Simkin, Tom and Fiske, Richard S., 1983, *Krakatau, 1883:* Smithsonian Institution Press, Washington, D.C.

Strahler, Arthur N. and Strahler, Alan H., 1973, *Environmental Geoscience: Interaction between Natural Systems and Man,* Hamilton Publishing Co., Santa Barbara, California.

Thurman, Harold V., 1985, *Introductory Oceanography:* Charles E. Merrill Publishing Company, Columbus, Ohio.

Ward, Roy, 1978, *Floods,* A Geographical Perspective: Halstead Press (A Division of John Wiley & Sons, Inc.), New York.

Wood, John A., 1979, *The Solar System:* Prentice Hall, Inc., Englewood Cliffs, New Jersey.

Other Sources

Vereenigin ter Bevordering van de Belangen des Boekhandels te Amsterdam, 1953, *de Ramp:* Amsterdam, Netherlands.

Chapter 5

Butler, Hal, 1976, *Nature at War:* Henry Regnery Company, Chicago.

Clark, Champ and Editors of Time-Life Books, 1982, *Planet Earth; Flood:* Time-Life Books, Alexandria, Virginia.

Coates, Donald R., 1981, *Environmental Geology:* John Wiley & Sons, Inc., New York.

Donahue, Roy L., Miller, Raymond W., and Shickluna, John C., 1977, *Soils,* An Introduction to Soils and Plant Growth: Prentice-Hall, Inc., Englewood Cliffs, New Jersey.

Edwards, Mike, 1982, "The Lash of the Dragon," in *Journey into China:* National Geographic Society, Washington, D.C.

Greenland, David and DeBlij, Harm J., 1977, *The Earth in Profile;* A Physical Geography: Canfield Press, San Francisco.

Gregory, K. J. and Walling, D. E., 1979, *Man and Environmental Processes:* Dawson-Westview Press, Inc., Boulder, Colorado.

Hollis, G. E., 1975, "The Effects of Urbanization on Floods of Different Recurrence Intervals," *Water Resources Research,* Vol. V: American Geophysical Union.

Hoyt, William G. and Langbein, Walter B., 1955, *Floods:* Princeton University Press, Princeton, New Jersey.

Moran, Joseph M., Morgan, Michael D., and Wiersma, James H., 1980, *Introduction to Environmental Science:* W.H. Freeman and Company, San Francisco.

Olson, Ralph E., 1970, *A Geography of Water:* Wm. C. Brown Company, Publishers, Dubuque, Iowa.

Perry, Ronald W., Lindell, Michael K., and Green, Marjorie R., 1981, *Evacuation Planning in Emergency Management:* D.C. Heath and Company, Lexington, Massachusetts.

Strahler, Arthur N. and Strahler, Alan H., 1977, *Geography and Man's Environment:* John Wiley & Sons, Inc., New York.

Ward, Roy, 1978, *Floods,* A Geographical Perspective: Halstead Press (Division of John Wiley & Sons, Inc.), New York.

Other Sources

NOAA, Department of Commerce, *Natural Disaster Survey Report,* 1994, "The Great Flood of 1993," U.S. Government Printing Office, Washington, D.C.

Wuerch, Donald E., MIC, *Monthly Report of River and Flood Conditions,* National Weather Service Office, Buffalo, New York, March 5, 1985.

Town of Amherst, Master Plan Update, Planning Office, Amherst, New York, October 30, 1986.

Special Publications, American Society of Civil Engineers, Report of the Committee on the Cause of the Failure of the South Fork Dam, *Transactions,* Vol. XXIV (January/June, 1891).

Chapter 6

Blair, Thomas A. and Fite, Robert C., 1957, *Weather Elements:* Prentice-Hall, Inc., Englewood Cliffs, New Jersey.

Calhoun, C. Raymond, 1981, *Typhoon: The Other Enemy:* Naval Institute Press, Annapolis, Maryland.

Critchfield, Howard J., 1960, *General Climatology:* Prentice-Hall, Inc., Englewood Cliffs, New Jersey.

Eagleman, Joe R., 1983, *Severe and Unusual Weather:* Van Nostrand Reinhold Company, New York.

Gross, M. Grant, 1971, *Oceanography:* Charles E. Merrill Publishing Company, Columbus, Ohio.

Hidore, John J., 1969, *A Geography of the Atmosphere:* Wm. C. Brown Company Publishers, Dubuque, Iowa.

Hubert, Lester F. and Lehr, Paul E., 1967, *Weather Satellites:* Blaisdell Publishing Company, Waltham, Massachusetts.

Lehr, Paul E., Burnett, R. Will, and Zim, Herbert S., 1957, *Weather:* Simon & Schuster, Inc., New York.

Miller, Albert, 1971, *Meteorology:* Charles E. Merrill Publishing Company, Columbus, Ohio.

Nash, Jay Robert, 1976, *Darkest Hours:* Nelson-Hall, Inc., Chicago.

Schaefer, Faith, "Sea Secrets," *Sea Frontiers,* Vol. 33, No. 1, Jan./Febr., 1987.

Taylor, George F., 1954, *Elementary Meteorology:* Prentice-Hall, Inc., New York.

Other Sources

NOAA, Department of Commerce, *Tornado Preparedness Planning:* U.S. Government Printing Office, Washington, D.C., 1978.

NOAA, Department of Commerce, *The Greatest Storm on Earth, Hurricane,* PA 76008: U.S. Government Printing Office, Washington, D.C., 1977.

NOAA, Department of Commerce, *Some Devastating North Atlantic Hurricanes of the 20th Century,* PA 77019: U.S. Government Printing Office, Washington, D.C., 1977.

NOAA, Department of Commerce, *Tornado,* PA 77027: U.S. Government Printing Office, Washington, D.C., 1978.

NOAA, Department of Commerce, *Tornado Safety,* PA 82001: U.S. Government Printing Office, Washington, D.C., 1982.

Chapter 7

Ahrens, C. Donald, 1982, *Meteorology Today:* West Publishing Company, New York.

Blair, Thomas A. and Fite, Robert C., 1957, *Weather Elements:* Prentice-Hall, Inc., Englewood Cliffs, New Jersey.

Clayman, Charles B., med,ed., 1989, *Home Medical Encyclopedia* (American Medical Association): Random House, New York.

Eagleman, Joe R., 1983, *Severe and Unusual Weather:* Van Nostrand Reinhold Company, New York.

Fless, Mary, "Thunderbolt Theory: Shedding Light on Lightning," *SUNY Research 82,* Vol. 2, No. 5, Nov/Dec, 1982.

Garelik, Glenn, "Different Strokes," *Discover:* Vol. 15, No. 10, October, 1984.

Gedzelman, Stanley David, 1980, *The Science and Wonders of the Atmosphere:* John Wiley & Sons, Inc., New York.

Hill, R. D., "TRIP Illumines Lightning," *EOS,* Transactions of the American Geophysical Union, Vol. 67, No. 20, May, 1986.

Krider, Philip E., 1986, "Lightning Damage and Lightning Protection," in *Violent Forces of Nature* (ed. Robert H. Maybury): Lomond Publications, Inc., Mt. Airy, Maryland.

Mooney, Michael M., 1975, *The Hindenburg:* Bantam Books, Dodd, Mead and Company, New York.

Riehl, Herbert, 1965, *Introduction to the Atmosphere:* McGraw-Hill Book Company, New York.

Taylor, George F., 1954, *Elementary Meteorology,* Prentice-Hall, Inc., New York.

Other Sources

Air Force Manual 105-5, *Weather for Aircrews,* Department of the Air Force, Washington, D.C., 1962.

NOAA, Department of Commerce, *Thunderstorms and Lightning,* PA 83001: U.S. Government Printing Office, Washington, D.C., 1985.

Letter to the author from Michael Cejka (Vice President of the Western New York Chapter of the American Meteorological Society, Buffalo, N.Y.), March 26, 1987.

Chapter 8

Ahrens, C. Donald, 1982, *Meteorology Today:* West Publishing Company, New York.

Allen, Oliver E., 1983, *Planet Earth;* Atmosphere: Time-Life Books, Alexandria, Virginia.

Bahr, Robert, 1980, *The Blizzard:* Prentice-Hall, Inc., Englewood Cliffs, New Jersey.

Bennett, Hugh H., 1939, *Soil Conservation:* McGraw-Hill Book Company, Inc., New York.

Blair, Thomas A. and Fite, Robert C., 1957, *Weather Elements:* Prentice-Hall, Inc., Englewood Cliffs, New Jersey.

Butler, Hal, 1976, *Nature at War:* Henry Regnery Company, Chicago.

Cressey, George B., 1960, *Crossroads,* Lands and Life in Southwest Asia: J.P. Lippincott Company, Chicago.

Eagleman, Joe R., 1983, *Severe and Unusual Weather:* Van Nostrand Reinhold Company, New York.

Editors of Encyclopedia Britannica, 1978, *Disasters!* When Nature Strikes Back: Bantam Books, Inc., New York.

Hedin, Sven, 1925, *My Life as an Explorer* (translated by Alfhild Huebsch): Garden City Publishing Company, Inc., Garden City, New York.

Moran, Joseph M. and Morgan, Michael D., 1986, *Meteorology,* The Atmosphere and the Science of Weather: Burgess Publishing, Edina, Minnesota.

Nash, Jay Robert, 1976, *Darkest Hours:* Nelson-Hall, Inc., Chicago.

Rossi, Erno, 1978, *White-Death, Blizzard of '77:* 77 Publishing, Port Colborne, Ontario, Canada.

Ryan, Paul B., 1985, *The Iranian Rescue Mission,* Why it Failed: Naval Institute Press, Annapolis, Maryland.

Smith, Guy-Harold (ed.), 1958, *Conservation of Natural Resources:* John Wiley & Sons, Inc., New York.

Whitaker, J. Russell and Ackerman, Edward A., 1951, *American Resources,* Their Management and Conservation: Harcourt, Brace and Company, New York.

Whittow, John, 1979, *Disasters.* The Anatomy of Environmental Hazards: The University of Georgia Press, Athens, Georgia.

Other Sources

American Medical Association, 1982, *Family Medical Guide:* Random House Publishing Company, New York.

Federal Aviation Agency and Department of Commerce (Weather Bureau), *Aviation Weather,* AC-006: U.S. Government Printing Office, Washington, D.C., 1965.

Israel Geography, Israel Pocket Library, 1973: Keter Publishing House, Ltd., Jerusalem.

NOAA, Department of Commerce, *Winter Storms,* PA 70018: U.S. Government Printing Office, Washington, D.C., 1975.

The Holy Scriptures, According to the Masoretic Text, 1955: The Jewish Publication Society of America, Philadelphia.

U.S. Army Corps of Engineers, 1977, *Operation Snow Go,* Blizzard of ÿ77: U.S.A.C.E., Buffalo, New York.

Chapter 9

Avery, Thomas E. and Berlin, Graydon L., 1985, *Interpretation of Aerial Photographs:* Burgess Publishing Company, Minneapolis, Minnesota.

Battan, Louis J., 1984, *Fundamentals of Meteorology:* Prentice-Hall, Inc., Englewood Cliffs, New Jersey.

Eagleman, Joe R., 1980, *Meteorology,* The Atmosphere in Action: D. Van Nostrand Company, New York.

Eagleman, Joe R., 1983, *Severe and Unusual Weather:* D. Van Nostrand-Reinhold Company, New York.

Goldstein, Avram, 1984, *Flying Out of Danger,* A Pilot's Guide to Safety: Airguide Publications, Inc., Long Beach, California.

Kimble, George H. T. and Good, Dorothy, 1955, *Geography of the Northlands:* The American Geographical Society and John Wiley & Sons, Inc., New York.

Maybury, Robert H. (ed.), 1986, *Violent Forces of Nature:* Lomond Publications, Inc., Mt. Airy, Maryland.

Miller, Albert, 1971, *Meteorology:* Charles E. Merill Publishing Company, Columbus, Ohio.

Mondey, David, 1983, *Airliners of the World:* Aerospace Publishing Ltd., London.

Moran, Joseph M. and Morgan, Michael D., 1986, *Meteorology,* The Atmosphere and the Science of Weather: Burgess Publishing, Edina, Minnesota.

Nash, Jay Robert, 1976, *Darkest Hours:* Nelson-Hall, Inc., Chicago.

Navarra, John G., 1979, *Atmosphere, Weather and Climate,* An Introduction to Meteorology: W.B. Saunders Company, Philadelphia.

Neiburger, Morris, Edinger, James G., and Bonner, William D., 1973, *Understanding Our Atmospheric Environment:* W.H. Freeman and Company, San Francisco.

Riehl, Herbert, 1965, *Introduction to the Atmosphere:* McGraw-Hill Book Company, New York.

Other Sources

Air Force Manual 105-5, *Weather for Aircrews,* Department of the Air Force, Washington, D.C., 1962.

Federal Aviation Agency and Department of Commerce (Weather Bureau), *Aviation Weather,* AC-006, U.S. Government Printing Office, Washington, D.C., 1965.

Federal Aviation Agency and Department of Commerce (Weather Bureau), *Aviation Weather,* AC-006-A (Revised), U.S. Government Printing Office, Washington, D.C., 1975.

Chapter 10

Barck, Oscar T., Wakefield, Walter L., Lefler, Hugh T., 1950, *The United States,* A Survey of National Development: The Ronald Press Company, New York.

Bark, L. Dean, 1978, "History of American Droughts," in *North American Drought* (ed. Rosenberg, Norman J.), AAAS Selected Symposium 15: Westfield Press, Inc., Boulder, Colorado.

Bennett, Hugh Hammond, 1939, *Soil Conservation:* McGraw-Hill Book Company, Inc., New York.

Brown, Lester R. and Wolf, Edward C., 1986, "Reversing Africa's Decline," in *State of the World,* Worldwatch Institute Report 1986: W. W. Norton & Company, Inc., New York.

Bryson, Reid A. and Murray, Thomas J., 1977, *Climates of Hunger:* The University of Wisconsin Press, Madison, Wisconsin.

Carter, George F., 1968, *Man and the Land:* Holt, Rinehart and Winston, Inc., New York.

Collinson, A. S., 1977, *Introduction to World Vegetation:* George Allen & Unwin (Publishers), Ltd., London.

Cressey, George B., 1963, *Asia's Lands and People:* McGraw-Hill Book Company, Inc., New York.

Dasman, Raymond F., 1984, *Environmental Conservation:* John Wiley & Sons, Inc., New York.

Durrell, Lee, 1986, *State of the Ark:* Doubleday & Company, Inc., Garden City, New Jersey.

Easterling, William E., 1987, "Drought as a Hazard: Can We Plan for It?", in *Natural Hazards Observer,* Vol. XI, No. 4, National Hazards Research and Applications Information Center, University of Colorado, Boulder, Colorado.

Ebert, Charles H.V., 1978, "El Nino: An Unwanted Visitor," *Sea Frontiers:* International Oceanographic Foundation, Vol. 24, No. 6.

Ehrlich, Paul R., Ehrlich, Anne H., Holdren, John P., 1977, *Ecoscience:* W.H. Freeman and Company, San Francisco.

Glantz, Michael H., 1977, *Desertification:* Westview Press, Inc., Boulder, Colorado.

Gross, M. Grant, 1995, *Principles of Oceanography:* A View of the Earth: Prentice-Hall, Inc., New Jersey.

Kobinsky, C.J., 1993, "Ocean Surface Topography and Circulation," Atlas of Satellite Observations Related to Global Change: Cambridge University Press, Canbridge.

Lydolph, Paul E., 1977, *Geography of the U.S.S.R.:* John Wiley & Sons, Inc., New York.

McGinnies, William G., Goldman, Bran J., Paylore, Patricia, 1968, *Deserts of the World:* University of Arizona Press, Tucson, Arizona.

Moran, Joseph M., Morgan, Michael D., Wiersma, James H., 1980, *Introduction to Environmental Science:* W.H. Freeman and Company, San Francisco.

Nixon, Richard M., 1982, *Leaders:* Warner Brothers, Inc., New York.

Raikes, Robert L., 1971, "Formation of Deserts of the Near East and North Africa," in *Health Related Problems in Arid Lands* (ed. Riedesel, M. L.), Desert and Arid Zones Research Symposium, Arizona State University, Tempe, Arizona.

Symons, Leslie, 1983, *The Soviet Union:* Barnes & Noble Books, Totowa, New Jersey.

Wagner, Richard H., 1971, *Environment and Man:* W.W. Norton & Company, Inc., New York.

White, Gilbert F., 1960, *Science and the Future of Arid Lands:* UNESCO, Paris.

Whittow, John, 1979, *Disasters, The Anatomy of Environmental Hazards:* The University of Georgia Press, Athens, Georgia.

Wijkman, Anders and Timberlake, Lloyd, 1984, *Natural Disasters,* Acts of God or Acts of Man?: International Institute for Environment and Development, London and Washington, D.C.

Other Sources

World Meteorological Organization, 1975, *Drought and Agriculture,* Technical Note No. 138 (WMO-No. 392), Geneva, Switzerland.

Chapter 11

Albini, Frank A., 1984, "Wildland Fires," *American Scientist,* Vol. 72, No. 6.

Borman, F. Herbert and Likens, Gene E., 1981, "Catastrophic Disturbances and the Steady State in Northern Hardwood Forests," in *Use and Misuse of the Earth's Surface* (ed. Skinner, Brian J.): William Kaufmann, Inc., Los Altos, California.

Butler, John E., 1976, *Natural Disasters:* Heineman Educational Australia, Richmond.

Critchfield, Howard J., 1960, *General Climatology:* Prentice-Hall, Inc., New Jersey.

Dasman, Raymond F., 1984, *Environmental Conservation:* John Wiley & Sons, Inc., New York.

Ebert, Charles H. V., 1963 "The Meteorological Factor in the Hamburg Fire Storm," *Weatherwise* (AMS), Vol. 16, No. 2.

Ebert, Charles H. V., 1963, "Hamburg's Firestorm Weather," *National Fire Protection Association Quarterly,* (NFPA), Vol. 56, No. 3.

Gale, Robert D., 1987, "Can Anything Be Done About Western Wildfires?" *Natural Hazards Observer,* Vol. XII, No. 2 [November, 1987:] Natural Hazards Research and Applications Information Center, University of Colorado, Boulder.

Holbrook, Stewart H., 1944, *Burning an Empire,* The Story of American Forest Fires: The Macmillan Company, New York.

Laubenfels, David J., 1970, *A Geography of Plants and Animals:* Wm. C. Brown Company, Publishers, Dubuque, Iowa.

Moore, Peter D., 1982, "Fire: catastrophic or creative force?" *Impact of Science on Society* (UNESCO), Vol. 31, No. 1.

Mullaney, Anthony J., 1946, "German Fire Departments Under Air Attacks," in *Fire and the Air War* (ed. Bond, Horatio): National Fire Protection Association, Boston, Massachusetts.

Odum, Eugene P., 1963, *Ecology:* Holt, Rinehart and Winston, Inc., New York.

Page, Jake and Editors of Time-Life Books, 1983, *Planet Earth; Forest:* Time-Life Books, Inc., Alexandria, Virginia.

Phillips, John, 1959, *Agriculture and Ecology in Africa:* Faber and Faber, London.

Pyne, Stephen J., 1982, *Fire in America; A Cultural History of Wildland and Rural Fire:* Princeton University Press, Princeton, New Jersey.

Wagner, Richard H., 1971, *Environment and Man:* W. W. Norton & Company, New York.

Wells, Robert W., 1968, *Fire at Peshtigo:* Prentice-Hall, Inc., Englewood Cliffs, New Jersey.

Whitaker, J. Russell and Ackerman, Edward A., 1951, *American Resources,* Their Management and Conservation: Hartcourt, Brace and Company, New York.

Wright, Henry A. and Bailey, Arthur W., 1982, *Fire Ecology; U.S. and Southern Canada:* John Wiley & Sons, Inc., New York.

Chapter 12

Balzer, Robert L., 1964, *The Pleasures of Wine:* The Bobbs-Merrill Company, Inc., New York.

Barker, Will, 1960, *Familiar Insects in America:* Harper Brothers, Publishers, New York.

Boorstin, Daniel J., 1983, *The Discoverers,* A History of Man's Search To Know His World and Himself: Random House, Inc., New York.

Brown, Lester R. and Wolf, Edward C., 1986, "Assessing Ecological Decline," in *State of the World,* Worldwatch Institute Report 1986: W.W. Norton & Company, Inc., New York.

Burke, James, 1978, *Connections:* Little, Brown and Company, Boston.

Carson, Rachel, 1962, *Silent Spring:* Houghton Mifflin Company, Boston.

Cressey, George B., 1960, *Crossroads,* Lands and Life in Southwest Asia: J.B. Lippincott Company, Chicago.

Durrell, Lee, 1986, *State of the Ark:* Doubleday & Company, Inc., Garden City, New York.

Ehrlich, Paul R. and Ehrlich, Anne H., 1981, *Extinction,* The Causes and Consequences of the Disappearance of Species: Random House, Inc., New York.

Essig, E. O., 1965, *A History of Entomology:* Hafner Publishing Company, Inc., New York.

Fleming, Walter E., "Soil Management and Insect Control," in *Yearbook of Agriculture, 1957,* U.S. Department of Agriculture: U.S. Government Printing Office, Washington, D.C.

Hobson, Richard, Jr. and Lawton, John, 1987, "New Battle in an Ancient War," *ARAMCO WORLD,* Vol. 38, No. 3 (May/June).

Koopowitz, Harold and Kaye, Hilary, 1983, *Plant Extinction; A Global Crisis:* Stone Wall Press, Inc., Washington, D.C.

Laycock, George, 1966, *The Alien Animal:* American Museum of Natural History, The Natural History Press, Garden City, New York.

Page, Jake and the Editors of Time-Books, 1985, *Planet Earth,* Grasslands and Tundra: Time-Life Books, Inc., Alexandria, Virginia.

Rudd, Robert L., 1964, *Pesticides and the Living Landscape:* The University of Wisconsin Press, Madison, Wisconsin.

Simkin, Tom and Fiske, Richard S., 1983, *Krakatau 1883,* The Volcanic Eruption and Its Effect: Smithsonian Institution Press, Washington, D.C.

Tuchman, Barbara W., 1978, *A Distant Mirror,* The Calamitous 14th Century: Alfred A. Knopf, Inc., New York.

Wagner, Richard H., 1971, *Environment and Man:* W.W. Norton & Company, Inc., New York.

White, Gilbert F., 1960, *Science and the Future of Arid Lands:* UNESCO, Paris.

Other Sources

Rainforest Action Network, Alert #12: San Francisco, March, 1987.

Chapter 13

Alverson, Dayton L., 1977, "Opportunities to Increase Food Production from the World's Oceans," in *Oceanography,* (ed. Pirie, R. Gordon): Oxford University Press, New York.

Arai, Eiko, 1994, "Buried in Garbage," *Pacific Friend,* Vol. 22, No. 4, Jiji Gaho Sha, Inc., Tokyo.

Bascom, Willard, 1982, "The Disposal of Waste in the Ocean," in *Life in the Sea,* Readings from Scientific American: W. H. Freeman and Company, San Francisco.

Blaikie, Piers, 1985, *The Political Economy of Soil Erosion in Developing Countries:* Longman, Inc., New York.

Bradley, P. N., 1986, "Food Production and Distribution—and Hunger," in *A World in Crisis?* (ed. Johnston, R. J. and Taylor, P. J.): Basil Blackwell, Ltd., Oxford.

Brookins, Douglas G., 1981, *Earth Resources, Energy and the Environment:* Charles E. Merrill Publishing Company, Columbus, Ohio.

Brown, Lester R., 1987, "Analyzing the Demographic Trap," in *State of the World,* Worldwatch Institute Report 1987: W.W. Norton & Company, Inc., New York.

Brown, Lester R., 1979, "Resource Trends and Population Policy: A Time for Reassessment," Worldwatch Paper No. 29: Worldwatch Institute, Washington, D.C.

Brown, Lester R., 1976, "World Population Trends: Signs of Hope, Signs of Stress," Worldwatch Paper No. 8: Worldwatch Institute, Washington, D.C.

Brown, Lester R. and Wolf, Edward C., 1984, "Soil Erosion: Quiet Crisis in the World Economy," Worldwatch Paper No. 60: Worldwatch Institute, Washington, D.C.

Coates, Donald R., 1981, *Environmental Geology:* John Wiley & Sons, Inc., New York.

Couper, A. D., 1972, *The Geography of Sea Transport:* Hutchinson & Company; (Publishers), Ltd., London.

Davis, Jr., Richard A., 1987, *Oceanography:* Wm. C. Brown Co. Publishers, Dubuque, Iowa.

Dorfman, Robert and Dorfman, Nancy S., 1972, *Economics of the Environment:* W.W. Norton & Company, Inc., New York.

Ebert, Charles H. V., 1965, "Water Resources and Land Use in the Qatif Oasis of Saudi Arabia," *The Geographical Review,* Vol. LV, No. 4, October, 1965.

Ehrlich, Paul R., Ehrlich, Anne H., and Holdren, John P., 1977, *Ecoscience:* W.H. Freeman and Company, San Francisco.

Flavin, Christopher, 1986, "Moving Beyond Oil," *State of the World,* Worldwatch Institute Report 1986: W.W. Norton & Company, New York.

Gross, Grant, 1987, *Oceanography;* A View of the Earth: Prentice-Hall, Inc., Englewood Cliffs, New Jersey.

Grun, Bernard, 1979, *The Timetables of History* (Based on Werner Stein's Kulturfahrplan): Simon & Schuster, Inc., New York.

Hausenbuiller, R. L., 1985, *Soil Science,* Principles and Practices: Wm. C. Brown Co. Publishers, Dubuque, Iowa.

Heyerdahl, Thor, 1973, "Ocean Pollution Seen from Rafts," in *Oceanography* (ed. Pirie, R. Gordon): Oxford University Press, New York.

Holdren, John P. and Ehrlich, Paul R., 1981, "Human Population and the Global Environment," in *Use and Misuse of the Earth's Surface* (ed. Skinner, Brian J.): William Kaufmann, Inc., Los Altos, California.

Keller, Edward A., 1979, *Environmental Geology:* Charles E. Merrill Publishing Company, Columbus, Ohio.

Leakey, Louis S. B., 1969, *Animals of East Africa:* National Geographic Society, Washington, D.C.

Lester, Stephen, 1987, "Garbage Incineration Makes No Sense at All," *Everyone's Backyard,* Vol. 5, No. 2, Citizen's Clearinghouse for Hazardous Waste, Inc., Arlington, Virginia.

Lutz, Wolfgang, 1994, "The Future of World Population," *Population Bulletin,* Vol. 49, No. 1, Population Reference Bureau, Washington, D.C.

Meadows, Donella H., Meadows, Dennis L., Randers, Jorgen and Behrens III, William W., 1972, *The Limits to Growth:* Universe Books, New York.

Meyer, David, 1986, "A Blueprint for World Population Stabilization," *Report:* The Population Institute, Washington, D.C.

Moran, Joseph M., Morgan, Michael D. and Wiersma, James H., 1980, *Introduction to Environmental Science:* W.H. Freeman and Company, San Francisco.

Odell, Peter R., 1986, "Draining the World of Energy," in *A World in Crisis?* (ed. Johnston, R. J. and Taylor, P. J.): Basil Blackwell, Ltd., Oxford.

Passerini, Edward, 1986, "Food for Everyone? Yes . . . From Trees," in *Agriculture and Human Values,* Vol. III, No. 3.

Pollock, Cynthia, 1987, "Realizing Recycling's Potential," in *State of the World,* Worldwatch Institute Report 1987: W.W. Norton & Company, Inc., New York.

Smith, R. M. and Newsom, T. M., 1976, *The Hyperion Process,* As Applied to Production of Natural Protein. (Prospectus; Revision I).

Stoddard, Robert H., Blouet, Brian W. and Wishart, David J., 1986, *Human Geography,* People, Places, and Cultures: Prentice-Hall, Inc., Englewood Cliffs, New Jersey.

Thompson, M. Louis and Troeh, Frederick R., 1978, *Soils and Soil Fertility:* McGraw-Hill Book Company, New York.

Thurman, Harold V., 1985, *Introductory Oceanography:* Charles E. Merrill Publishing Company, Columbus, Ohio.

Wenk, Jr., Edward, 1977, "The Physical Resources of the Ocean," in *Ocean Science,* Readings from Scientific American: W. H. Freeman and Company, San Francisco.

Whipple, A. B. C., and The Editors of Time-Life Books, 1983, *Planet Earth; Restless Ocean:* Time-Life Books, Inc., Alexandria, Virginia.

Woods, Robert, 1986, "Malthus, Marx and Population Crises," in *A World in Crisis?* (ed. Johnston, R. J. and Taylor, P. J.): Basil Blackwell, Ltd., Oxford.

Zimmermann, Erich W., 1964, *Introduction to World Resources:* Harper and Row, Publishers, New York.

Other Sources

Shell Oil Company, The National Energy Outlook 1980–1990, *Special Report* (August, 1980), Houston, Texas.

U.S. Department of Defense, *The Effects of Nuclear Weapons,* published by U.S. Atomic Energy Commission: U.S. Government Printing Office, Washington, D.C., 1962.

Chapter 14

Brodigan, Amy, Lipsett, Brian, Gibbs, Lois M., 1986, *Hazardous Waste Fact Pack:* Citizens Clearinghouse for Hazardous Waste, Inc., Arlington, Virginia.

Brown, Lester R., Jacobson, Jodi, 1987, "Assessing the Future of Urbanization," in *State of the World,* Worldwatch Institute Report, 1987: W. W. Norton & Company, Inc., New York.

Calder, Ritchie, 1962, *Living with the Atom:* University of Chicago Press, Chicago.

Carson, Rachel, 1962, *Silent Spring:* Houghton Mifflin Company, Boston, Massachusetts.

Chalaby, Abbas, 1981, *Egypt:* Casa Editrice Bonech, Florence, Italy.

Clark, David, 1985, *Post-Industrial America;* A Geographical Perspective: Methuen, Inc., New York.

Coch, Nicholas K., 1995, *Geohazards:* Natural and Human: Prentice-Hall, Inc., Englewood, New Jersey.

Eagleman, Joe R., 1991, *Air Pollution Meteorology:* Trimedia Publishing Company, Lenexa, Kansas.

Ebert, Charles H. V., 1980, "Love Canal: An Environmental Disaster," *TRANSITION,* Vol. 10, No. 3 (Socially and Ecologically Responsible Geographers).

Echols, James R., 1981, "Population vs. Environment: A Crisis of Too Many People," in *Use and Misuse of the Earth's Surface* (ed. Skinner, Brian J.): William Kaufmann, Inc., Los Altos, California.

Edwards, Mike, 1987, "Chernobyl—One Year After," *National Geographic,* Vol. 171, No. 5, May, 1987.

Ehrlich, Paul R., Ehrlich, Anne H., Holdren, John P., 1977, *Ecoscience:* W. H. Freeman and Company, San Francisco.

Foth, Henry D., 1984, *Fundamentals of Soil Science:* John Wiley & Sons, Inc., New York.

Hayes, Denis, 1979, "Pollution: The Neglected Dimensions," Worldwatch Paper No. 27: Worldwatch Institute, Washington, D.C.

Hohenemser, Christoph, 1986, "Chernobyl: The First Lessons," *Natural Hazards Observer,* Vol. XI, No. 1 (September 1986): Natural Hazards Research and Applications Information Center, University of Colorado, Boulder.

Levine, Adeline Gordon, 1982, *Love Canal;* Science, Politics, and People: D.C. Heath Company, Lexington, Massachusetts.

Likens, Gene E., Wright, Richard F., Galloway, James N., and Butler, Thomas J., 1979, "Acid Rain," Reprint from *Scientific American,* Vol. 241, No. 4 (October 1979): W.H. Freeman and Company, San Francisco.

Lydolph, Paul E., 1977, *Geography of the U.S.S.R.:* John Wiley & Sons, Inc., New York.

Mathews, Samuel W., 1990, "Is Our Planet Warming?" *National Geographic,* Vol. 178, No. 4 (October 1990): National Geographic Society, Washington, D.C.

Menard, H. W., 1974, *Geology, Resources, and Society;* An Introduction to Earth Science: W.H. Freeman and Company, San Francisco.

Mikesell, Marvin M., 1974, "Geography as a Study of Environment: An Assessment of Some Old and New Commitments," in *Perspectives on Environment,* Publication No. 13: Association of American Geographers, Washington, D.C.

Milbrath, Lester W., 1989, *Envisioning a Sustainable Society:* State University of New York Press, Albany, New York.

Monastersky, Richard, 1992, "Haze Clouds the Greenhouse," *Science News,* Vol. 141, No. 15 (April 11, 1992).

Navarra, John Gabriel, 1979, *Atmosphere, Weather and Climate;* An Introduction to Meteorology: W.B. Saunders Co., Philadelphia.

Neiburger, Morris, Edinger, James G., Bonner, William D., 1973, *Understanding Our Atmospheric Environment:* W.H. Freeman & Company, San Francisco.

Plass, Gilbert N., 1959, "Carbon Dioxide and Climate," Reprint from *Scientific American* (July 1954): W.H. Freeman & Company, San Francisco.

Postel, Sandra, 1986, "Altering the Earth's Chemistry: Assessing the Risks," Worldwatch Paper No. 71: Worldwatch Institute, Washington, D.C.

Postel, Sandra, 1984, "Air Pollution, Acid Rain, and the Future of Forests," Worldwatch Paper No. 58: Worldwatch Institute, Washington, D.C.

Schmid, James A., 1974, "The Environmental Impact of Urbanization," in *Perspectives of the Environment,* Publication No. 13: Association of American Geographers, Washington, D.C.

Solomon, Susan, 1990, "Progress Toward a Quantitative Understanding of Antarctic Ozone Depletion," *Nature,* Vol. 347 (October 1990).

Stoddard, Robert H., Blouet, Brian W., Wishart, David J., 1986, *Human Geography;* People, Places, and Cultures: Prentice-Hall, Inc., Englewood Cliffs, New Jersey.

Symons, Leslie, 1983, *The Soviet Union:* Barnes & Noble Books, Totowa, New Jersey.

Thurman, Harold V., 1990, *Introductory Oceanography:* Charles E. Merrill Publishing Company, Columbus, Ohio.

Wagner, Richard H., 1971, *Environment and Man:* W.W. Norton & Company, Inc., New York.

Whittow, John, 1979, *Disasters;* The Anatomy of Environmental Hazards: The University of Georgia Press, Athens, Georgia.

Zuesse, Eric, 1981, "Love Canal: The Truth Seeps Out," *REASON* magazine (February 1981): Reason Foundation, Santa Monica, California.

Other Sources

Canadian Embassy, *Report:* Fact Sheet on Acid Rain (Undated); Canadian Embassy, 1771 N. Street, N.W., Washington, D.C.

CCHW, *Report:* Deep Well Injection; An Explosive Issue (August, 1985): Citizens Clearinghouse for Hazardous Wastes, Inc., Arlington, Virginia.

Ebert, Charles H.V., *Report No. 2* (Unpublished): Comments on the Love Canal Pollution Abatement Plan (November 13, 1978), Department of Geography, SUNY/Buffalo, New York.

EPA 600/8-79-028, *Research Summary;* Acid Rain (October 1979); U.S. Environmental Protection Agency, Washington, D.C.

Intergovernmental Panel of Climate Change (IPCC), 1991, *Climate Change:* The IPCC Scientific Assessment; Press Syndicate of the University of Cambridge.

Proceedings of a Joint Symposium by the Board on Atmospheric Sciences and Climate, and the Committee on Global Change, Commission on Physical Sciences, Mathematics, and Resources, National Research Council: National Academy Press, Washington, D.C., 1989.

United States-Canada Research Consultation Group, *Second Report* (Unpublished): Long-Range Transport of Air Pollutants (November, 1980).

Chapter 15

Anderton, David, 1979, "Jet Bombers," in *Air Power:* Phoebus Publishing Company/BPC Publishing, Ltd., London.

Berton, Pierre, 1986, *Vimy:* McClelland and Stewart, Ltd., Toronto, Canada.

Bronowski, Jacob, 1973, *The Ascent of Man:* Little, Brown and Company, Boston, Massachusetts.

Brunswig, Hans, 1978, *Feuersturm "über Hamburg:* Motorbuch Verlag, Stuttgart, W. Germany.

Buckingham, Jr., William A., 1982, *Operation Ranch Hand;* The Air Force and Herbicides in Southeastern Asia, 1961–1971: Office of the Air Force History, USAF, Washington, D.C.

Burke, James, 1978, *Connections:* Little, Brown and Company, Boston, Massachusetts.

Caidin, Martin, 1960, *The Night Hamburg Died:* Ballantine Books, Inc., New York.

Calder, Ritchie, 1962, *Living with the Atom:* University of Chicago Press, Chicago.

Clark, Robin, 1968, *The Silent Weapons:* David McMay Company, Inc., New York.

Dornberger, Walter, 1954, *V-2:* Ballantine Books, Inc., New York.

Eagleman, Joe R., 1991, *Air Pollution Meteorology:* Trimedia Publishing Company, Kansas.

Earle, Sylvia A., 1992, "Assessing the Damage One Year Later," *National Geographic,* Vol. 181, No. 2, National Geographic Society, Washington, D.C.

Ebert, Charles H. V., 1963, "Hamburg's Firestorm Weather," *National Fire Protection Association Quarterly,* Vol. 56, No. 3.

Foth, Henry D., 1984, *Fundamentals of Soil Science:* John Wiley & Sons, Inc., New York.

Frankland, Noble, 1970, *Bomber Offensive,* The Devastation of Europe: Ballantine Books, Inc., New York.

Friedman, Norman, 1992, *Desert Victory;* The War for Kuwait: Naval Institute Press, Annapolis, Maryland.

Friedrich, Otto and Editors, 1991, *Desert Storm: The War in the Persian Gulf,* Time Books: Little, Brown and Company, Boston.

Frogett, Steve, 1992, "Tomahawk in the Desert," *U.S. Naval Institute Proceedings* (January 1992), U.S. Naval Institute, Annapolis, Maryland.

Gräff, Siegfried, 1955, *Tod im Luftangriff:* H. H. Nolke Verlag, Hamburg, W. Germany.

Grun, Bernard, 1979, *The Timetables of History* (Based on Werner Stein's Kulturfahrplan): Simon and Schuster, Inc., New York.

Hecht, Selig, 1955, *Explaining the Atom:* The Viking Press, New York.

Hersh, Seymour M., 1968, *Chemical and Biological Warfare,* America's Hidden Arsenal: The Bobbs-Merrill Company, New York.

Hogg, Ian V., 1970, *The Guns: 1939–1945:* Ballantine Books, Inc., New York.

Hollander, Edwin P., 1976, *Principles and Methods of Social Psychology:* Oxford University Press, New York.

Irving, David, 1963, *The Destruction of Dresden:* Holt, Rinehart and Winston, New York.

Jukes, Geoffrey, 1968, *Kursk,* The Clash of Armour: Ballantine Books, Inc., New York.

Keegan, John, 1993, *A History of Warfare;* Vintage Books, Random House., New York.

Kissinger, Henry, 1994, *Diplomacy:* Simon & Schuster, New York.

Klöss, Erhard, 1963, *Der Luftkrieg über Deutschland, 1939–1945:* Deutscher Taschenbuch Verlag, München, W. Germany.

Lorenz, Konrad, 1966, *On Aggression* (Translated by Marjorie Kerr Wilson): Harcourt, Brace & World, Inc., New York.

Maple, Terry and Matheson, Douglas W., 1973, *Aggression, Hostility, and Violence;* Nature or Nurture?: Holt, Rinehart and Winston, Inc., New York.

Menard, H. W., 1974, *Geology, Resources and Society;* An Introduction to Earth Sciences: W.H. Freeman and Company, San Francisco.

Meselson, Mathew and Robinson, Julian P., 1980, "Chemical Warfare and Chemical Disarmament," *Scientific American,* Vol. 242, No. 4 (April 1980).

Morgenthau, Hans J., 1949, *Politics Among Nations;* The Struggle for Power and Peace: Alfred A. Knopf, Inc., New York.

Musgrove, Gordon, 1981, *Operation Gomorrah:* Jane's Publishing Company, Ltd., London.

O'Connell, Robert L., 1989, *Of Arms and Men:* A history of war, weapons, and aggression: Oxford University Press, New York.

Reid, William, 1986, *Weapons Through the Ages,* Crescent Books, New York.

Salisbury, Harrison E., 1978, *The Unknown War:* Bantam Books, Inc., New York.

Salisbury, Harrison E., 1969, *The 900 Days;* The Siege of Leningrad: Harper and Row, Publishers, Inc., New York.

Sanborn, Forrest J., 1946, "Fire Protection Lessons of the Japanese Attacks," in *Fire and the Air War* (ed. Bond, Horatio): National Fire Protection Association, Boston, Massachusetts.

Saundby, Robert, 1961, *Air Bombardment:* Harper and Brothers, New York.

Simkin, Tom and Fiske, Richard S., 1983, *Krakatau, 1883;* The Volcanic Eruption and Its Effects: Smithsonian Institution Press, Washington, D.C.

Stagner, Ross, 1965, "The Psychology of Human Conflict," in *The Nature of Human Conflict* (ed. McNeil, Elton B.): Prentice-Hall, Inc., Englewood Cliffs, New Jersey.

Stone, George Cameron, 1961, *A Glossary of the Construction, Decoration and Use of Arms and Armor:* Jack Brussel, Publisher, New York.

Sulzberger, C.L. and Editors of American Heritage, 1966, *Picture History of World War II,* Vol. I: American Heritage Publishing Company, Inc., New York.

Wallace, Robert and Editors of Time-Life Books, 1978: Time-Life Books, Inc., Alexandria, Virginia.

Westing, Arthur H., 1984. *Herbicides in War:* Taylor and Francis Publ. Co., Philadelphia, Pennsylvania.

Williams, John, 1976, *Atlas of Weapons and War:* Aldus Books Limited, London.

Wyden, Peter, 1985, *Day One: Before Hiroshima and After:* Warner Books, New York.

Zentner, Kurt, 1963, *Illustrierte Geschichte des zweiten Weltkriegs:* Südwest Verlag, München, W. Germany.

Ziemke, Earl F., 1968, *Stalingrad to Berlin;* The German Defeat in the East: Office of the Chief of Military History, U.S. Army, Washington, D.C.

Other Sources

Hiroshima-Nagasaki Committee, 1985, *The Impact of the A-Bomb:* Iwanami Shoten, Publishers, Tokyo.

U.S. Department of Defense, 1962, *The Effects of Nuclear Weapons:* U.S. Atomic Energy Commission, U.S. Government Printing Office, Washington, D.C.

Chapter 16

Blaikie, Piers; Cannon, Terry; Davis, Ian; Wisner, Ben, 1994, *At Risk:* Natural Hazards, People's Vulnerability, and Disasters: Routledge, New York.

Booth, Basil and Fitch, Frank, 1979, *Earthshock:* Walker and Company, New York.

Brown, Lester R., 1995, "Nature's Limits," in *State of the World,* Worldwatch Institute Report 1995: W.W. Norton & Company, Inc., New York.

Bryant, Edward A., 1991, *Natural Hazards:* Cambridge University Press, New York.

Ebert, Charles H.V., 1982, "Consequences of Disasters for Developing Nations," Impact, Vol. 32, No. 1, UNESCO, New York.

Geipel, Robert, 1982, *Disaster and Reconstruction:* George Allen & Unwin, Boston, Massachusetts.

Hollander, Edwin P., 1976, *Principles and Methods of Social Psychology:* Oxford University Press, New York.

Levine, Adeline F., 1982, *Love Canal:* Science, Politics, and People: D.C. Heath and Company, Lexington, Massachusetts.

Nash, Jay R., 1976, *Darkest Hours:* Nelson-Hall, Inc., Chicago.

Noji, Eric K., 1987, *The Public Health Consequences of Disasters.* Oxford University Press, New York.

Pararas-Carayannis, George, 1982, "The Effects of Tsunami on Society," *Impact,* Vol. 32, No. 1, UNESCO, New York.

Roberts, Neil, 1994, "The Global Environmental Future," in *The Changing Global Environment:* Blackwell Publishers, Cambridge, Massachusetts.

Rogers, Peter, 1982, "The Social and Economic Impact of Tropical Cyclones," *Impact,* Vol. 32, No. 1, UNESCO, New York.

Simmons, I.G., 1989, *Changing the Face of the Earth:* Culture, Environment, History: Basil Blackwell, Inc., Cambridge, Massachusetts.

Slovic, Paul, 1994, "Perception of Risk," in *Environmental Risks and Hazards,* ed. Susan L. Cutter: Prentice-Hall, Inc., Englewood, New Jersey.

Wijkman, A. and Timberlake, L., 1984, *Natural Disasters:* Acts of God or Acts of Man?: Institute for Environmental Development, London and Washington, D.C.

Other Sources

Natural Hazards Observer Report, Vol. XVII, No. 3, January 1994: University of Colorado, Boulder, Colorado.

Introduction

McCall, G.J.H., Laming, D.J.C., Scott, S.C., 1992; *Geohazards:* Natural and Man-made: Chapman & Hall; New York

INDEX